Lecture Notes in Social Networks

More information about this series at http://www.springer.com/series/8768

Iyad Muhsen AlDajani

Internet Communication Technology (ICT) for Reconciliation

Applied Phronesis Netnography in Internet Research Methodologies

Springer

Iyad Muhsen AlDajani
Jena Center for Reconciliation Studies
Friedrich Schiller University Jena
Jena, Germany

ISSN 2190-5428 ISSN 2190-5436 (electronic)
Lecture Notes in Social Networks
ISBN 978-3-030-41202-9 ISBN 978-3-030-41203-6 (eBook)
https://doi.org/10.1007/978-3-030-41203-6

This Springer imprint is published by the registered company Springer Nature Switzerland AG.
The registered company address is: Gewerbestrasse 11, 6330 Cham, Switzerland

Abbreviations

CAQCAS	Computer Assistant Qualitative Content Analysis Software
E-Democracy	Electronic Democracy
E-Government	Electronic Government
E-Service	Electronic Services
HoF	Hear of Flesh Not Stone Project
ICT	Internet communication technology
ICT-MR	Internet communication technology for the advancement of moderation, reconciliation, and democracy research
ICT for reconciliation	Internet communication technology for reconciliation
IRW	Interactive Reconciliation Workshops
NGO	Non-governmental organization
OIRA	Online Interactive Reconciliation Articles
PMMN	Phronetic Mixed-Method Netnography approach
PhMM	Phronetic Mixed-Method Tradition
QLCA	Qualitative Content Analysis
QNCA	Quantitative Content Analysis
QD	Qualitative Descriptions
QDS	Qualitative Data Structure
WASATIA NGO	Non-governmental organization in Palestine
Wasatia	Moderation as an Islamic concept

Contents

Part IV The Empirical Communication Research in Practice "The Techne Approach"

List of Figures

List of Tables

Part I
Internet Communication Research Overview

Chapter 1
Internet Communication Research Overview

1.1 Preface

This study is an interdisciplinary research that applies Phronesis in Social Sciences illustrated in Internet research methodologies in an experimental intervention research design to investigate the impact of the usage of internet communication technologies on the reconciliation process, applied in a case study in the Palestinian –Israeli context.

The research is part of the trilateral project "Heart of Flesh Not Stone," financed by the German Science Foundation, Deutsche Forschungsgemeinschaft (DFG). It involves Friedrich-Schiller University, presented by Martin Leiner, Director of the Jena Center for Reconciliation Studies (JCRS), Nikolaus Knoeppfler, and assisted by Martin O'Malley from the Ethics Centre; the University of Tel-Aviv presented by Ari Nadler; the University of Ben-Gurion, presented by Shifra Sagy; and Wasatia in Palestine, presented by Mohammed Dajani Daoudi.

The research develops a scientific approach to conduct a scientific investigation and develop a research methodology on whether internet communication technologies can instigate social change towards the reconciliation process in the midst of a protracted conflict. It includes the integration of disciplines such as philosophy, sociology, religion, ethics, political science, communication science, and internet information technologies. It explores the hypothesis that Internet Communication Technology (ICT) can promote social change toward moderation, reconciliation, and democracy in the midst of conflict; thus to be referred to as the Reconciliation Process in the research.

Invigorating that the reconciliation process is much needed to generate social change in the stagnant peace process between Israelis and Palestinians, engaged in a violent conflict for more than seven decades. It requires sincere efforts to overcome social prejudices, the clash of narratives, feelings of victimhood, denial of rights, and historical attachments of the other, that keep perpetuating the conflict.

I. M. AlDajani, *Internet Communication Technology (ICT) for Reconciliation*, Lecture Notes in Social Networks, https://doi.org/10.1007/978-3-030-41203-6_1

The research illustrates the intrinsic value of adhering to a culture of moderation, reconciliation, and the advancement of democracy in the middle of the conflict. The reconciliation process is an essential requirement to achieve a just and lasting settlement in combining the present 'Top-down approach' with the 'bottom-up approach' using the new expanding social communications technologies of the internet to push for social change toward reconciliation process. It is done by using digital social networks as a tool to lay the foundation for creating the environment conducive for moderation, reconciliation, and democracy, to commit both communities in conflict to adopt the reconciliation process.

The Reconciliation process is more achievable when the bottom-up approach affiliated with stakeholders convince the political leadership that this is what the people want; and that they should adopt a reconciliation agenda to respond positively to what the people demand. In this way, deliberative democracy is in progress where instead of people voting to express their voices, they are debating out loudly for politicians to hear them. ICT can help to build trust and put an end to dehumanization, delegitimization, and stereotyping of others and to usher mutual recognition and respect for the narrative of the other and acceptance for their aspirations, interests, and attachments with their history.

The communication research demonstrates that using the online digital social network as a tool to spread moderation, reconciliation, and democratic values which are vital to initiating the process of reconciliation. Utilizing online social networks as a tool to spread peaceful concepts within both communities helps to prepare them to reconcile by breaking the psychological barrier of hatred between them.

The conclusion offers a proper orientation and comprehensive knowledge of the theoretical framework of the research utilizing theories and concepts helpful to understand the reconciliation process. Such theories and concepts include the "*Hölderlin Perspective,*" referencing the German poet Friedrich Hölderlin (1770–1843) for (Leiner and Flämig 2012, p. 16), which reveals that reconciliation can be present in the middle of the conflict. Another is the Aristotelian concept "*Phronesis*" in Applied Phonetic Social Science as used in internet communication technology for social transformation, those theories are the theoritcal fundation of the research.

The research implements a new method to develop an internet research methodology by utilizing the Internet communication technologies applied in phronetic social science, to expose social change in the middle of conflicts. Moreover develops the phenomenology of Internet Communication Technology in Applied Phronetic Social Science, which can be adopted as an internet research methodology for other internet research, described more in detail in Chap. 2. In Chap. 3, the communication research researches reconciliation in the middle of conflict, methods, practices, and strategies by using ICT in different aspects. In Chap. 4, the research introduced how ICT is considered a haven for social, cultural, and economic transformational change.

In Chap. 5, communication research introduces the Wasatia movement in Palestine as an agent for social change. The case study in this research is the non-governmental non-political Palestinian organization "*Wasatia*" established in

Palestine in January 2007 by Mohammad Dajani Daoudi.[1] The research focuses on how this movement is using Internet Communication Technologies and the social networks to expand its Reconciliation Process to compensate for its lack of funds to promote its moderation values. It has become an essential agent for social transformation advancing the moderation process in the middle of conflict for that purpose.

The study explores the Wasatia Circle by Dajani Daoudi in several of his publications and how It can be exploited and understood using social media and illustrate it within the expansion of big-data. The study describes how to develop a cycle of change using the evolvement of the Wasatia Circle into a Wasatia Cycle using ICT, paving the way for the reconciliation process and thus ushering in conflict resolution and reconciliation processes. Thus this approach is described in more detail in Chap. 5. The study explores online deliberative democracy using online social networks and ICT applications, to within the impact of an online deliberative democracy dispute, this approach is more introduced and explored in Chap. 6.

The research applies "*Netnography*" approach to developing the analytics for the research, which was introduced by Kozinets (2015) and explored in Chap. 7. In addition to ethical conduct in Internet research to be considerd, accredit to the association of internet researchers (AOIR) standards in research which is introduced in the chapter. For analyzing and interpreting the analytics, qualitative content analysis methodologies are introduced depending on methods introduced by Krippendorff, which are more described in Chap. 8. The empirical work for qualitative content analysis the combines the methods with the procedures are demonstrated in an analytical form on the bases of theories and methods of Krippendorff and other experts in the field of qualitative content analysis, they are illustrated in Chap. 9.

The study explores, collects, analyzes, investigates, interprets the behaviors and engagements of participants in online social media and their influence towards social change in the public and community providers, who affects and is affected by a problem of concern within the community. It studies the influence of Internet communication technologies, such as online social networks, for developing an ego-centered social network that evolves to adopt in its core the values of moderation, reconciliation, and democracy as part of the Israeli-Palestinian reconciliation process. It also evaluated the effects of a change of intentionality from hatred and animosity toward more acceptance and change.

The study helps to understand ICT applications as a tool for the advancement toward moderation in Islam, as a tool for reconciliation and online deliberative democracy. The main finding is that ICT is proving to be a useful analytic tool for measuring and examining the advancement of the community towards the Reconciliation Process; also, for researching the potential of online deliberative democracy ingrained in digital social networks in the online community that can develop social change towards the reconciliation process. The findings and results affirm that ICT can foster and bridge the reconciliation process in communications

[1] Mohammed Dajani Daoudi, is the founder of the Wasatia movement in Palestine, also a founder and former director of the American Studies at Alquds University.

between communities in conflict and has the potential to build bridges of understanding between foes as well as partners in peace and to establish peaceful and cooperative relations. Also, ICT can influence social, cultural, and economic transformational change.

The innovation in this research is the creativity in developing internet research methodologies to research the internet and how to analyze "Big-Data"; data develops from the use of the online application for internet researchers. This kind of innovation is new and is still under development. The study introduced a theoretical synthesis to be adopted as a discourse to conduct internet research. The synthesis is extracted from applying Phronetic social science in mixed-method experimental intervention design applied in a Netnography approach and examined by qualitative content analysis methods and procedures.

The empirical work illustrates that it made social change in the online social media; the percentage is quite significant; the social change illustrates that the participants can adapt to change in the middle of the conflict. The perspective to involve participants actively in the research and bringing about change for individuals or communities using internet communication technology is a new tool, and it must be used.

ICT is part of the global human interaction that may help expose ethical virtues, such as trust, empathy, moderation, reconciliation, democracy, coexistence, and it is time to develop online social networks that encourage those virtues in developed countries. The communication research adheres to a standard of academic research methodology design, the theoretical framework was adapted from integrating theories researched and explored by different professors in the field of qualitative data analysis, the empirical work is new and innovative, and this kind of research is still evolving in all disciplines, as the study is an interdisciplinary research.

Today the internet is used as a tool for all kinds of research; this research can be a reference resource for other doctoral students in the field of qualitative data analysis and Internet research methodologies.

Here are some advantages of using Internet research as a tool in applying academic research:

1. The researcher can explore different studies from different academic fields that are related to his own field of inquiry.
2. Data collection is enormous, as the Internet generates datasets of hundreds of thousands of data that can be collected in different academic methods.
3. The Internet is a sustainable system; it can also be easy to research specific phenomena of a wide spectrum of participants.
4. Internet research is ultra-dynamic; data keeps changing according to social environments that support the evolvement of the datasets.
5. No bridges or boundaries, open space for all participants to interact, and no distinctions between participants, but it also can be restricted as required by the researcher.

1.2 Definition of Terms and Concepts

Big-Data All datasets evolved from conducting the research, including the online social network datasets and online materials such as images, videos, posts, articles that are included in the social website (www.wasatia.info) and the Facebook Groups (Wasatia.info & Wasatia Vision) and all kind of observations and behavioral conducted by online participants in the research.

Mixed-Method Research Design This is a research design defined by Creswell "*as a methodology, it involves philosophical norms that guide the direction of the collection and analysis of data and the mixture of qualitative and quantitative approaches in many phases in the research process*" (Creswell 2014, p. 244).

Netnography This term refers to a study conducted on an online community. According to Kozinets (2015),

> *it is a name given to a specific set of related data collection, analysis, ethical representation research practice, where a significant amount of data collected and participant-observational research conducted originates in and manifested through that data shared freely on the Internet, including mobile applications.* (Kozinets 2015, p. 79)

Phronetic Social Science This refers to the philosophical framework developed by Bent Flyvbjerg (2001). It is used to develop a Phronetic social practice that progresses a discourse for researching better practice and practical wisdom of outcomes. It is derived from what Aristotle highlighted as '*Phronesis,*' which is the ability to make well-informed judgments in the specific and often original context (Ess and Jones 2004).

Phronetic Mixed-Method Netnography Approach A discourse of analysis by applying Complex Mixed-Method Experimental Intervention Design and applying the Netnography approach for analysis by applying qualitative content analysis, this approach is in more detail in Chap. 7.

Practice Theory Refers to a new method to research the practice of human conduct, especially the influence of taken-for-granted assumptions (Ritzer 2008); the significant theorists that contributed to this theory are Grafinkel, Bourdieu, Giddens, and Foucault. Bourdieu's Theory of Practice can be identified as a Grand Theory that illustrates the conduct and behaviors of human nature (Walther 2014, p. 1). A Grand Theory is an "*abstract and normative theory of human nature and conduct*" (Skinner 1985, p. 1).

Qualitative Content Analysis According to Krippendorff (2013), the qualitative content analysis is an "*analysis of the manifest and latent content of a body of communicated materials through classifications, tabulations, and evaluation of its keys symbols and themes to ascertain its meaning and probable effect*" (p. 1).

Reconciliation Process In the context of this research, the process is referring to advancement for social change towards moderation, reconciliation, and democracy in the middle of conflict by the use of ICTs for that purpose.

Theoretical Synthesis These are theories that are outside sociology and brought into sociological theory; this type involves a wide range of theories from two or more theories, which involves structural functionalism and symbolic interactionism (Ritzer 2008, p. 72).

Wasatia NGO A Non-Governmental Organization in Palestine, working to expose moderation from a political-religious perspective, in Palestine. The word "*Wasatia*" refers to Moderation in the Arabic Language.

Wasatia Cycle The correlation developed from the influence of ICTs on the Wasatia Circle explored in Chap. 5. They cycle illustrate a Phronetic approach towards applying social change more for moderation, reconciliation, and democracy towards a cycle of transition, explored in detail in Chap. 5.

The social networks are rapidly becoming an indispensable part of the everyday lives of a majority of people around the globe. People perform various political, social, economic, and religious activities on the Internet and the so-called online social networks such as Facebook, Twitter, You-Tube. The research presents an overview of the empirical study evaluating the impact of social networks and online forums on adherence to the reconciliation process in the middle of the conflict.

The Reconciliation process is related to ethical values such as peace, truth, justice, love, trust, empathy, and the investigation of the moral issues in practices that dignify the process toward social change towards moderation, reconciliation, and democracy. Therefore, the communication research is concerned with the analysis of particular moral issues in private and public spectrum, using Internet communication technologies for that purpose. Therefore, we can consider that the reconciliation process is part of applied ethics, combining the use of the datasets generated from offline & online interactions to explore, collect, code, analyze and interpret the social change from utilizing ICT in exposing the reconciliation process. As part of the state of the art, I signify this scientific method as **Applied Ethics in Big Data**.

The Reconciliation Process is essential for social change that inhibits the peaceful development relations between entities in prolonged conflicts. Social change requires enormous efforts to overcome the lack of charismatic leadership, religious bigotry, backward cultural habits, and dire economic conditions, in a protracted conflict; all of these are insurmountable obstacles perpetuating the conflict. Consequently, the research studies how to overcome present obstacles with a stagnated conflict and pave the way towards reconciliation.

1.2.1 The Communication Research

The research develops an interdisciplinary scientific approach to study whether internet communication technology (ICT) can initiate social change to impact the reconciliation process in the Palestinian-Israeli conflict. It combines different disciplines such as philosophy, sociology, religion, political science, communication science, information internet technologies, and information sciences. Therefore, the research explores the hypothesis that Internet Communication Technology (ICT) can advance social change toward moderation, reconciliation, peace, and democracy in the midst of a protracted conflict. The research also aims to bring a better understanding of ICT as a tool for the advancement toward Reconciliation Process in the Palestinian community. It investigates how to create a culture of moderation, reconciliation, and democracy starting bottom-up approach by using the new advancement in social communication networks to push social change toward a reconciliation process. It argues that using digital social networks as a tool would bring better understanding and more empathy for the human factor to help commit the two nations that are in conflict with the reconciliation process, developing a culture of peace.

The research utilizes different theoretical approaches, such as the Applied Phronetic Social Science. It introduces different concepts to advance to the reconciliation process using ICT. One such concept is the "*Hölderlin Perspective*" accredit to Leiner (Leiner and Flämig 2012, p. 16). Another is the Aristotelian concept "*Phronesis*" in Applied Phonetic Social Science for Flyvbjerg, Landman, and Schram. It introduces the phenomenology of Internet Communication Technology in applied Phronetic Social Science as a methodology for internet communication research. The study also introduces the religious "Wasatia" moderation concept developed by Dajani Daoudi (Barakat 2017).[2] This research describes the evolvement of the "Wasatia" Circle[3] of moderation to reconciliation to democracy by using internet communication technology and develops the "Wasatia Cycle," that would evolve the community to social change towards the reconciliation process.

As for the methodological aspects, the communication research constructed theoretical synthesis by integrating several types methodological traditions, which are the Mixed-Method Experimental Intervention Research Design for Creswell and Clark, (Creswell and Plano Clark 2018), and "*Netnography*" as introduced by Kozinets (2015) and deploying the Qualitative Content Analysis methodologies as conducted by Krippendorff and other experts in the field of qualitative content analysis, to explore, collect, analyze, interpret and investigate the analytics. The empirical work was conducted on the non-governmental organization called "*Wasatia*" established in Palestine in 2007. It explores how Wasatia utilizes the online social networks in advancing the cause of reconciliation in the midst of conflict through

---

[2] Barakat, Z. (2017). From Heart of Stone to Heart of Flesh: Evolutionary Journey from Extremism to Moderation. München: Herbert Utz Verlag GmbH.

[3] Wasatia: A term used to identify Moderation in Islamic theology.

the promotion of moderation to affect social change using internet communication technologies. It researches how Wasatia uses the Internet technology systems, such as Facebook and its online social network, to compensate for its shortage of funds in calling upon both people to adopt its core values of moderation, reconciliation, and peace to put an end to the Israeli-Palestinian conflict.

In conclusion, communication research explores some concepts and methodologies for a more comprehensive approach to online ICTs for social transformation and social change. It includes procedures for the analysis of analytics of online social networks and investigates different results from the use of analytics.

So far, the reconciliation process was conducted through a top-bottom approach of political leaders reluctant to make the sacrifices needed to reach a peace agreement between the two communities involved in the conflict. It concludes that peace is more achievable, adopting a bottom-up approach than a top-bottom traditional political leadership approach, with the use of ICTs for this purpose.

The current study reached significant results and improved several concepts to be adopted, that are of high importance for social change in the middle of the conflict. First, the online social networks do have the potential to promote mutual trust, mutual recognition, acknowledgment, economic, political relations, and respect for the sensitivity and consideration of the other and of the enemy. Second, moderation, reconciliation, and democracy are essential elements vital for peacebuilding where uninhabited digital social media can play a significant role in paving the way for the reconciliation process. Third, utilizing online social networks is a powerful tool to spread the culture of moderation, reconciliation, and democracy in a community suffering from protracted conflict to a community willing to reconcile using the bottom-up approach. Forth, ICT can play a significant role in online deliberative democracy as a useful analytic tool for promoting the advancement of the community towards the reconciliation process within the use of ICTs.

Several concepts were introduced as a result of conducting the communication research, such as the phenomenology of ICT in applied Phronetic social science, ICT can be a tool for social transformation, the transformation of Wasatia Circle into the Wasatia Cycle by the use of ICTs to empower social change. This communication research introduces the applications of ICTs in exposing and implementing online deliberative democracy. In addition to other concepts, such as moderation (Dajani 2009), ICT for the development of cultural, social, and economic capital (Bekerman and Horenzsyk 2009; Brantmeier 2009; Lenhardt and Nicholsen 1990); deliberative democracy (Elster 1998; Gutmann and Thompson 2004), and deliberative polling (Fishkin and Habermas 1995; Luskin 2003). Furthermore, the theories of deliberative democracy and online deliberative democracy are used to explain how the social networks became the outlet for people to express their views freely, including other theories that advance the methods for internet research.

The following practiced methods were applied: netnography for studying communities online (Basset and O'Riordan 2002; Kozinets 2015), qualitative content analysis methods for qualitative data analysis with NVivo (Bazeley and Jackson 2013; Kohlbacher 2012; Penna 2013; Saldana 2013a, b; Sandelowski 2000; Schreier 2012; Schulz 2012), and methods for qualitative content analysis (Krippendorff 2013). The research aimed to explore and inquire data by applying qualitative content analysis for collecting datasets to examine the primary hypothesis and measure social changes using the interpretation provided by the analytics results. The communication research developed the theoretical synthesis of applying mix-method Netnography and Qualitative contact analysis for online social network analysis.

Keywords Applied Ethics, Big-Data, Internet Communication Technologies (ICT), "Hölderlin Perspective" , "Wasatia" in Palestine, Phronetic Social Science, Applied Phronesis, Mixed-Method Research Design, Conflict Transformation, (Wasatia)-Moderation, Reconciliation, ICT for reconciliation, E-Democracy, Online Deliberative Democracy, Netnography, Qualitative Internet Research, Content Analysis, Qualitative Content Analysis (QLCA), Quantitative Content Analysis (QNCA). Complex Mixed-Method Explementary – Intervention Research Design, Phronetic Mixed-Method Netnography Approach

1.2.2 Audience

This study was intended to help students and researchers in the field of social and behavioral sciences, with some knowledge of research design methods and some Internet technological information background in communication science. The study is sought to help researchers implement the mixed-method experimental-intervention research design and deploy qualitative-quantitative traditions for an online social network adhering to the Netnographic approach.

The research was intended for Netnographers, researchers who have wanted to conduct Internet communication research that implements the mixed-method approach applied to a netnography framework. Moreover, it is for researchers who wanted to deploy a social structure for netnography design and apply social network analysis, from Facebook, YouTube, and datasets that develops "*Big-Data*," utilizing qualitative-quantitative content analysis as the method of analysis and interpretation to develop a more understating for the changing phenomenon.

1.2.3 The State of the Art

The research is applied amidst a stagnated conflict in the Middle East between Israelis and Palestinians; this conflict has detained people with hatred and animosity between two nations for over 60 years. One nation does not believe in the existence of the other or creates propaganda for having a rightful claim over the other. This kind of protracted conflict existed due to the complex abhorrence for the narrative of the other.

The research is an activist for reconciliation and an expert in ICT for over 15 years, he was intrigued to develop ICT platforms, which can bridge the divide of hatred between two nations facing a deteriorated conflict. Also, thought of developing a community that adheres to a mutual understanding and acceptance of the narrative of the other and is willing to coexist and acknowledge and recognize one another and tries to develop an innovative way to develop a culture of peace. At the start, the researcher studied concepts for social change toward the reconciliation process in the middle of conflict using ICT platforms. For that reason, he explored theories and concepts that advanced the use of ICT for understanding, discovering, and exploring the notions of moderation, reconciliation, and democracy and research that ICT can develop social change towards social transformation.

he studied the integration of theoretical syntheses of different approaches. Such approaches include the Phronetic approach, mixed-method research design, netnography, and qualitative content analysis. The empirical work consists of developing an online social network site and publishing materials online on the social website, to boost posts in the Facebook pages to integrate interaction and participation into the social network site, record impressions and behaviors as well as to develop analytics of shared views, comments, likes, and posts from visitors, participants, and members. he tried to encourage participants to acknowledge moderation and reconciliation as the path to be taken to develop a peaceful culture. The empirical work involved two Facebook pages: Wasatia.info, Wasatia Vision, and http://www.wasatia.info, an online social network website.

One of the core frameworks for the research was to identify and review other works completed on those different theories, concepts, and approaches to develop a skeptical understanding for the use of those methods in various studies. he wanted to advance more understanding of the innovation and the contribution that ICT could achieve from the integration of those approaches in the field of online social networks. The opportunity was in demonstrating ways in which the work extended research and researchers, whether other research was empirically examined, and drew implications for practice and a future perspective by theorizing methods and procedures. The empirical work included the study of the innovations of other researchers, which were emphasized in the review section in more detail. Therefore, the research and the empirical work attempted to address whether ICT could adhere to the values of moderation, reconciliation, and democracy in the middle of the conflict, influences from an online community. The study examined if the impact of social change towards the reconciliation process was possible and attainable, then the study examined the ambivalence of the content that resulted from utilizing ICT amid conflicts.

he refers to the empirical work for the study as Internet communication research (ICT-MR). The study explores the integration of different approaches, as mentioned above, to examine the hypothesis in question; ICT can pave the way for social change towards reconciliation process (moderation, reconciliation, and democracy) in the middle of the conflict. Because ICT-MR applied to interdisciplinary research, the approaches were presented and researched separately. For example, the Figure below indicates the rational model applying the discourse for implementing the integration of three theoretical syntheses, which are explained thoroughly in Chap. 2. This kind of integration can assist in analyzing and exploring online social networks and is referred to as the Phronetic Mixed-Method netnography approach, as discussed in Chap. 7 (Fig. 1.1).

However, he researched different concepts and theories, such as moderation in Islam, reconciliation, e-democracy, deliberative democracy, and online deliberative

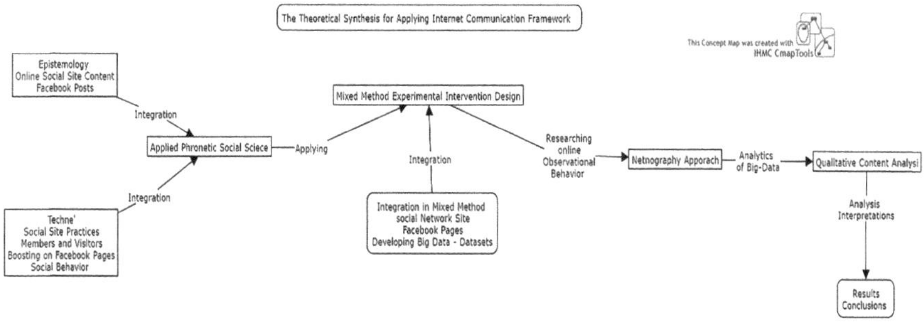

Fig. 1.1 Rational model for the communication research discourse design ICT-MR

democracy. Those concepts were broad in various studies, and these did not include the innovation that ICT-MR contributed to the research. For that reason, these were not covered in the State of the Art, as these were also widely researched in Chaps. 5 and 6. Those theories and concepts were defined according to different scholars and experts in the field of moderation, reconciliation, and deliberative democracy in the research work. The order is according to their contribution in the consequences of chapters, (Flyvbjerg 2001; Flyvbjerg et al. 2012; Creswell 2014; Creswell and Plano Clark 2018; Dajani 2014; Gutmann and Thompson 2009; O'Malley et al. 2017; Kozinets 2015; Krippendorff 2013; Leiner and Flämig 2012; Schreier 2012). Those scholars are the leading academics that the research evolved from in the review of pertinent literature for their research with their co-researchers.

The primary objective of the Internet communication research (ICT-MR) was to encounter the transformation from the intentionality of hate and animosity to the intentionality of moderation, reconciliation, and democracy within the Palestinian community, which encourages social change toward reconciliation process. This ICT research also explored ways in which social networks were used to break psychological barriers of fear by helping ordinary citizens to connect and share information and data to build confidence, trust, and hope toward understanding moderation and reconciliation in the midst of conflict. **The hypothesis, also researchers whether ICT can facilitate ways for healing and restoring the physical, psychological, cultural, and social psychological damage that occurs in conflicts, whether these are direct violence or indirect violence,[4] advanced in the Israeli-Palestinian conflict**. Hence, the empirical work adapted a practical research plan based on (24 months) of experience and testing of an ICT platform. The expectation from the findings of the Internet communication research was that ICT could be efficiently directed toward the reconciliation process under conditions or several conditions, as applied through the online social network communication technologies, which demonstrated proof of the hypothesis in question.

This ICT research observed the stimulation that Internet communication research had for attempting to change society toward moderation, reconciliation, and

[4] cf. **Vrasidas et al.** 2009, **p. 233.**

democracy in an online Palestinian community. This kind of empirical research was called *Netnography or ethnography on the Internet is a "new research design that adapts ethnographic research techniques to study of culture and communities emerging through computer-mediated communication"* (Kozinets 2002, *p. 2). Based on this,* he illustrated that others in the community and on the other side of the wall yearned for peaceful coexistence, economic prosperity, social tranquility, and reconciliation. The purpose of the ICT research was to understand the role of new platforms, as well as to explore and discover ways in which the use of ICT could have a part in the transformation process of changing citizens to inspire social change towards healing the bitterness in the community.

The research was innovative and new; he aslo found it challenging to seek past studies that used the interdisciplinary approach, as the scope of each study varied within each discipline. For that reason, he applied theoretical synthesis for combining and researching each discipline and its correlations between one scope of discipline with the other and integrate these to achieve the best outcome that could explore and examine the hypothesis, as explored in the rest of the communication research.

1.2.4 Identifying and Researching the Relevant Literature

This section includes developing a literature map with relevant literature explored and studied in the research, such as books, journals, and articles for present studies and past research, on ICT that relate to the central theme of the research.

1.2.4.1 Literature-Map

The literature map contains books, journals, and articles, which relate to the hypothesis, theories, and methodology used in the research. The literature map has two categories (see Fig. 1.2):

1. *Literature books*: These are based on the topic of hypothesis and theories, applied Phronesis in Mixed-Method research design, Netnography in social science, and analytics used in the qualitative content analysis.
2. *Journals and articles*: These are on the topic of hypothesis and theories, qualitative content analysis, Mixed-Method research, Phronetic research plan, and the Mixed-Method research design.

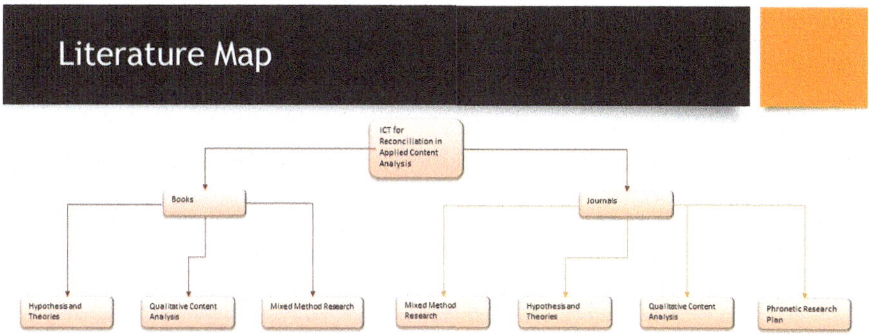

Fig. 1.2 Literature map sample research

1.2.4.2 Relevant Literature – Books

Some of the literature used for the research that relates to the topic of the hypothesis and theories adopted in the research are illustrated in this section. Flyvbjerg (2001) developed the Phronetic approach, and he aided colleagues in developing applied phronesis (Flyvbjerg et al. 2012). Also studied the concept of applied phronesis more clearly in his next two books. Green (2007) studied mixed-method design to describe the inquiry of when to implement the mixed-method approach. Creswell and Clark's (2011) book was about designing and conducting mixed-method research. Creswell's (2014) book was famous for its research approaches; the book indicated methods for qualitative, quantitative research approaches, and the mixed-method research approach. This ICT-MR study applied the complex experimental intervention mixed-method research design (Creswell and Clark 2011) as an approach to examining the research through the sequential explanatory multi-phase mixed-method, as defined by Creswell (2014).

Some literature about ICT and its role in the development of society, education, and endorsement for reconciliation included Pool (1983), Lenhardt and Nicholsen (1990), Couldry (2012), Vrasidas et al. (2009). Researchers have explored the endorsement of ICT for education and reconciliation in conflict areas (Vrasidas et al. 2009). Studies of different cases of implementing ICT for reconciliation in the world, such as ways in which ICT could aid education, justice, and the development of the society (Nussbaum 2012), to explain what school principles and leaders need to know about digital technologies and social media to aid in education development in schools (Cuban 2003; Gastil and Levine 2005a, b). The researchers have strived to foster deliberative democracy and implement it within the society, as well as having conducted some case studies in cities in the United States using ICT to apply deliberative democracy using online applications (Cuban 2003; Gastil and Levine 2005a, b).

I also researched literature on the concept of Wasatia (Dajani 2007) and about Wasatia as a theory and a vision for the future (Dajani 2010), which was written in the Arabic language. Dajani (2014) explained Wasatia as a concept of centrism and moderation in Islam in the English language, as well. In addition, research was completed concerning Dajani's (2014) work on moderation which is also funded by the Heart of Flesh Not Stone Project.

Leiner and Flämig discussed the concepts for reconciliation (2012, p. 16), where they explained the *Hölderlin Perspective* derived from a notion written by the German poet Friedrich Hölderlin (1770–1843). Hölderlin wrote in his last section of his novel Hyperion, "Versöh*nung is mitten in Streit und alles Getrennte findet sich wieder*" (p. 760). In English, it is translated as "*reconciliation is in the middle of the strife and all that was separated finds each other again.*"

Researching some deliberative democracy literature included Cronin (2008), who studied concepts of deliberative democracy and how it should be implemented between *naturalism* and *religion*; Cohen (1998) explored democracy and *liberty in a deliberative democracy*; Fishkin (2009), who explained deliberative democracy and how to implement deliberative polling; and Gutmann and Thompson (2004), who described and defined deliberative democracy with all its characteristics and principles. The book described kinds of democracy and illustrated why deliberative democracy was a unique method to implement in society. More literature that aided the research such as, Kozinets (2015), redefined the Netnography approach to explain concepts, principles, and guidelines to conduct a Netnography approach to any designated research. Boellstrorff et al. (2012) wrote a useful handbook for researchers who wanted to study communities online.

Researching analytics qualitative content analysis literature included Bernhard's (1952) content analysis in communicative research; Sandelowski (2000), Schulz (2012), Schreier (2012), who described the procedure of coding and illustrates examples for types of conducting for content analysis procedure; and Kohlbacher (2012), who studied using qualitative content analysis in case study research. I also studied Saldana (2013a, b), Penna (2013), Bazeley and Jackson (2013), and Krippendorff (2013). Figure 1.3 illustrates how the literature map was developed with some of the books used for the research on the topics of qualitative-quantitative content analysis, hypothesis, theories, and Mixed-Method research design that relates to the central theme of the research.

Figure 1.4 illustrates a literature map for some of the books used for research on the subject of mixed-method Intervention Experimental research design.

1.2.4.3 Relevant Journals and Articles

The literature map that was developed for journals contains some of the journals on the themes of the hypothesis and research questions of the communication research. The journal topics are related to the Phronetic research plan, mixed-method research, hypothesis and theories, netnography, and qualitative content analysis. Journals

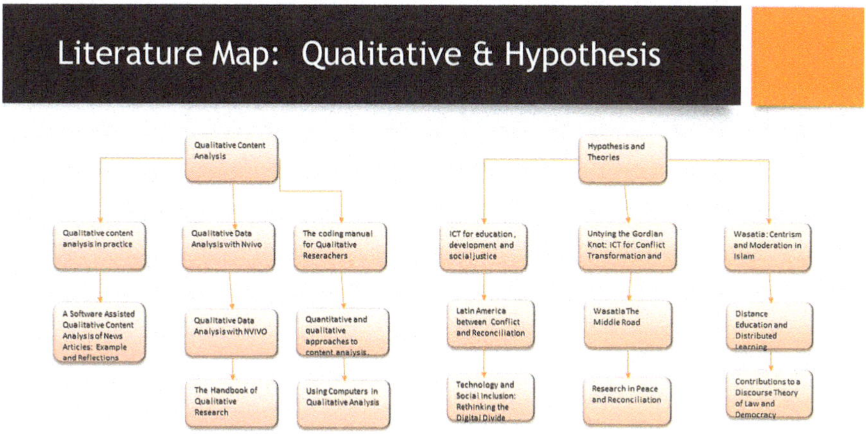

Fig. 1.3 Literature map for researching the hypothesis

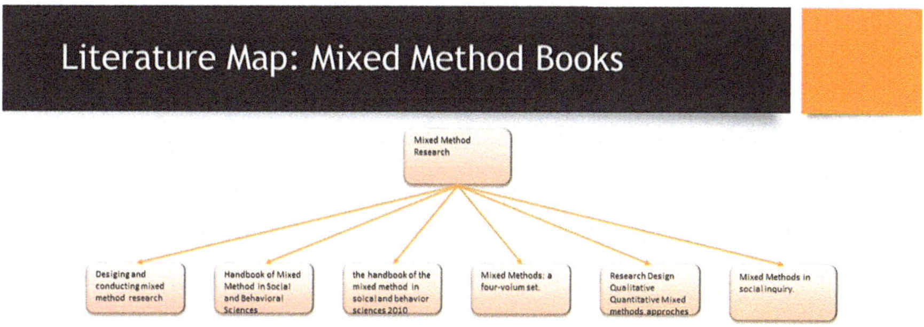

Fig. 1.4 Literature map for researching mixed-method Intervention Experimental Design

were about Mixed-Method research, quality and quantity, field methods, and multiple research approaches. In addition, numerous published research studies in social and human sciences used the mixed-method approach, such as occupational therapy (Lysack et al. 1994), interpersonal communication in studying the use of emails for personal relationships (Boneva et al. 2001), AIDS prevention (Jentz and Zimmerman 1996), and dementia caregiving (Weitzman and Levkoff 2000). In addition, journals were about the implementation of Kozinets (2002) version of netnography and Flyvbjerg (2006). In addition, some journals described the topic of the hypothesis of the research (Botes 2001; Hattotuwa 2004; Jorgic 2013; Karefe

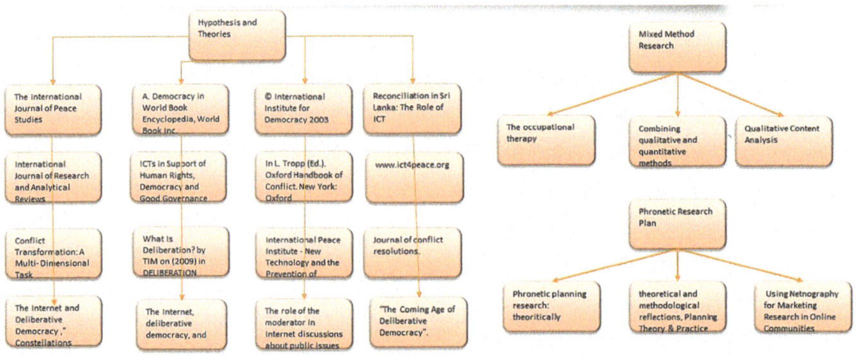

Fig. 1.5 Literature map for researching the hypothesis research journals

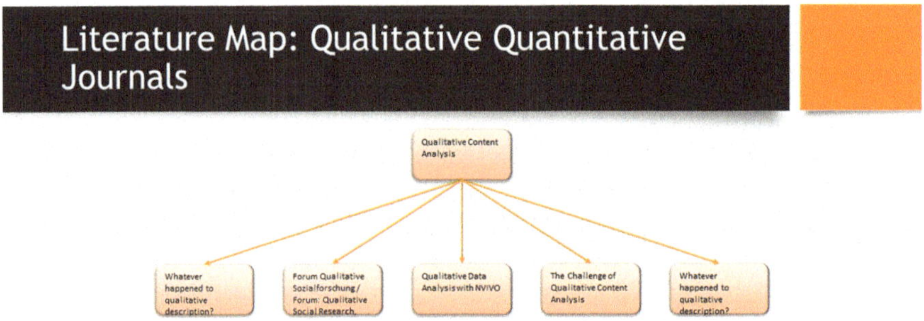

Fig. 1.6 Literature map for researching the qualitative, quantitative procedures and methods journals

2012; Kozinets 2002; Larrauri 2013; Miall 2003; Selian 2002; UNDP 2008). Other journals included Gibson (2006), Perera (2015), Cole and Crawford (2007), Larrauri (2013), van der Meer (2016), Boneva et al. (2001), Dangwal and Srivastava (2015), Leiner (2015a, b). Figure 1.5 illustrates some of the journals used for this ICT.

Figure 1.6 illustrates some of the journals used for qualitative content analysis.

1.2.5 Chapter Synopses

Having introduced the research project, the figure below serves to build up the theoretical foundations for the Internet communication research design (Fig. 1.7).

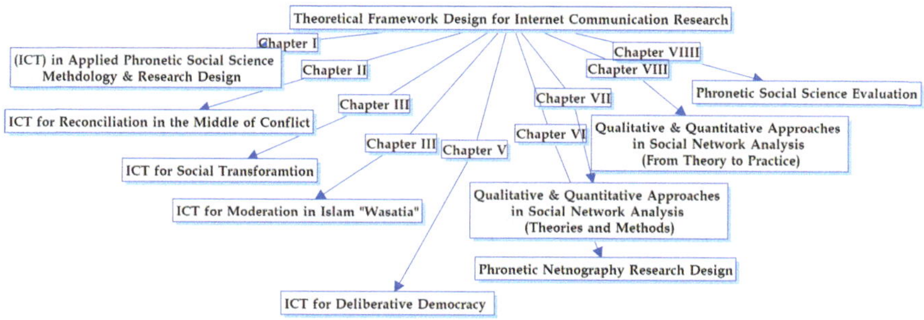

Fig. 1.7 Internet Communication Research Design for ICT-MR

1.2.5.1 Chapter 2: The Phenomenology of Internet Communication Technology (ICT) in Phronetic Social Science (Methodology and Research Design)

This chapter introduces some new innovative approaches for conducting Phronetic social science in Internet communication technologies. The chapter explores the methodological approach to conduct Internet communication research (ICT-MR). The chapter illustrates how to **apply the mixed-method experimental-intervention research design, according to** (Creswell and Clark 2018), in Phronetic social science. The chapter presents how to conduct Phronetic social science research by using some applications for practical judgment and reasoning, attentive to the specific context in conducting the research; this is what "*Aristotle*" identified as "*Phronesis.*" (Kozinets 2015). The chapter explores a **new phenomenology for applying ICTs for social change adopted in Applied Phronetic social science** and introduces it as a theoretical framework for the research. In addition, the chapter illustrates a new methodological theory to be adapted to conduct Internet communication research and refers to it as the **Applied Phronetic Mixed-Method Experimental Intervention research design**. The theoretical framework and the methodological aspects of the communication research are evolved from the theories and methodologies researched by experts in the field, such as Flyvbjerg, Landman, Schram, Creswell, and Clark. Finally, the chapter illustrates an epistemological theory behind planning, conducting, validating, analyzing, and interpreting the Internet communication research.

1.2.5.2 Chapter 3: Reconciliation in the Middle of Conflict Exposed in Internet Communication Technology (ICT)

This chapter introduces the influence of ICT as a tool for the advancement of reconciliation and a tool for social change. It also explores the ambiguous use of ICT as a means for hatred and animosity to an instrument of reconciliation and moderation. The chapter researches the aid of ICT for the evolvement of reconciliation in the middle of the conflict, and it introduces a brief, straightforward introduction to the notion for the importance **of why ICT can be used for reconciliation in the middle of the conflict**.

Furthermore, the chapter introduces methods for reconciliation from experts in the field, and **explores ICT for reconciliation, methods for reconciliation, ICT strategies for reconciliation, reconciliation as a term in the Arabic language**.

1.2.5.3 Chapter 4: Internet Communication Technology (ICT) for Social Change

This chapter explores the hypotheses for the research, as well as describes philosophies and theories for implementing Internet communication technologies as these are being adopted into societies to affect transformation for social change. The chapter researches the influence of ICT on culture, society, and the economic capitals of the society. Moreover, it demonstrates the effect of ICT platforms on the characteristics for the adaptation of transformation and accentuates more on the role of ICT in the advancement for moderation, reconciliation, and democracy in the Internet communication research.

The chapter introduces **ICT in transforming content, structure, profiles, groups,** and the transformation of underlying issues in societies towards social capital changes. Also, the chapter explores the ambiguous use of ICT in developing societies, such as Palestine, and the disadvantages of the use of ICTs.

1.2.5.4 Chapter 5: Internet Communication Technology (ICT) for Moderation "*Wasatia*" as Concept and a Movement in Palestine

The chapter describes research done on "*Wasatia*" as a concept and a movement in Palestine. It uses terminologies outlined by the NGO Wasatia in Palestine founded in 2007, under the supervision of Dajani Daoudi,[5] who is a scholar and an advocator for moderation in Islamic studies, he is also the founder of the **Wasatia Movement** in Palestine. Also, the chapter explores in depth the role of ICTs in the evolvement of the **Wasatia Circle**[6] for advancing social change in the middle of the conflict, the evolving term is identified as the **Wasatia Cycle.**

[5] Professor Mohammad Dajani Daoudi, is the founder of Wasatia Movement in Palestine in 2007and the Former Dirctor of the American Studies Institute at Al-Quds University.

[6] The term is form a research done by Barakat, Z. (2017). From Heart of Stone to Heart of Flesh: Evolutionary Journey from Extremism to Moderation. München: Herbert Utz Verlag GmbH. That defines Wasatia Circl as a transfromation from reconciliation to moderation that leads to democracy.

1.2.5.5 Chapter 6: Internet Communication Technology (ICT) for Deliberative Democracy

This chapter explores the definitions of **deliberative democracy and online deliberative democracy**. The chapter describes the role of ICTs in the advancement of democracy in societies developing what is referred to as e-government to understand and implement democratic principles as part of the social life in societies. The chapter also introduces deliberative democracy with the full nation for its characteristics and its utilization in different aspects that includes cultural social and economic capitals. The chapter also highlights the use of ICT as a tool to foster the process of deliberative democracy and evolve deliberation to be an online deliberative democracy. In addition, it explores the advantages and disadvantages of using deliberative democracy over the Internet. Finally, the chapter introduces a case study of an experimental study to conduct an online deliberative democracy. The experimental study was conducted by Pietro Speroni di Fenizio (Coimbra University of Portugal), Alois Paulin (Faculty of Information Systems, Novo Mesto, Slovenia), and Cyril Velikanov (foundation Euractiv/Politech, Brussels). The chapter introduces ICT-MR for the online deliberative process and the advancement of peacebuilding from utilizing ICTs. The chapter also introduces e-democracy and its deployment for e-government and e-services and illustrates how ICT can play a role in the online democratic process.

1.2.5.6 Chapter 7: Phronetic Netnography Research Design

This chapter explores an online ethnography **using netnography approach** for the social network site developed in the empirical work. The online social network site is http://www.wasatia.info. This type of practice was introduced by Robert Kozinets (2015). The chapter discusses the development of an online social network site for Internet communication research ICT-MR. The empirical work included the developing of the following online website and social pages, and they are **http://www.wasatia.info and two Facebook pages: Wasatia.info and Wasatia Vision**. The chapter conducts an inspection and explores strategies for learning about the effect of ICT on participants' behaviors that are active in the social network and describes the development of an online community in conflict. The chapter explores and examines the exposure for the reconciliation process in the online community developed from the netnography approach. The chapter introduces the development of an ego-centered social network in Palestinian society. Finally, it explains in detail the new term, **Phronetic Mixed-Method netnography approach**, with traditions and methods of the research using the netnography approach.

1.2.5.7 Chapter 8: Internet Research Analytics (Theories and Methods)

This chapter researches qualitative and **quantitative content analysis** and develops new methods and procedures to analyze Internet communication research in theory. The aim is to research how to explore, examine, and analyze communities that are online from theories developed through the discourse of the study. The chapter illustrates different ways to develop empirical datasets, which eventually examines the hypothesis of this study.

The chapter describes in-depth and sequentially ways in which the methods and theories of content analysis procedures are applied to online social networks. The chapter elaborates on the content analysis framework for the research components and introduces the computer assistant software for qualitative analysis **(CAQCAS) NVivo for Windows**. Also, explains the qualitative content analysis procedure for mixed-methods experimental design for analytics. The chapter also illustrates the Internet communication research framework design in qualitative content analysis, and presents a combinations for work done by major professors in the field of qualitative content analysis, **such as** Sandelowski (2000), Schulz (2012), Schreier (2012), Kohlbacher (2012), Saldana (2013a, b), Penna (2013), Bazeley and Jackson (2013), and Krippendorff (2013).

1.2.5.8 Chapter 9: Applied Internet Research Analytics, from Theory to Practice

The chapter explores hands-on practical methods of qualitative and quantitative content analysis of online social networks. This chapter describes the procedures for developing analytics for social networks in practice. It includes the practical qualitative content analysis techniques for the online social network site, http://www.wasatia.info, and the two Facebook pages in the communication research. It illustrates the method for exploring, collecting quantitative and qualitative datasets, analyzing qualitative and quantitative data, and interpreting content from the online social network site (http://www.wasatia.info) and the datasets from two Facebook pages: Wasatia.info and Wasatia Vision.

The chapter illustrates using (CAQCAS) which is NVivo 10 for Windows as one of the tools to apply the technique and methods to explore the analytics introduced from the datasets, thereby used for interpreting the results of the Internet communication research using methods from researchers (e.g., Krippendorff 2013), including other professors who are experts in conducting analyses.

1.2.5.9 Chapter 10: Phronetic Social Science Evaluation

The chapter introduces the evaluation method from conducting the Phronetic social science in Internet communication research. Describing the aims and purpose and the philosophical and theoretical foundation in the research. The chapter introduces a new innovative process referred to it as **ICT Mechanism in the reconciliation process**. The chapter shows how the social network analysis methodology process was conducted in the research, also exploring the mixed-method netnography integration. It also illustrates the experimental discussion phase. The chapter introduces the consequence of conducting internet research, referring to **it as a synthetic**

theory approach. The chapter also introduces the empirical work conducted and illustrates results from conducting the internet research ICT-MR.

1.2.5.10 Chapter 11: Research Findings and Recommendations

The chapter introduces the conclusion from conducting internet communication research and concludes the role of ICT for social change towards the reconciliation process in the middle of the conflict. Also, introduces the experimental research findings, research questions, and answers, and a summary of each chapter. It also introduces a further perspective for a study to be explored by different researchers and describes how ICT develops social change towards the reconciliation process in a different aspect of the social conflict. The chapter also includes conclusion remarks and exploration for applied ICT in Phronetic Social Science and other future perspectives in conducting Applied Phronetic Social science in different studies.

1.2.6 Reviewing Related Literature with an Individual Reflection

This section reviews specific literature used in conducting the communication research. Also, the section includes whether the literature reviewed strengthens the work being done, as well as the gaps between studies, including the good and the bad. In addition, it introduces reflection to discuss differences between methods approached in the literature, which was used as a scientific source for conducting the communication research.

1.2.7 Alternative Approaches in Conflict Resolution

Leiner & Schließer (eds.) (2018), the book introduces different approaches to conflict resolution and tackles a new approach of mechanisms towards reconciliation process, from experts in the field of conflict resolution and reconciliation. The authors describe their experience in applying reconciliation in different disciplines and introduce their experiences and findings. It introduces four distinct approaches to conflict resolution, respectively; negotiation, religion and gender, reconciliation and forgiveness, and the arts. In addition, it is an excellent reference for understanding conflict resolution and the transformation towards reconciliation. Also, it explores the reconciliation process as a definition, a process, and an outcome. The new mechanism that is missing in the book, the approach of using ICT for applying reconciliation mechanisms in the middle of the conflict, and that's what would be illustrated in the communication research.

1.2.8 Applied Phronesis, Real Social Science

Flyvbjerg, B., Landman, T., and Schram, S. (eds) (2012), the book presents a new, hands-on approach to the social inquiry to implement mega projects. The theoretical and methodological ideas behind the book, inspired by Aristotelian phronesis, represent an original perspective within the social sciences and to empower the research design to implement forever. The book illustrates different examples that describe what applied phronesis looks like in practice. The reflexive analysis of values and power gives new meaning to the impact of research on policy and practice. This book will benefit scholars, researchers, and students who want to make a difference in practice.

1.2.9 Asia-Pacific Between Conflict and Reconciliation

Tolliday, Palme, and Kim (eds.) (2016) discussed the Asian-Pacific and East Asian conflicts by introducing concepts and strategies for dealing with the past. They expressed a practice of reconciliation to build a future based on peace rather than conflict. They addressed reconciliation efforts in different case studies and areas of the world. They included comparative case studies analysis from the Asia-Pacific, the United States, and Germany.

1.2.10 Content Analysis: An Introduction to Its Methodology

Krippendorff (2013) illustrated the history of the evolution of content analysis, which was considered an extensive source for understanding content analysis. Also, the book aided in the understanding of the methodology for conducting qualitative content analysis for the research. Krippendorff provided definitions, methods, and procedures for QCA in theory and in practice that helped me conduct the research. The researcher also studied various methods theoretically and illustrated these visually to the researcher, which helped researchers understand different theories for different methods.

I used the book to develop new procedures to conduct communication research. Also, it aided the theoretical practice in the research, as adapted in Chaps. 8 and 9. Also, it introduced the computer assistant qualitative content analysis software, CAQCAS NVivo, which became the depended software to use in the research for conducting QCA procedure. The book was considered one of the significant assets for conducting ICT-MR research in theory and practice.

1.2.11 Classical Sociological Theory

Ritzer (2008) provided a rich, clear understanding of philosophical, theoretical work; also, Ritzer provided a major classical overview for theorists and their sociological thought. One of the essential objectives of this book was that Ritzer presented theorists with their biographical sketches; their theories were presented in a historical and intellectual context. The book was inspiring, as it defined critical theories in sociology, two of which were necessary for this research, which involved theoretical synthesis and the practice theory. These two theories showed the philosophical, sociological theory for the empirical work done with a sociological background. This book was suitable only for scholars who wanted to understand the sociological theory. However, no empirical work was completed as part of this book.

1.2.12 Communication Rights and Democratic Media Roles

Cammaert and Carpentier (2007) illustrated ways in which the media could play a role in democratizing citizens, using technologies, and gave a thorough understanding of media influencing political agendas by exemplifying the influence of the media in a democracy. The primary hypothesis of this book was whether the contemporary media conjuncture offered enough to democracies to evolve in societies. Also, Cammaerts and Carpentier elaborated on how democratic investment could be deepened within society by using media and how communication rights could be expanded using technologies.

Cammaert and Carpentier (2007) included four directions in the areas of a structure for the opportunities of democratizing society using the media. The first section was devoted to citizenship and the public spheres, elaborating on the general theme of communication rights. The second section elaborated further on a notion central to communication rights, namely that of participation of participants. The third section returned to the traditional representational role regarding democracy and citizenship, scrutinizing and criticizing the democratic efforts of contemporary journalism that democracies in society should protect as a power. The fourth section moved outside of the traditional media system and dealt with the diversity of media and communication strategies of activists, such as using media tools for the democratization of the society.

Cammaert and Carpentier (2007) did not discuss the notion of the type of democracy that influenced society by utilizing media as a tool for that purpose. Cammaerts and Carpentier strengthened the concept of the hypothesis in question, but this ICT-MR illustrated more on the type of democracy influencing the society by using ICT as a tool for democratizing society. ICT-MR brought more understanding about the kind of democracy used in ICT, illustrated it as deliberative democracy, and elaborated on different ways to influence society using this kind of democracy, as illustrated in Chap. 6.

1.2.13 Digital Media Ethics

Ess and Jones (2004) discussed the central ethical issues of digital media, from computers and the Internet to mobile phones. Ess discussed it from a global perspective, which was also useful for readers who were interested in learning new ways of thinking about principles and values that guided Internet interactions and research. Ess also acquired newly global perspectives issues of digital media, such as privacy, copyright, pornography, violence, and the ethics of cross-cultural communication online. As part of conducting the ICT-MR, the research applied an ethical consideration, which was described and defined by Ess as consequentialist ethics, utilitarianism, and deontological ethics. Ess did not present how that ethical consideration was implied on digital media. The ICT-MR considered the ethical review and applied consequentialist ethical conduct by the association of Internet research regulations, as discussed in Chap. 7. Ess and Jones (2004) also described the capacity of ethical reflections from what Socrates and Aristotle defined as phronesis: "*It is the ability to make a well-informed judgment in specific and often novel context*" (p. xiv).

1.2.14 From Conflict Resolution to Reconciliation

Bar-Siman Tov (ed.) (2004) provided one of the best reference books to understand reconciliation as a method, a term, and a process. It was a "*Master of Art*" for reconciliation studies and also introduced the "*Intractable conflict*" theory. The book consisted of essays for the best minds in conflict resolution and studies about reconciliation processes. It introduced a framework to understand reconciliation in different concepts and terms, as applied to different areas of this study. Bar-Siman Tov, with other professionals in the field of reconciliation, addresses methods and ways to understand how the reconciliation process could occur to develop a culture of peace, using different studies in different fields.

1.2.15 ICT for Education Development and Social Justice

Vrasidas et al. (2009) studied how ICT could be used to promote sustainable development and social justice. The paper aimed to research how new technologies could be directed toward the ambitions of societies to explore the implementing of ICTs as a tool in social transformation. Vrasidas et al. researched how change might occur from using ICT as a tool, and they showed how the world might learn more from using ICTs in education, as well as how ICT could make a change in social capitals of societies. The authors' objective was to show the role of ICT in education, development, and social justice around the world.

Vrasidas et al. (2009) aided this research's hypothesis, as it explored how ICTs could be turned to a sacred or profane purpose and how it transformed social capitals in societies. This ICT-MR research illustrated, in theory, and practice, how ICTs made a change for social transformation. Also, this study illustrated some theories of how ICTs influenced social change from different scholars in the field. Vrasidas et al. illustrated examples from all over the world but did not illustrate the change of influencing ICTs on societies. The communication research showed how ICT played a role in social transformation in theory and how it was being implemented in practices using ICTs for that purpose. The theories in the book did not evolve from an empirical work; Vrasidas et al. illustrated theories and depended on experiences of different scholars in the field of ICTs in social capitals change, but did not discuss these thoroughly.

1.2.16 Latin America Conflict and Reconciliation

Leiner and Flämig (eds.) (2012) illustrated theories that should be anointed in research for the reconciliation process. They argued that reconciliation did not only happen after the conflict but also in the middle of the conflict. They argued that reconciliation must be used to embrace the present, past, and future issues. They addressed issues from a transdisciplinary perspective as a method towards the reconciliation process in the middle of conflicts. In addition, they addressed the past that was shaped by fierce conflicts and the idea of fostering reconciliation in the middle of the disputes in different areas of the world. The book consisted of the case and comparative studies about global perspectives of different conflicts around the world.

1.2.17 Making Social Science Matter

Flyvbjerg (2001) studied understanding the concept of phronesis. He explored phronesis as an Aristotelian concept. He also defined it to give a more clear understanding of the philosophical methodology behind it. He explored works from Bourdieu, Habermas, Foucault, and other philosophers to connect the rational power for knowledge, as he stated, "*Knowledge leads to power and power leads to knowledge*" (p. 125). Flyvbjerg (2001) explained the relation of power in theory. Moreover, in practice. Also, Flyvbjerg et al. (2012) described applied phronesis in projects to provide real examples of projects completed around the world concerning phronesis. The book was an excellent example to follow when applying projects that had a Phronetic approach as the philosophical methodology.

1.2.18 Netnography Redefined

Kozinets (2015) wrote an essential book for social, cultural research; it contained procedural guidelines for conducting netnography research. The book was used extensively in Chap. 7 to research the development of the online Palestinian community. Kozinets also introduced ethical conduct for carrying out netnography research. One of the main topics of this book included a discussion on how to research big data and social media analytics. This ICT adopted these netnography guidelines to research the online Palestinian community development in the discourse of the study, and it was an excellent guide for online ethnography research.

1.2.19 Networks: An Introduction

Newman (2010) gave a profound introduction to networks and discussed the empirical study of social networks, which was necessary for the empirical study in conducted research. Also, Newman introduced and explored technology networks, social networks, networks of information, and the fundamentals of network theory. The book was an asset for the research, as it defined one of the leading concepts that evolved in the research: the development of the egocentric network, known as the ego-centered network. In the research, the base of the ego-centered network was the notion that the NGO Wasatia was defined on, which was moderation in Islam, a notion defined by the founder of the NGO Wasatia in Palestine Mohammad Dajani Daoudi.

1.2.20 Projects and Risks

Flyvbjerg (2013) provided more understanding of how megaprojects were built, considering the Phronetic social science approach. He described a mega project as one that costs billions of dollars, such as the Channel Tunnel Project between France and Britain, and the bridge between Sweden and Denmark. Phronesis was an approach applied to make those projects sustainable.

1.2.21 Qualitative Data Analysis with NVivo for Windows

Bazeley and Jackson (2013) provided a clear overview of planning and conducting qualitative data analysis with NVivo; they explored mixed-method techniques using NVivo methods for analysis. Beazley and Jackson taught how to handle literature reviews, to develop visual data, and to work with the mixed-method on the practical level. The authors also illustrated in practice how to use the book to analyze social media using tools provided by NVivo. However, Beazley and Jackson did not explore theories that a scholar should understand before using the software. In the research, before using the software, the researchers investigated in theory how to study, inquire, analyze, examine data, and then illustrate these in practice. The best method was to use this book after reviewing the theories for analysis and then study this book to deploy empirical work for the conducted research.

1.2.22 Qualitative Content Analysis in Practice

Schreier (2012) introduced qualitative methods for research and the process of conducting qualitative research. This book was also a hands-on technical guide for conducting any qualitative research. Schreier described significant steps to conduct qualitative content analysis (QCA) and included examples from other researchers in the field of QCA. The book was an asset in conducting the communication research, as it helped define and understand QCA, and assisted in developing coding frames for ICT-MR to study how the analysis was done theoretically. Also, the book was used as a guiding tool to develop the theories to conduct social network analysis, as illustrated in Chap. 8.

1.2.23 Reconciliation Justice, and Coexistence (Theory and Practice)

The book was published in 2001, and edited by Mohammed Abu-Nimer, a Professor in dialog and conflict resolution. The book is a good guide to understand the concept of reconciliation, justice, and coexistence, as it introduces works for different professionals in the field of reconciliation and conflict resolutions. The book also introduces theories and concepts of the reconciliation process as a guide in theory and in practice. The book explores different reconciliation work done in different conflicts in different countries. The book is master peace for promoting the understanding of theories to conduct reconciliation theories in practice in different conflicts. The book introduces an example from different countries that had to work for reconciliation, justice, and coexistence, some that were successful, and some had failed.

1.2.24 Research Design Qualitative Quantitative, and Mixed-Method Approach

Creswell (2014) authored this book as guidance for researchers to develops their research design on an academic level. The book is an excellent source to understand how research design is done using qualitative and quantitative measures, and how to conduct academic research that has the standard of a master and a doctoral degree. The book illustrates an advanced framework for the new process that helps to design proposals for qualitative, quantitative, and Mixed-Methods and conduct the empirical work for the research that is specialized in human and social sciences.

1.2.25 Research Design Qualitative, Quantitative, and Mixed-Methods Approaches

Creswell (2014) studied how to develop a framework for scientific research; he applied approaches to design a study for qualitative, quantitative, and mixed-methods research for social sciences. Creswell also addressed critical elements for developing any investigations to assist in developing the research process. Creswell's suggestions for creating a mixed-method design were applied to this ICT, as discussed in Chap. 2. Creswell also discussed ways in which to conduct the qualitative, quantitative, mixed-method design in practice. He covered issues in research, such as ethical issues for conducting the research. The only problem was that Creswell was not thick enough in the coverage of theories for analysis. He also only pinpointed what the researcher should do, and then it was on the researcher's conscience to figure out how it was to occur.

1.2.26 The Coding Manual for Qualitative Research

Saldana (2013a, b) described the coding and sub-coding procedure for QCA. Saldana also illustrated steps in practice for the coding method. The book had 32 methods of coding examples. I used the book to assist in writing the details of the procedure for coding and analyzing the communication research, as discussed in Chap. 9. The book was used in different studies internationally. It was a guide to conduct qualitative data analysis for communication research.

1.2.27 Thüringen: Braucht das Land Versöhnung? Kritisches Jahrbuch der Philosophie, Band 17

The Book is in Germany Language, still there is no issue with it in English, the article that was in English in the book was for scholar who assisted me in my research, Martin O'Malley, who illustrates the Jena Approach from his point of view as a scholar in the Heart of Flesh Not Stone Project. He introduces principles and priorities for reconciliation studies, giving a more understanding of the

"Hölderlin Perspective," for Leiner, and Schiller Correlative in reconciliation perspectives, reconciliation priority rules encountering ethical, political and local priorities. The article is more about a philosophical perspective and critical approaches for the Jena Approach for reconciliation done in Jena Center for Reconciliation Studies in Jena, Germany. The writer in his article did not address practical reconciliation perspectives in daily life. It is always excellent to have a philosophy of reconciliation, but how can it become part of our thoughts and daily habits. This kind of notation I found in the researches done by Flyvbjerg (2001), where he tries to make social science matter again by presenting the philosophy and then applied philosophy or what I would suggest applied ethics, is essential to develop a social change in societies. The important aspect of this book is about justice as a virtue of a person who does the right thing; this concept is very related to starting the process of reconciliation in the middle of the conflict.

1.2.28 The Media, Society, World Social Theory, and Digital Media Practice

Couldry (2012) discussed the media's contribution to the social organization, as well as described the practice theory and social theory to show both evolved to consider digital media. Couldry researched the study of media, contributed it to the social theory, and defined the Internet as part of the media as an institution. He researched ways in which the media influenced the transformation to capital change in society.

Couldry (2012) illustrated the ways in which social change occurred in and around the media. He studied the media's influence on different conflicts in the world, such as the Arab spring and what happened in Abu Ghraib jails in Iraq, after the war. The book showed the influence of the media on social change. Couldry's findings strengthened the hypothesis in the study but did not illustrate that the practice theory could manage and influence change. He also did not discuss ways in which the media could be directed toward reform, such as intentionalities as in ICT-MR research. The gap was that Couldry did not illustrate how the practice theory could change people in practice; he only demonstrated the theories' existence and provided examples of these.

In this ICT-MR research, those theories were shown in practice to examine ways in which participants changed and the change regarding the increase of members in the social network site. This ICT-MR research showed the influence of change from hatred and animosity to empathy and accepting the other. The ICT-MR also indicated ways in which members from virtual society became active members in real life. Couldry (2012) added to this study by illustrating ways in which media influenced daily life and the power it had on ritual, capital, and social order.

1.2.29 Technologies of Freedom

Pool (1983) exposed the role of technology in the freedom of speech. The pool provided a technological tour in his critical review of the history of technology. He contributed to an understanding of how ICT evolved in the media and information about ICT exposing free speech to the world. When reading this book, one understands how the role of the media came to power. Pool *(*1983*)* stated, "*Networked computers will be the printing presses of the twenty-first century... Soon most published information will be disseminated electronically*" (p. 224). This aided the communication research and the hypothesis in question; this ICT-MR research sought to understand the role of ICT in society and if it could be managed as a tool for moderation, reconciliation, and democracy in the middle of the conflict. The book upheld that ICT had a considerable role to play in society, which was in the past and in the future. However, the book had no theories to discuss or methods to elaborate, and Pool (1983) did not assure the role of ICT in societies. The purpose of the book was to uphold that technology could bring a free speech, which could be acknowledged today. I sought more understanding of the role of ICT in conflicts to advance ICT as a tool to bring peace and prosperity to conflict areas, as presented in Chap. 4.

1.2.30 The Social Media Gospel

Gould (2013) described a broad view of how the Internet social network was being deployed but attempted to illustrate how it could expose religious and faithful views using the Internet. The author used social media as a platform to talk about the Gospel online and develop a discussion of religious views. Gould provided an excellent example of how Internet social networks could be directed depending on the topic of issue. This idea was also reflected in the Internet communication research, such as adopting social network to discuss views about moderation and reconciliation and democracy in a society that was online.

1.2.31 What School Leaders Need to Know About Digital Technologies and Social Media

Mcleod and Lehmann (2012) introduced using ICTs as tools in education and presented a description for online applications of ICTs, such as Blogs, Wiki's, Podcasts and Webinars, RSS and RSS readers, Digital Video, and Virtual Schooling. McLeod & Lehmann introduced ways in which social media changed the way society lived and how it changed the way that society learned. The importance of this book was that it defined the usage of an online application that could be a real asset to advance education in schools. The book assisted in the research by providing the types of applications that the social network site needed to be operational online. Also, it helped to provide the ICT tools for assisting interactions between online members.

1.2.32 Why Deliberative Democracy?

Gutmann and Thompson (2004) described the definitions and methods of deliberative democracy. Also, they introduced different conceptions of deliberative democracy, the process, and what was beyond that process. Gutmann and Thompson assisted in understanding deliberative democracy and other types of democracy, as supported in Chap. 6. Lastly, the book illustrated some samples of the deliberative process and its methods, which were helpful in understanding the ways in which deliberative democracy had a role in ICT. What was missing in the book, is the online deliberative democracy, I had to research more Russian scholars, and I was amazed that Russian scholars work more on online deliberative democracy and its implication and how to practice it in real life.

1.2.33 When the People Speak

Fishkin (2009) illustrated deliberative democracy in different types of conduct, such as deliberation day and deliberative polling. The researcher also illustrated deliberative as a normative, empirical, and practical work within different democratic theories. In this research, the book aided me in understanding the ways in which deliberative democracy work aided online deliberative democracy. The book also assisted in understanding deliberative democracy and implementing it in the research.

Part II
The Phenomenology of Internet Communication Technology (ICT) in Applied Phronesis

Chapter 2
The Phenomenology of Internet Communication Technology (ICT) in Applied Phronetic Social Science

2.1 Introduction: The Methodology and Research Design

The research addressed the problem of stagnated hate and animosity in conflict by utilizing ICT to influence intergroup intentionality—from enmity and hatred to moderation and reconciliation during conflicts to lead toward social change that would form a path for the reconciliation process. This phenomenology introduced an interdisciplinary methodology that used Internet communication technology applications to engage the disciplines of social psychology, philosophy, politic science, and informatics, as applied in conflict to lead the community for the aspiration to moderation reconciliation and democracy. Having introduced the research project, the figure below serves to build up the theoretical foundations for the research design in researching the phenomenology of ICT in Applied Phronetic Social Science in this communication research (Fig. 2.1).

This communication research develops a scientific methodology that was interdisciplinary, according to the following definition of the National Science Foundation Academies'[1]:

> *Interdisciplinary research is defined as mode of research by teams or individuals that integrates information, data, techniques, tools, perspectives, concepts, and/or theories from two or more disciplines or bodies of specialized knowledge to advance fundamental understanding or to solve problems whose solutions are beyond the scope of a single discipline or area of research practice.*

Guiding this interdisciplinary endeavor was Phronetic social science developed by Flyvbjerg (2001) and applied by different scholars (Flyvbjerg et al. 2012),

[1] National Science Foundation: Committee on Facilitating Interdisciplinary Research, Committee on Science, Engineering, and Public Policy (2004). *Facilitating interdisciplinary research*. National Academies. Washington: National Academy Press, p. 2. Also check https://www.nsf.gov/od/oia/additional_resources/interdisciplinary_research/definition.jsp

I. M. AlDajani, *Internet Communication Technology (ICT) for Reconciliation*, Lecture Notes in Social Networks, https://doi.org/10.1007/978-3-030-41203-6_2

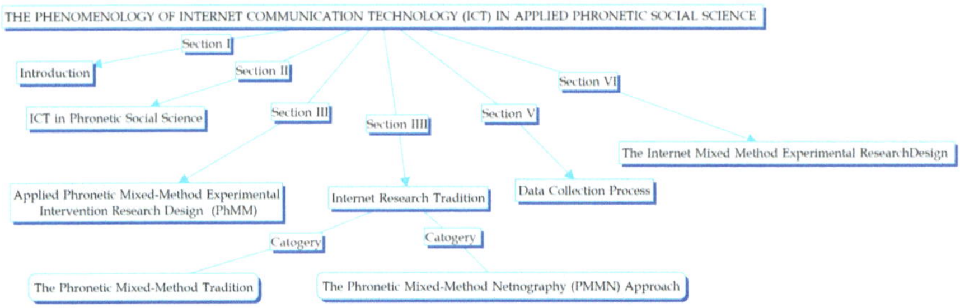

Fig. 2.1 Internet Research Design for the Phenomenology of ICT in Applied Phronetic Social Science

Flyvbjerg provided a teleological and result-oriented framework to discern, prioritize, and utilize scientific concepts through philosophical interpretations, theories, perspectives, and tools. The communication research also includes other tools and concepts such as the integration of the sequential explanatory mixed-method research design (Bryman 2006; Creswell 2014; Creswell and Clark 2011) and applies the mixed-method Experimental Intervention design (Creswell and Clark 2018).

These concepts and methodologies aided in interdisciplinary research to study the influence of ICT to promote virtues such as moderation, reconciliation, and democracy in the Palestinian/Israeli conflict. The figure below illustrates the framework and theoretical foundation for the research; the concepts and theories are discussed in detail within the discourse of the communication research (Fig. 2.2).

The method designed for this research allowed the examination of ICTs' influence on the reconciliation process (moderation, reconciliation, and democracy) in the midst of conflict. The method dealt with the high complexity of themes of social and political intentions, as well as the role of ICTs, utilized to encourage social change. One aspect of complexity provided a significant amount of relevant statistical data. Another was the random variation of group intentionality due to the ongoing, politically relevant events, and further technological advances in social, digital media. This type of method was used to develop patterns, principles, theories, methods, and new techniques to help other researchers with using the applied Phronetic approach in implementing the Mixed-Method experimental intervention research design. Also, the emerging for results from conducting such a method was relatively stable, sustainable, and reproducible, as well as validated and reliable for any future researchers.

2.2 The Philosophical Framework

The communication research aimed to develop an ICT-MR platform to develop an approach for social change towards the reconciliation process within a community that was in a conflict state. A research strategy was developed according to the ancient Aristotelian principle, "*Phronesis*," as developed by Flyvbjerg, to address

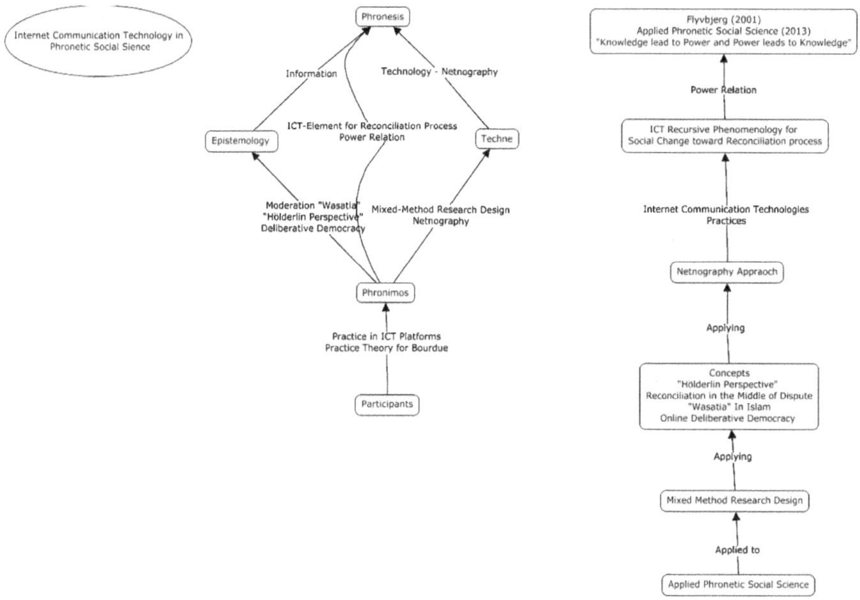

Fig. 2.2 Framework and theoretical foundation for the communication research

contemporary social science-related projects, namely the Phronetic research plan (pp. 283–306). Flyvbjerg recognized that contemporary social science issues lent poorly to strict single-discipline methodologies, therefore requiring specific methodologies that were tailored to achieving specific goals. Choosing appropriate methods required practicality or prudence, as described well in Aristotle's ancient writings on ethics. Flyvbjerg defined a Phronetic Research Plan as a "*to the study of planning based on a contemporary interpretation of the classical Greek concept phronesis, variously translated as practical wisdom, practical judgment, common sense, or prudence*" (p. 284). Moreover,

> *We may grasp the nature of prudence [phronesis] if we consider what sort of people we call prudent. Well, it is thought to be the mark of a prudent man to be able to deliberate rightly about what is right and advantageous…But nobody deliberates about things that are invariable…So…prudence cannot be a science or art; not science [episteme] because what can be done is variable (it may be done in diverse ways, or not done at all), and not art [techne] because action and production are generically different. For production aims at the end other than itself, but this is impossible in the case of action because the end is merely doing well. What remains, then, is that it is a real statement, reasoned, and capable of action about things that are good or bad for man. We consider that this quality belongs to those who understand the management of households or states.* (N.E., VI pp. 1140a24–b12). Phronesis is prudent to attain the best judgment for an argument

2.2.1 The Phronetic Research Plan (Flyvbjerg)

The Phronetic research plan, as developed by Flyvbjerg (2004), guided methodology development that remained dynamic as research work progressed. Early research findings were subject to continual evaluation regarding the research goals, and the method might incorporate learning in a conscious and rigorous way to integrate learning regarding newly available evaluative software and findings to reveal effective practices. Phronetic was an element of the methodology that required a regular methodological assessment to integrate it into the method from the beginning.

This discussion was developed in Flyvbjerg's *Theoretical and Methodological Reflections*. Moreover, "*the Phronetic classical interpretation of phronesis is strong on values but weak on issues of power" (p. 284).*

For a social organization to be successful, it should have intellectual capacities and systems for achieving the right ends. This condition is known as the *Phronetic condition,* according to an interpretation of Aristotle, whereby planning research emphasizes continuing consequentialist practicality captured in the term phronesis (as cited in Flyvbjerg 2004, pp. 283–306). Aristotle defined phronesis as an intellectual virtue that *was "reasoned and capable of action with concerning things that are good or bad for man*" the italic is formed the original (E.N.VI, pp. 1140a24–b12, 1144b33–1145a11), also cited in (Flyvbjerg 2004, p. 284).

Phronetic research planning concerns reflection about values and reflecting interests with power (Flyvbjerg 2004). Aristotle indicated that planning had three philosophical terms: "*episteme," "techne," and "phronesis." These were used* to understand the differences considering Flyvbjerg's (2006) basic understanding of those words: *"episteme concerns theoretical know why and techne denotes technical know-how, phronesis emphasizes practical knowledge and practical ethics"* (p. 287).

Flyvbjerg (2006) also discussed that theoretical knowledge was more valuable compared to practical knowledge; one could not generalize from a single study, and his study did not contribute to scientific development. This ICT-MR demonstrated there was a way to moderation and reconciliation through the Phronetic social practice using ICT as a tool for the implementation discourse. This ICT-MR studied the different phases of the research. Flyvbjerg (2006) described the Phronetic research plan as "*a scientific discipline without a large number of thoroughly executed case studies is a discipline without systematic production of exemplars, and discipline without exemplars is an ineffective one.*"(p. 219) Therefore, this ICT-MR implemented the Phronetic approach to the research through many applications and phases of actions using the ICT as a tool to approach a change from theory to practice for endorsing the reconciliation process in the midst of conflict. According to Flyvbjerg et al. (2012), *"Phronetic social science is ultimately about producing knowledge that can challenge power not in theory but in ways that inform real efforts to produce change*" (p. 20).

2.2.2 *Critical Questions for Conducting Phronetic Social Science by Means of ICT*

According to Flyvbjerg (2001), to develop social science research that matters in society, the power relation must be present and introduced by the element of power in a Phronetic approach. This aspect must occur because the reason for the power relation in social science is what makes research efficient within society. In this research, I introduced ICT as the element of power in the Phronetic approach to social change. Flyvbjerg (2004) argued that some issues guided the development of social research to keep it focused on the practical goal, and this research engaged the Phronetic process, as stated below:

What Is the Purpose of This Study? The aim was to develop an efficient ICT platform that provided the understanding and measurement tools capable of positively influencing social groups participating in or affected by political conflict, explicitly focusing on reconciliation in the Palestinian/Israeli conflict. Questions arose, including the following: How can the reconciliation process be fostered even in the midst of continuing political conflict; what are ICT practices, theoretical principles, and tools that are effective in measuring the influence on society for social change; who gains and who loses; and by which mechanisms of power does one implement the Phronetic research plan?

Who Gains by Reconciliation, Moderation Using ICT? Present and future generations of Israelis and Palestinians would gain from the prosperity of reconciliation such development of economic reconciliation that would increase motives towards the reconciliation process that to a stagnated conflict. Introducing moderation as the path for reconciliation was an aim that was measured by real standards of competent governing, rule-of-law access and effectiveness, and well-being achievements, such as health-care access, education resources, and employment opportunities. Implementing or even promoting moderation and reconciliation—and not just cessation of conflict—fostered democratic principles as understood in this communication research in a deliberative manner. For this end, moderation and reconciliation had great potential for fostering democratic governance and civil society institutions using ICT as the element for that purpose.

Who Loses? Some groups would not benefit from a situation of peace, including those who benefited from the lack of adequate governing institutions and the legal oversight of democratic values and government. These groups have an interest, whether rational or not, in continuing situations of hate, grief, and war between both nations of the Israelis and the Palestinians. These groups have less interest in peaceful solutions to conflict because they gain more from the persistence of conflict.

Flyvbjerg (2001) described power as an essential aspect to gain knowledge, according to Francis Bacon's Theory of Ideology and Culture" "*knowledge is power*[2]" Flyvbjerg address "*that knowledge and power, truth and power, rationality and power are analytically inseparable from each other; power produces knowledge and knowledge produce power*" (p. 132). The element of power between *episteme* and *techne*, develops a power relation between both, that can develop a wider power relation between entities that are in contact with each other, such as participants in online social networks, participants gain knowledge and interact with other participants, this develop a power relation that would affect a social change towards the better judgment or the force of a better argument, as Foucault stated "*if one tries to erect a theory of power one will always be obliged to view it as emerging at a given place and time and hence deduce it to reconstruct to its genesis*" (p. 119).

This ICT-MR included the knowledge of understanding moderation, reconciliation, and influencing these using ICT as the element of power for social change. According to (Flyvbjerg 2001), "*the structure and functioning of many societies, got their issues on the public agenda no primary by rational consensus but through the power of struggles and conflicts characteristic of activism and social change*" (p. 105), This ICT-MR could represent a solution to influencing Palestinians and Israelis in the conflict toward social change for moderation reconciliation and democracy in the midst of the power struggle in conflicts.

Is ICT a Favorite Tool to Conduct ICT-MR Research? ICT is a tool that affects both in-group beliefs and feelings, as well as is related to out-of-group relationships. The communication research shows how ICT could be an instrument for influencing the reconciliation process between groups that were in protracted conflicts, such as the Israeli/Palestinian one. ICT could help one develops a knowledge base of theories, principles, approaches, and measuring tools relevant to influencing moderation, reconciliation, and democratic principles and definitions. ICT can influence the public and generate interactions of reasonable, practical reasoning for change within the public and more; also, ICT presents a doorway to solving the Palestinian/Israeli conflict. In this study, ICT was introduced as the element of power in the Phronetic research approach to implementing persisting changes in the research towards the reconciliation process.

What Can ICT-MR Research Achieve? ICT-MR was the only appropriate, potentially moderation-reconciliation-promoting tool, and its efficient use depended on the political context within which it was used. Nevertheless, the research showed that as a tool, there were practical implications for developing the ICT-MR platform to influence a community positively and negatively from an online presence. This resource could influence social intentionality regarding moderation and reconciliation in the midst of Israeli/Palestinian conflict for a social change that the internality for a stagnated protracted conflict.

[2] Eugenio-Enrique Cortes-Ramirez, "Knowledge is Power. Francis Bacon's Theory of Ideology and Culture". Via Panorâmica: Revista Electrónica de Estudos Anglo-Americanos/An Anglo-American Studies Journal. Série 3, Número Especial (2014): 25–42. ISSN: 1646-4728. Web: http://ler.letras.up.pt/

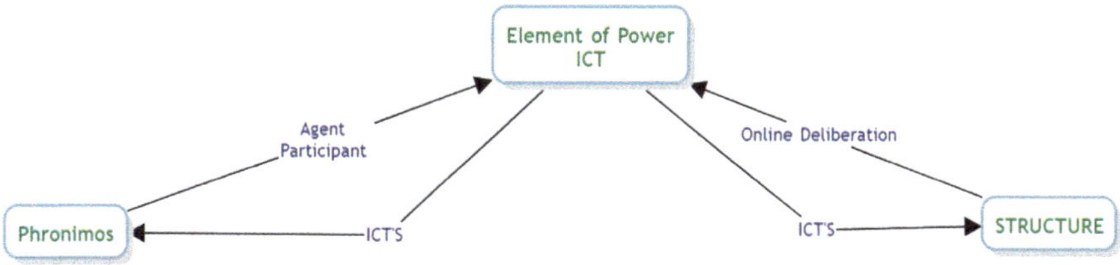

Fig. 2.3 ICT as the element of power in developing the online deliberation

2.2.3 Methodological Guidelines (Flyvbjerg 2001, p. 130)

Focus on Values The Phronetic research plan, used for developing the ICT-MR platform, integrated ethical principles and the values of the Wasatia Circle, namely, moderation, reconciliation, and democracy in the Palestinian society. In conducting the ICT-MR, the research develops various interpretations, and each interpretation was built on claims of validity, which produced reliability for the research and interpretations that were valid from having comparison between different social networks that are involved in the research, mainly the social network site www.wasatia.info, the Facebook Pages, Wasatia.info and Wasatia Vision.

Flyvbjerg (2001) introduced the logical values in questions, such as the following: "*Where are we going? Is it desirable? What should be done?*" (p. 130). Flyvbjerg (2001) applied these to phronesis to "*balance instrumental rationality with value-rationality by increasing the capacity of individuals, organizations, and society to think and act in value rational terms*" (p. 130). This was demonstrated by applying netnography, as described in Chap. 7, for more details (Fig. 2.3).

Place Power at the Core of the Analysis The power at the heart of the analysis was understood through the perspective of the Wasatia definitions and concepts. As such, the core was to influence social change in the community toward the reconciliation process that would lead to the understanding of reconciliation in the middle of the conflict. In this case, the power element was ICT, which can be characterized in the following six features, according to Nietzschean-Foucauldian's interpretation of power (as cited in Flyvbjerg 2001, pp. 131–132):

1. Power is productive and positive. As a tool, ICT can illustrate both measures.
2. Power is viewed as a dense net of omnipresent relations. ICT is a global domain that uses its sphere of billions of participants, which is not part of a center or institution.
3. The concept of power is known as "*ultra-dynamic.*" ICT is ultra-dynamic, as it "*appropriates and exercises in a constant back and forth movements in relations with strength, tactics, and strategies*" (Flyvbjerg 2001, pp. 132).
4. ICT can be a means that produces knowledge that leads to power with the conduct of accruing more knowledge. "*Through knowledge and power, truth and power, rationality and power, power produce knowledge and knowledge produces*

power" (Flyvbjerg 2001, pp. 132); therefore, ICT can be the domain and the element that reduces or increases the power in this ICTs platform, such as this ICT-MR.

5. Power focuses on process, in addition to the structure; ICT as an element of power focuses process and structure together while influencing the forward shift of power and the backward shift between participants.
6. Power delivers a point of departure in small questions, small statements, and develops empirical practice; as an element of power, ICT can provide a departure point to return with different results and re-establish a new and different departure point that keeps on routing until it reaches the cautious Phronetic approach.

The design above describes the power relation between ICT as an element of power between the structure of the design which is the social network sites, and the Phronimos who is the agent that develops the kind of practice. In the ICT-MR they are the phronimos are the participants, and the power relationship between them is ultra-dynamic, described later in this chapter. The idea also presented by Hans-Georg Gadamer, that dialogue is initiated to get a better judgment of the argument, "*the dialog moves to a better understanding of the subject under discussion*" (p. 345). ICT produces dialogues to get a better understanding of the subject in discussion (Gadamer et al. 1992).

Get Close to Reality Getting close to reality refers to having the researcher as part of the practice; therefore, I was always exposed to the reactions of the participants, both negative and positive. The Phronetic researcher becomes part of the phenomena studied without any biased attractions. This type of Phronetic approach advances the study of phenomena in question; Flyvbjerg (2001) expressed this concept by "*fusion of horizons*" (p. 132) originally developed by Hans-Georg Gadamer.

ICT-MR, as developed in this research, was a value-laden approach that fostered moderation, reconciliation, and democracy amid conflict. This approach built on the assumption that Palestinians would benefit from intentions open to moderation and reconciliation because such openness could better counter stigmatized conflict between Israelis and Palestinians. This aspect could lead to immense prosperity through reconciliation. I led the research as the *Phronimos;* a label described "a *person possessing practical wisdom*" by Flyvbjerg (2001, p. 57).

Searching the Great Within the Small Part of the Phronetic approach involved researching little questions while conducting the research, which illustrated the departure of change; in this case, the departure toward understanding moderation led to emphasis about reconciliation in the middle of the conflict. Due to aspiration from Nietzsche and Foucault, Flyvbjerg (2011) emphasized that the fundamental approach in the Phronetic approach be that small questions often led to essential answers. I used the ICT-MR platform to conduct small question posts on the social network site and in the digital social media, such as Facebook and YouTube videos, where participants interacted to answer problematic situations. Flyvbjerg (2011) called this approach the "*patience and a knowledge of details*" *an emphasis by Nietzsche and Foucault* (p. 133).

The Practice Before Discourse According to Flyvbjerg (2011), the Phronetic approach was intertwined with practical activity and knowledge in the research. This ICT-MR focused on both aspects, which developed a discourse for functional analysis, as described in Chap. 8. This kind of practice focused on actual practices, which constituted to a given hypothesis. Flyvbjerg (2011) emphasized the researcher could inquire to understand the practices studied regarding the hypothesis of the study.

Researching Cases and Context Flyvbjerg (2011) emphasized that the phronesis approach must have the discourse of practical rationality and judgment. It was best understood through old cases and that judgment derived from practices that were cultivated from communication. This ICT-MR applied the approach through the social network site and the Facebook pages, as illustrated in Chap. 7 within the conduct of Netnography and phases of actions.

Move Beyond Agency and Structure According to Flyvbjerg (2011), the structure was found as part of actors and the actors as part of the structure. The concept adapted Bourdieu's (1977) practice theory "*Habitus*," as I was part of the practice and part of searching for information to explore and inquire about the hypothesis in question. Also, it researches the relationship between the actors and the structure to understand what influences the structure (Flyvbjerg 2001, p. 138). This ICT-MR developed a platform for the Palestinians to interact with each other to create more understanding of the concepts of moderation, reconciliation, and democracy. I observed and collected data to concentrate more focus on the deliberation about the hypothesis in question.

Participate in Dialog with Pluralistic Values, Beliefs, and Voices Using ICT applications endorsed dialog between the participants within the community. This ICT-MR provoked guidelines using the ICTs in its platform. The practice and interaction of participants produced input to the ongoing social dialogue and praxis in society (Flyvbjerg 2001, p. 139). This ICT-MR concentrated on influencing social change toward reconciliation, moderation, and democracy by influencing an online community. One of the most critical aspects of this Phronetic research involved interpretation, which was supported by valid claims, and I had the responsibility of checking the validity claim for interpretation. Flyvbjerg (2001) stated, "*Phronetic researcher thus recognize a human privilege and basic conditions: meaning dialogue in context.*" (p. 139).

Asking How and Why the Narrative Is Essential Flyvbjerg (2001) stated that the researcher should work on historical inquiries, then develop descriptions and interpretations of the phenomenon from the perspective of the participants, research, and perspective materials. Flyvbjerg (2001) stressed that "*the analytical and instrumental rationality which lie at the core of both episteme and techne, the practical rationality of phronesis is based on social conditioned, intersubjective "between reasons*" (p. 139). The relation between "*episteme*" and "*techne,*" the first becomes a reason to apply the other, and they become consequently between reasons for applying each other to advance a status. This lets us to the premise that *ICT can foster moderation, reconciliation, and democracy in the midst of conflict.* The approach involved applying the Mixed-Method netnography approach, as seen in Chap. 7. This approach developed new themes that emerged because of the narrative of the participants in the study. Those results related to examining the primary hypothesis of the communication research and whether the hypothesis contrasted the results or not (Flyvbjerg 2001, pp. 137–138).

Figure 2.4 illustrates the ways in which this ICT-MR research plan used the Mixed-Method netnography approach to conducting the theoretical synthesis of acquiring the study. Creswell (2018) introduced a Mixed-Methods intervention research design, and Kozinets (2015) introduced a netnography approach for online ethnography study. This approach included the agent, which was me, and the structure, which was introduced in the research as the Phronetic Mixed-Method netnography approach (PMMN), using both presents the Phronetic social science in practice.

ICT-MR Research plan applies the Mixed-Method integrated with the Netnography approach; the research plan is the first model illustrated by the research, to develop the cycle in conducting the research.

2.2.4 The Phenomenology of ICT in Applied Phronetic Social Science

The concept was inspired by the work of Flyvbjerg[3] (Flyvbjerg et al. 2012) who was stirred from the Aristotelian phronesis to apply phronesis in practice. According to Flyvbjerg et al. 2012), phronesis can be applied "*as a transformative methodology that builds into its foundation a critical spatial dimension and relational dynamics*" (Flyvbjerg et al. 2012, p. 266).

The Phronetic approach was applied in practice for problematic situations in social science. In the communication research, the case study was to influence social change, within the community, from hate and animosity to less hate and animosity by using ICTs. This case study was defined by (Flyvbjerg et al. 2012) as paradigmatic because the selections included samples from groups who were information oriented in ICTs and the intention was to develop a school for the domain (i.e., develop moderation and reconciliation of understanding and practice in the Palestinian community; Flyvbjerg 2001, pp. 78–79). Figure 2.5 illustrates the phenomenology of ICT for utilizing the Phronetic practice for a social change.

[3] Bent Flyvberg: Chair of Major Programme Management at Oxford University's Saïd Business School and Director of the Oxford Centre for Major Programme Management. He works for better planning and management of megaprojects and cities, plus he writes about phronetic social science and case study research.

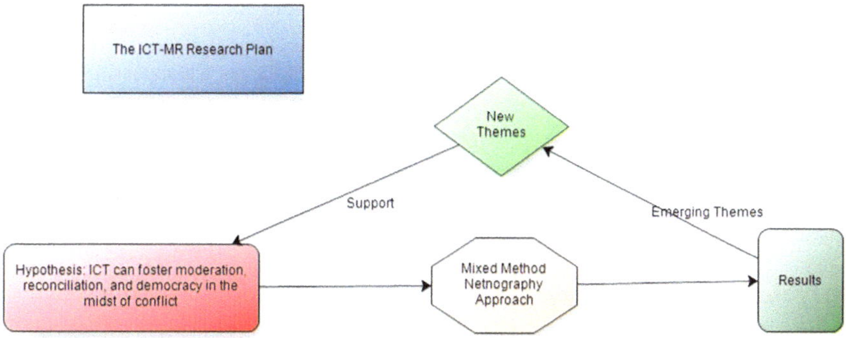

Fig. 2.4 Research plan to apply Mixed-Method Phronetic Netnography Approach

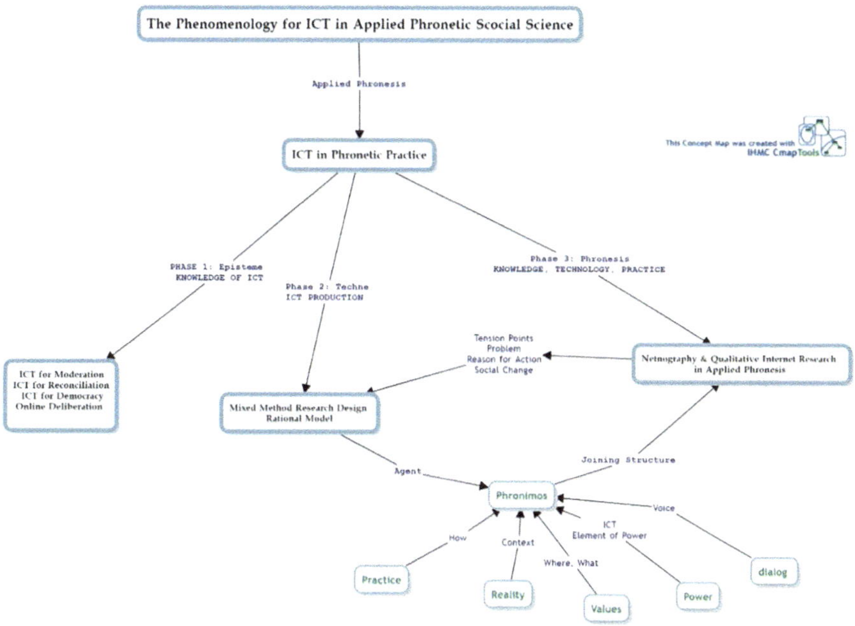

Fig. 2.5 The phenomenology of ICT in Applied Phronetic social science

To explain the concept map in the figure, one should acknowledge the different concepts that need definition. These are defined as follows according to (Flyvbjerg 2011, p. 57):

- *Phronetic social science* refers to practical science that "*can help people involved in ongoing political struggle question the relationships of knowledge and power and thereby work to produce change*" (Flyvbjerg et al. 2012, p. 19). The purpose of the case study was to lead toward social change within the community from hatred and animosity to less hatred and animosity and social change towards the reconciliation process.

- *Episteme* refers to epistemic, scientific knowledge, and it was based on general analytical rationality (Flyvbjerg 2001; i.e., epistemology).
- *Techne*, which is a craft, context-dependent, and oriented path to production, is known today as technical (Flyvbjerg 2001, p. 57).
- *Phronesis*, according to Flyvbjerg (2001, p. 57), is the ethical deliberation about values concerning the practice in a specific context, and it is oriented toward actions.
- *Phronimos* refers to the person who is processing the knowledge applied in practice and knows ways in which to behave in different circumstances in a way that his or her actions cannot be equated with or reduce knowledge of the general truth (Flyvbjerg 2001, p. 57).

Flyvbjerg (2001) described the phronimos as a researcher who operated in accordance with the following premises (p. 115):

1. Researchers are both involved in, and partially produced data, the same cultural practice that they study; the researcher cannot stand and be an outsider, and also the researcher is not identical with that which he or she studies.
2. Practice refers to the researcher, with his or her knowledge of the culture and the people he or she studies, can provide a way of acting and think of once, without being a crucial part of the deliberation process.
3. To have meaning for the discourse of the research, only the researcher can make the discourse understood and meaningful as part of society's history.

According to Fig. 2.3, the hypothesis was that ICT could foster social change towards the reconciliation process (moderation, reconciliation, and democracy) in the middle of the conflict. To apply the Phronetic social science, I began with Phase 1, which was the episteme or the knowledge, which used moderation as a concept and theory, ICT for reconciliation, ICT for democracy, and the understanding of online deliberative democracy. Phase 2 applied the Phronetic Mixed-Method. Phase 3 was phronesis, given that phronesis was somehow techne but with a brain, "*techne with a head*"(Flyvbjerg 2001, p. 168), where actions and practices were for the benefit of testing the hypothesis. Applying phronesis meant applying practice with deliberation. To start to engage in this practice, one should need power; the power must be of need always and in continuous modes. In this case, the element of power was ICT, which acted as an open domain for participants deliberate matters. The power of ICT kept participants engaged in matters of deliberation continuously "*episteme*" and "*techne*," which placed ICT at the core of analysis and element of power to ignite the practice and develop the power relations between "*episteme*" and "*techne*." Flyvbjerg (2001, p. 57), applied power to phronesis in accordance to Foucault's rational power theory: *"If one tries to erect a theory of power one will always be obliged to view it as emerging at a given place and time and hence deduce it, to reconstruct its genesis"* (Flyvbjerg 2001, p. 119). The (*phronimos*) applied tools to produce input for an ongoing online deliberation in matters of moderation,

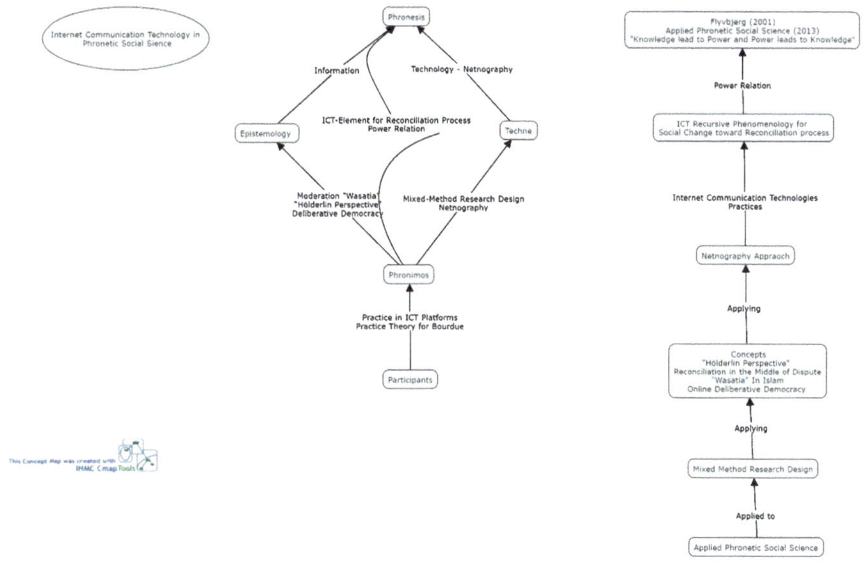

Fig. 2.6 Applied Phronetic internet research theoretical framework

reconciliation, and democracy within the domain of ICT. Thus, I delivered interpretations, which must be based on absolute validity, and thus made the study reliable (Flyvbjerg 2001, p. 115).

2.2.5 *Applied Phronetic Social Science for Social Change Using ICTs*

Figure 2.4 illustrates the development of power relationships between the "*Episteme*" and "*Techne*" within the use of practice theory to emerge to "*phronesis*" best practice for gaining social change (Fig. 2.6).

Figure 2.4 shows the element of power as the phronimos (participants) applying ICT platforms with its discourse of the *habitus* practice theory from Bourdieu (1977). This evolved from having a problem to solve, such as spreading concepts of moderation, reconciliation, and democracy in the middle of the Israeli/Palestinian conflict. In the discourse of applying ICT platforms for that purpose stated before, a concurrent discourse evolved between the "*episteme*" and the "*techne*" to try to solve the problem that occurred, which led to developing tensions between the participants. The positive social change would lead toward better intentions toward moderation and reconciliation, and the negativity led social change toward hatred and animosity, ICT-MR communication research wants to lead for social change towards the reconciliation process.

2.2.6 ICT-MR Expectations from Implementing the Phronetic Social Science

"phronetic approach (close readings, sequencing, intersubjectivity, interactions), its final advantage is central to phronesis; namely, it allows the social scientist to uncover perceptions, experiences, and feelings about power, power relations and institutionalized constraints as they are confronted (or not) through social and political engagement." (Flyvbjerg et al. 2012, p. 32). As ICT was the power for change between the praxis of moderation and reconciliation, the effect of ICT would be high at the start of the influence. According to Newman (2012), ICT develops a snowball sampling effect for a significant number of people within the community; in this case, they would have fewer intentions to change at the start of the snowball sampling effect. However, at the core of the snowball effect, a community develops who has an ego-centered network that affected the social network. In this case, it was the intention for social change toward moderation, reconciliation, and democracy in the midst of conflict, which was the hypothesis. Snowball sampling. Snowball sampling "is a technique for studying hidden populations that rely on social networks for its operation. It is essential to keep this distinction clear. To judge by the literature, some professional social network analysts do not, and the results are often erroneous conclusions and lousy science". (Newman 2010, p. 58). Figure 2.7 illustrates the effect of snowball sampling to develop the ego-centered networked community for social change. The change starts with high less intentionality for change, and by continuing snowball sampling the more the intentionality changes to high intentionality for change, but less affecting less in the community. The snowball sampling takes effect in phases, as illustrated in the figure below 4 phases for change. The start of phase one, affect a large spectrum of the community but less intentionality for change, by the start of phase 4, the figure introduced less affected but more intentionality's for change, phase 4 is always at the core of change.

This ICT-MR targeted groups who were multipliers in the Palestinian society, such as educators (university professors and schoolteachers), facilitators, journalists, religious leaders, youth, and women community leaders. This ICT-MR focused on getting those members of the ICT-MR social network site and social network applications. This ICT-MR facilitated using Internet dialogues within a forum; publishing workshops online, providing public online lectures; and using e-learning for the empowerment of moderation, reconciliation, and democracy. The ICT-MR activities were mainly online events, such as conducting an online forum for members and developing a social network website for the NGO Wasatia. Facebook pages were developed for the advancement of core democratic values of moderation, reconciliation, and democracy to undermine trends of radicalism and extremists in the community. Such social intentionality for Palestinians, open to pluralism, was a path away from the current climate of religious, social, and political radicalism and extremism that was escalating fear, enmity, and insecurity.

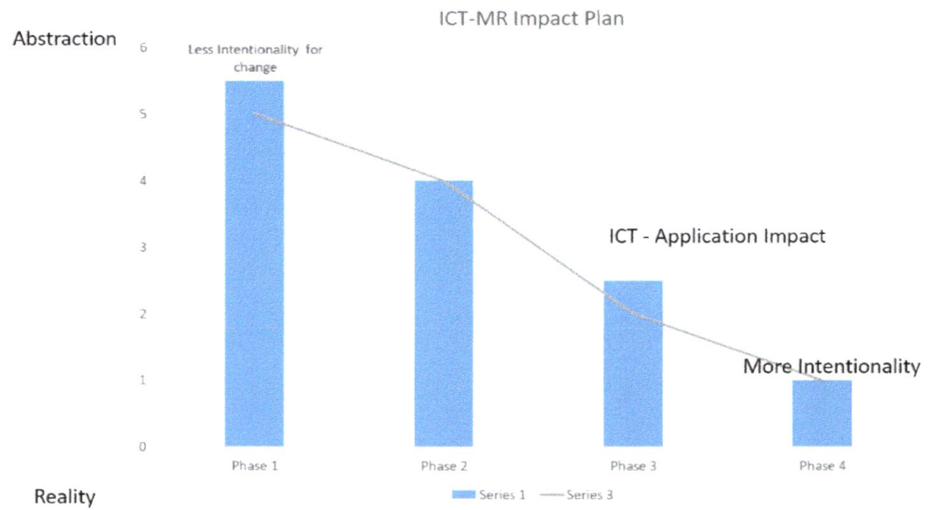

Fig. 2.7 Expectations to change from least to more intentionality, as drawn up by Iyad Dajani & Martin O'Malley (Martin O'Malley, PhD (Boston College), Research Scholar at the Center for Applied Ethics, EthikZentrum, at the Friedrich Schiller University, Jena. Research Scholar at the Jena Center for Reconciliation. Current Project (coordinator): DFG Trilateral "Hearts of Flesh-Not Stone" (2013–2018).)

2.2.7 Developing Principles for ICTs in Applied Phronetic Social Science

ICT-MR Internet communication research addressed the influence of ICT for social change towards reconciliation process. I utilized the Phronetic mixed-method netnography approach through phases of actions, in more detail in Chap. 7. By applying Phronetic social science in the ICT-MR Internet communication research, principles emerged that are discussed in the following subsections.

Basis: An Actual Presence Within the Community The reason ICT-MR Internet communication research can influence the community because the base of the ICT-MR Internet communication research was an actual NGO in Palestine, known as Wasatia, which worked toward reconciliation more detail about the NGO in Chap. 5. ICT-MR Internet communication studies in those phases of action became an instrument for the elevation of the reconciliation process in the Palestinian community.

Boundaries: Having a Significant Target Group I needed to have a definite target group, such as the targeted groups in the ICT-MR who were Palestinians who have tendencies to accept the message of peace and reconciliation, which the NGO Wasatia portrayed in its activities and workshops. The ICT-MR created a virtual community for those people who wanted to understand the concepts of peace and reconciliation through the online activities developed by the ICT-MR.

Clear Telos The goals of this ICT-MR was to promote Wasatia/moderation, according to the Wasatia circle model (see Chap. 5), reconciliation, and democracy, to advance the reconciliation process. Action planned to reach the central phenomena and be ready to change the stages of the plan based on experience (Phronesis), thereby developing best practice.

Back to Real Physical Activities Ideally, ICT-MR would form first in online or virtual communities, and then strive to bring participants to nonvirtual physical contact, in whatever form of action that might take (workshops, activities, etc.). This goal was to "*make it real,*" which would theoretically have a positive feedback-looping effect on the virtual community insofar as it could lend legitimacy and roundedness to the virtual community. In any case, the goal was to develop groups within the community who had tendencies to learn more about reconciliation and the process of reconciliation. ICT-MR developed new members, who wanted to participate in physical activities done by the NGO Wasatia on reconciliation and moderation. However, due to the lack of finances and other resources, the ICT-MR activities and measurements were limited to digital forums.

2.3 ICT in Applied Phronesis

This section discusses applied the Phronetic approach in Mixed-Method referring to it as the Phronetic Mixed-Method approach in the Internet communication research ICT-MR. The approach was a result of a combination of integrating the work of researchers (Creswell and Clark 2018; Flyvbjerg et al. 2012). The argument to integrate Phronetic approach with mixed-method was applied in many studies that are within the combination of social science research. Researchers stated, "*A diversity of data collection methods to best inform attempts to promote change related to the issues being studied. Depending on how they are conducted, mixed-methods projects are entirely consistent with Phronetic social science*" (Flyvbjerg et al. 2012, p. 20). The importance of conducting the approach was to advance the research from *theory-driven* to *problem-driven* research, as based on suggestions by Shapiro (2005). Therefore, this ICT-MR communication research was a problem-driven study, which proved "*better able to deliver on the founding promise of the social sciences to connect science and democracy, leading to the improvement of both, not just in theory but in practice as well*" (Flyvbjerg et al. 2012, p. 24).

2.3.1 Definitions: Qualitative, Quantitative Research, Mixed-Method Research

Qualitative Research Method Qualitative research means to take notes according to the behavior and activities of individuals that affected the group at the research site to record observations (Creswell 2014, p. 264). Qualitative research method doctrines the following (Creswell 2002, p. 58):

- Explores a central phenomenon as one key statement for the research.
- Asks participants broad, broad questions, open-ended question, and sub-questions that are on the topic of the research.
- The researcher must collect detailed views of participants in the form of words, images, and videos.
- Analyses and codes the data for the description to develop themes for exploring the central statement.
- Interprets the meaning of the information depending on personal reflections and past research when writing the final report that includes personal biases and a flexible structure for valid interpretations of findings.

Quantitative Research Method Quantitative research is an experimental design for testing scientific theories through a statistical procedure that analyzes data in some numerically measurable way. Moreover, the final report has a set structure consisting of an introduction, literature, and theory, methods, results, and discussion (Creswell 2014, p. 265). The basis of using this approach is to test the influence of behavior on a group by controlling all other factors that may influence the group, such as assigning individuals to groups and exposing those groups to an experimental statistical test. The researcher can isolate the groups to the experimental treatment and explore the factors that influence the outcome (Creswell 2014, p. 265). This method is best used for surveys and questionnaires, depending on the level of interpretation the experimental test will target and was used in the Heart of Flesh Not Stone Project.

Mixed-Method Intervention Research Design The mixed-method research combines both quantitative and qualitative methods of research design. It involves philosophical assumptions and using qualitative and quantitative approaches to mixing or integrating these together, systemically and in an exploratory method (Creswell 2014, p. 244). According to Creswell and Clark (2018),

> *Mixed-Methods research is the type of research in which a researcher or team of researchers combines elements of qualitative and quantitative research approaches (e.g., use of qualitative and quantitative viewpoints, data collection, analysis, inference techniques) for the purposes of breadth and depth of understanding and corroboration.* (p. 123)

The Mixed-Method research design involves the following elements (Creswell 2014, p. 217):

1. The collection of both qualitative data, such as open-ended data, whether these were the data or the questions of the research and quantitative closed-ended data, such as the hypotheses or research questions.
2. The analysis included both qualitative and quantitative data analysis.
3. The procedure involved the collection of both procedures, quantitative and qualitative procedures. Examples included sampling, discovering the source of information, data collection, and data analysis.
4. It involved the integration of results from both pursuits in the final analysis, such as merging the data, connecting the data, or embedding the data.

2.3.2 The Objective of the ICT-MR Internet Communication Research

The ICT-MR research aimed to study the influence of ICT on the community for the promotion of moderation, reconciliation, and democracy; the area of study was the Israeli/Palestinian conflict. This ICT-MR Internet communication research objective was to develop a social network site that would advocate the achievement of reconciliation and prosperity via the promotion of a culture of moderation that would lead to leaving the current climate of religious and political extremism that was escalating hatred, animosity, and violence in the Palestinian community. I wanted to study whether ICT could be a bridge for knowledge and a tool to promote awareness of moderation, reconciliation, and democracy in the community, such as on the Palestinian/Israeli conflict. Questions included the following:

- Can ICT create a community of tolerance within two communities that are in conflict?
- Can ICT empower individuals and groups to analyze and criticize the effects of injustice in conflict areas?
- Can ICT create an online community, which achieves the knowledge of moderation, reconciliation through ICT?
- Can ICT develop empathy and the understanding of the narrative of the other in conflict?
- Can ICT develop respect for the narrative of the other in the middle of conflict.?

Moreover, ICT adhered this community to transform from an online community to an active community that followed moderation, reconciliation, and democracy in its daily activities.

2.3.3 Applied Ethics in Big-Data

The communication research applies ethical values and virtues in consideration of conducting Internet communication research by the researcher, and to examine the practices of moral values by participants in the research. The internet communication research provides a massive number of datasets that assist in investigating the hypothesis of the research. Therefore, the researcher must include some ethical approaches such as the *deontological* approach; this approach is desired more when the researcher is conducting research that would articulate what is morally right irrespectively to the consequences of the actions (Hoven 2000, p. 135). Usually when using ICTs as a tool affects the participants by different factors, but the morally right would be the outcome of conducting the research. The other approach is to research the discourse of actions that consequently maximizes the good, referring it to the *teleological*, where the researcher frames what is good for the participants, and expose what is wrong with the actions of participants from gaining the good from research (Hoven 2000, p. 135).

The third is consequential ethics, Consequentiality *is of all the things a person might do at any given moment; the morally right action is the one with the best overall consequences. If there is no one best action because several actions are tied for best consequences, then, of course, any of those several actions would be right*" (Haines 2018). Kozinets explores the discourse of consequentiality in conducting research that studies the behavioral of online participants which are the consequence of actions that directs towards a better outcome. (Kozinets 2015).

In conducting the internet communication research, social interaction might lead to behavior of violations of ethical conduct, and sometimes those conducts might be unforgivable for some reason. Classical consequentialism is a theory which evaluates actions depending on the consequences and not by the rules or intentions like other theories. In the internet communication research, consequentiality is considered an act that depends on ethical consequences in achieving the results in the internet communication research (Kozinets 2015, p. 139) therefore, the internet communication research must have some guiding rules for the participants and members of the online social network.

Any internet research must expose participants to guiding rules in conducting internet research, Internet communication research is considering the rules stated by Virginia Shea (1999), and they are as follows:

(i) Remember the human behind the internet.
(ii) Adhere to the same ethical standards online that you as a participant follow in real life.
(iii) Acknowledge where you are in cyberspace.
(iv) Respect other participants, their time, and their online speed of interaction.
(v) Make yourself respectable and of moral slandered and of reliable information.
(vi) Share expert knowledge
(vii) Keep conversations under control.
(viii) Respect the privacy of other people.
(ix) Do not abuse the knowledge you get out of participating in the internet.
(x) Don't abuse your power.
(xi) Be forgiving of other people mistake.

The Internet communication research must also consider some moral issues in practice, according to (Gret and Moor 1998), who elaborated systems of ethics and applied those ethical conducts as moral principles in practice to conduct Internet research, and they are as follows:

1. Do not Kill or harm any participants
2. Do not cause pain in any factors
3. Do not disable your opponent.
4. Do not deprive of freedom
5. Do not deprive of the pleasure of conduct.
6. Do not deceive participants.
7. Keep your promises in conduct.
8. Do not cheat others and do not lie.
9. Obey the law, be a good participant online by obeying the laws.
10. Do your duty, respect your role as a participant.

According to Robert Goodin's concept of *negative task responsibility*, every participant in the online social network has a responsibility to prevent from expressing harm in internet communication, direct, indirect or in the virtual space (Hoven 2000, p. 153).

Applied ethics in Big-data, has an important factor, which is the Epistemic factor, this factor influence fact-finding, fake news, inquiry, and information collecting from online social networks, web pages, online articles, online videos, online forums, online chats, and online databases, online polls, and other ICT applications.

According to Pual Thagard (1999), the Epistemic factor can provide important practices when using ICT platforms, the first practice is *the power of practice*, assists participants find true knowledge among debating with other participants online, this makes a participant distinguish between fake and true knowledge, and also gains a lot of knowledge from the practice of being exposed to a lot of information. The second practice is *Fecundity* the ability of ICTs to lead a large number of participants and expose them to many true beliefs and values. The Speed of practice, the ability to gain knowledge in the fasted spectrum, no need to hassle in researching books, today we have a saying called "Ask *Google*," Google is an online search engine that has a great database of information. The efficiency of practice, participants are gaining true answers and can gain true knowledge and learn at a minimum cost and easily accessed. The last practice is Reliability, *is the ratio of truths to the total number of beliefs fostered by the practice* (p. 144). The Reliable epistemic practice empowers users and researchers to upgrade their capacity of reasoning and assist them making correct decision by means of online computational devices, and online social networks (Hoven 2000, p. 144).

2.3.4 The Significance of the Study

This study needed to be conducted because the ICT-MR Internet communication research would stimulate moderation, reconciliation, and democracy in the Palestinian community. The Internet communication research developed an online community, which developed a virtual society who understood moderation, reconciliation, and democracy. Also, it created a community willing to develop a society of tolerance and understanding, as based on the knowledge of moderation, reconciliation, and democracy.

Researchers stated, *"Peace starts with facing the harsh reality of hate"* (Maroun 2016a); moreover, the best phrase to write is reconciliation starts with confronting the harsh reality of animosity. The Palestinians and Israelis are in a stagnated warfare despite the UN resolutions 1948, the Oslo accords in 1993, and the Oslo II Accords, signed in Taba in 1995. One of the problems that the Israeli/Palestinian conflict faces is the hate and animosity between the two nations. Also, this includes the historical abhorrence that is inciting hate and war against each other, including the religion and faith of the two nations. One of the challenges that face the two nations in this war is the ethical conditions that may be adopted by either of these two nations.

Nevertheless, neither of those two nations has achieved the goal of winning the moral right that proves one has a legitimate ethical reason for being at war with the other. The problem was that neither armies nor one nation occupying the other could win ethical consideration of righteousness. The only way to win the ethical consideration of righteousness was to adopt the path of moderation, reconciliation, and democracy, as it was the only path not taken for the last 60 years between those two nations. This ICT-MR Internet communication research helped bridge that road for moderation, reconciliation, and democracy between the two nations using ICT to inspire the minds of the people, who were in a war, to help future generation to ascend to a better future.

2.3.5 The Purpose Statement of the Study

The topic of the study explored and examined ICT for reconciliation, using netnography in the applied content analysis. The ICT-MR evolved through a mixed-method experimental intervention research design applied to a netnography and examined by qualitative content analysis on the datasets of results acquired to develop a validated interpretation for the datasets arising from the research. This study involved implementing the *Hölderlin Perspective* theory by (Leiner and Flämig 2012, p. 16), Leiner implied that as conflict persisted between two nations, reconciliation remained possible in the midst of conflict. The purpose of Internet communication research was to research, understand, describe, develop, and discover moderation, reconciliation, and democracy in the Palestinian community by using ICTs. ICT-MR Internet communication research was used to define moderation and reconciliation in practice through articles, forums, and ICT programs; news feeds and web links, and other social media applications.

2.3.6 Preliminary Considerations

The Internet communication research was part of The Heart of Flesh Not Stone Project, which was a transdisciplinary study of intergroup reconciliation that focused on the Israeli/Palestinian conflict. The Internet communication research emphasized more on the practical part of conflict transformation for moderation and reconciliation through the use of Internet communication technologies.

Palestinians have the tolerance for peace, as they are trying to have a just peace between them and the Israelis. The Palestinians have signed peace accords with Israelis that would end with them having their state, and they have strived to reach that goal. With both parties not respecting those accords, most Palestinians have lost hope, ICT may enhance discourse to evolve moderation and reconciliation in the Palestinian community. Some of those who lost hope in coexistence and peace might reconsider that hope and come again within the area of moderation and reconciliation, leading to democracy.

Using ICT in enormous numbers within the Palestinian population, such through Facebook, Twitter, YouTube, and other social media showed that ICT could work as a tool to improve the understanding of moderation reconciliation and democracy within the Palestinian community. According to Ghassan Khatib (2015), about half of the Palestinians in the West Bank and Gaza use the Internet, and among these are young people; the percentages have grown higher every day, as researchers have estimated that 1.5 million Palestinian use Facebook, Electronic Media, Twitter, and YouTube (ESTRIN 2015). Therefore, ICT development has evolved in Palestinian cities. For example, the West Bank now boasts about 300 firms operating in the information technology industries (The Economist 2014). This aspect encouraged the development of an ICT Internet communication research to influence the Palestinian community and work on promoting moderation, reconciliation, democracy.

2.3.7 Expected Outcomes

I expected to develop an online social network site that develops social change towards the reconciliation process in a virtual society that influences the real community. The social network site debated the definitions and used terminologies of moderation, reconciliation, and democracy within the community. Integrating social networks, such as a Facebook page, a YouTube Channel, with a social media site to enhance acknowledgment of moderation, reconciliation, and democracy. Therefore, the outcome of the research is expected to encourage an online society to develop a virtual community that wants to implement moderation and reconciliation as a way of life. One of the expected outcomes was also to develop a society that believed and adhered to the concepts of moderation, reconciliation, and democracy and pave the way for the reconciliation process. In adopting the social network for social change towards reconciliation, empathy and the respect of the narrative for the other is a crucial expected outcome from the research.

2.3.8 The Hypotheses and Research Questions

The research questions were the following:

1. Why reconciliation in the middle of the conflict.
2. What does reconciliation mean in the Palestinian-Israeli conflict?
3. Would the Palestinian community accept the terminologies of moderation, reconciliation, and democracy or it would be considered as a taboo?
4. How would Palestinians conduct be, in regards to reconciliation and moderation within the use of the ICT?
5. What levels would the Palestinians engage themselves in the Internet communication research; would they stay an online participant or an online bystander or start to participate within the communities in real life activities to promote for the reconciliation process.
6. Can ICT be used as a tool for education, culture development, and social justice?
7. Can ICT be an element of social change towards the reconciliation process in the middle of the conflict?
8. Is there any ambiguity from using ICT as a tool for social change?

The central hypothesis was the following: ICT is a powerful tool for encouraging social change towards the reconciliation process in the middle of the conflict.

2.4 Internet Research Traditions

In Palestine, the ICT industries are obtaining much wealth, and most ICT networks in Palestine are inflaming hatred and animosity within the Palestinian community against Israeli communities and vice versa. Radicals from both sides have used Facebook, YouTube, Twitter, or any other social website networking to spread their hateful ideologies of dehumanizing the other. The tendency to find social networks or website that are encouraging hatred and animosity has grown, and these are further supported by radicals from both sides: Israel and Palestine. There is a need to develop a social network site that would encourage moderation, reconciliation, and democracy in the Palestinian society, and the need for social change towards the reconciliation process in the Israel Palestinian conflict made it significant to have a social network site that can develop social change towards reconciliation process. The ICT-MR helps develop the social network site for the NGO Wasatia and measures the social change towards reconciliation process in the middle of conflict by means of qualitative content analysis, explored in the discourse of the research.

2.4.1 The Phronetic Mixed-Method Tradition

As stated by Creswell and Clark (2014), there are five qualitative traditions:

- *Narrative research*, which involves studying the life of an individual and analyzing the implications of his theories that affected social change.
- *Ethnography*, which describes and interprets a cultural or social group, the study of participants and their social behavior in accordance with a specific situation.
- *Grounded theory*, which means the group has developed a theory grounded from data in the field and can be valid an attainable in different situations.
- *Case study*, which involves developing an in-depth analysis of a case or multiple cases, and comparing between cases to emerge with a theory that can define the effect of the case studying.
- *Phenomenology*, which involves studying and understanding the principle of experiences about a phenomenon, that happens in time and place.

This ICT-MR Internet communication research added another tradition: the Phronetic Mixed-Method netnography approach, which involves describing, exploring, examining, and interpreting an online community. This type of research for the online community was called netnography by Kozinets (2002, pp. 61–71). The Mixed-Method research design depended on the term of Phronetic mixed-method netnography approach, as discussed in previous sections and in practice in Chap. 7. Netnography qualitative tradition was a new tradition to evolve from the research, which calls to study an online community by using mixed-method approach to apply qualitative content analysis method to explore, examine, and analyze the culture of an online society to develop a validated interpretation for the research and examine the hypothesis of the study.

2.4.2 *The Phronetic Mixed-Method Netnography (PMMN) Approach*

The ICT-MR communication research went beyond ethnography to integrate the netnography approach. However, the Phronetic research stays true to the Phronetic principles by allowing for the diversity of data collection methods to help produce contextualized knowledge to explore and examine the hypothesis of the study (Flyvbjerg et al. 2012, p. 25; see Chap. 7).

According to the (Flyvbjerg et al. 2012), Phronetic research then, along with digital ethnography "*Netnography*" for (Kozinets 2015), is dialogical in the sense that it includes and (if successful in conduct) is included in a polyphony of voices, with no one voice, including that of the researcher, claiming final authority. It is also potentially liberator or empowering, in creating the space for people to tell their own stories (Freire 1970). The goal is to produce an input into the ongoing social dialogue and praxis of a society, rather than to generate ultimate, unequivocally verified knowledge. (Kozinets 2015, p. 143).

Conducting the Phronetic mixed-method netnography approach occurred through interpreting and studying the culture of an online community, thereby analyzing their engagements in real life. Then, data were obtained from initial observations of the digital social media from online applications, such as social network sites of Facebook pages and YouTube channels. In addition, I collected these datasets using the online application for an extended time in the open online domain (e.g., 6 months to 1 year). Then, I applied the data analysis technique by developing descriptions for the online community and applying analytical procedures to the online collected material. In addition, I made interpretations of the dataset collected. Finally, I studied the narrative form in the online community to develop descriptions of the social behaviors of a group or an individual in the online community, which led to examining the hypothesis in question. This ICT-MR integrated the Phronetic mixed-method approach to netnography with the process for data collection, data analysis, and narrative analysis, according to researchers (Creswell and Clark 2004), which included the integration of quantitative and qualitative data to develop a rational modal.

2.5 Data Collection Process

Sites studied included http://www.wasatia.info and Facebook pages: http://www.facebook.com/wasatia.info and http://www.facebook.com/WasatiaVision. I studied members on those websites. Every individual who registered on the site, http://www.wasatia.info, authorized me to explore the ideas of the online community that he or she represented. Data derived from website content, such as articles, images, news articles, videos, members' applications, posts, and comments from the sites listed above. I collected all forms generated on the website and all application results generated through online application software, such as the polls and the forums on the site. The ethical considerations that relate to participants in the research were explained in Chap. 7 under the Rules and Regulations for the site, as from the perspective of Wasatia NGO in Palestine. There were principles and guidelines for conducting the ethical consideration of an ethnography research. Boellstrorff et al. (2012) considered some principles when conducting research on an ethical level such as "*the principle of care, informed consent, mitigating institutional and legal risk, anonymity, deception, sex, and intimacy, doing good and compensation, taking leave, accurate and empathic portrayal*" (Boellstrorff et al. 2012, p. 129). Since we are exploring and examining social change, the research is clearer within the ethical consideration for the participants; the more important is to measure social change towards the reconciliation process, and not the participants themselves. This makes it clear that there is no "*body harm*" to any participant, this aspect of ethical consideration is more explored in Chap. 7.

2.6 The Mixed-Method Experimental Research Design

This section illustrates the conducted telos that was considered when developing a research design applying mixed-method. The research design focused on developing a Mixed-Method research design for ICT-MR; the telos were as follows:

- Why use mixed-method experimental intervention design (Creswell and Clark 2018)?
- Consider the theory; theory informs practice and practice inform theory.
- The purpose of Internet communication research.
- The Mixed-method implemented in the Internet communication research.
- The researcher explores the Mixed-Method netnography approach in the Internet communication research.
- The coding procedure and the analysis process for Internet communication research.
- The validation process for Internet communication research.
- The researcher develops the rational model for Phronetic mixed-method netnography design.

2.6.1 Applying Mixed-Method in the Communication Research

There are some fundamental differences between qualitative, quantitative, and mixed-method approaches. To choose which method to conduct the Internet communication research, I explored all those methods; according to Creswell (2014, p. 19), the critical factors for those differences were as follows: The quantitative approach *is "post-positivist, experimental design, and pretest and posttest measures of attitudes"* (Creswell 2014, p. 19). The quantitative method approach involves testing a theory from a narrow hypothesis with experimental design (Creswell 2014, p. 19).

Conversely, the qualitative method approach is *"constructivist, ethnographic design, and observation of behavior, transformative, narrative design, and open-ended interviews*" (Creswell 2014, p. 19). In this ICT-MR Internet communication research, I applied netnography[4] research, transformation, and a confirmatory on experimental design to understand the behavior and conduct of the online community, where I observed and collected data in the beginning and the end.

Finally, the Mixed-Methods approach is *"pragmatic, and the collection of both quantitative and qualitative data sequentially in the design"* (Creswell 2014, p. 19). It involves an inquiry into the assumption of collecting various data that best provide more understanding of a research problem than either quantitative or qualitative data can offer. It involves exploring the meaning of a phenomenon. Therefore, I chose to utilize the mixed-method research approach for the following reasons (Creswell 2014, pp. 18–19):

- Pragmatic knowledge claims mean that themes are sequential, concurrent, and transformative.
- I used both open-ended questions and closed-ended question to develop emerging and predetermined approaches, as well as to explore and interpret both quantitative and qualitative data. Moreover, I analyzed both qualitative and quantitative data using CAQCAS, such us NVivo 10 for Windows.
- I collected both quantitative and qualitative data.
- The mixed-method developed mixed rational data.
- The mixed-method integrated the data at various stages of inquiry.
- I used it to explore and interpret different visual pictures in the research.
- The mixed-method research implemented both practices of qualitative and quantitative approaches, by collecting the various types of data that best provided a complete understanding of the research problem than either quantitative or qualitative method.
- The mixed-method was an experimental design, as well an ethnographic design; this ICT-MR used the mixed-method as an experimental design to develop quantitative data, where themes emerged, and then use those themes for qualitative understanding for the data. The design was a *netnography design of an online community.*
- I developed a complete knowledge of the research by converging quantitative and qualitative data and comparing the two data at various times.
- I used it to understand the collated data at a more detailed level by using qualitative follow-up data to help explore and interpret the quantitative data.

[4]Netnography: term develop by Robert Kozinets to research communities that are online.

2.6.2 From Theory Into Practice, and from Practice Develops a Theory

The theory is the framework that the researchers have utilized in their design; it informs many aspects of the study from the issue to the problem, to the findings and final suggestions for proving it (Creswell 2014, p. 284). This Internet communication research ICT-MR adopted the Phronetic research plan by Flyvbjerg (2001) by applying a netnography Mixed-Method approach to the research plan. Then, utilizing the methods of qualitative content analysis of the research to examine, evaluate, and validate the research according to the principles and methods of qualitative content analysis.

2.6.3 The Purpose of the ICT-MR Internet Communication Research in a Netnography Mixed-Method Approach

The netnography mixed-method approach addressed the utilization of ICT for reconciliation in the midst of the Israeli/Palestinian conflict. Moderation, reconciliation, and democracy represented the right path to break the chains of stagnation. Therefore, I expose moderation, reconciliation, democracy terminologies, and understanding within the Palestinian community. An explanatory sequential of a mixed-method design was utilized in the research and then developed more to a Mixed-Method intervention experimental design, which I found in the new book of Creswell and Clark (2018), which was a type of design in which qualitative, quantitative data were collected sequentially, and then analyzed separately and had an intervention experiment, that when using the social network sites including Facebook Pages, to develop social change. Themes emerged by explaining the quantitative results with exploring in depth a follow up for the qualitative data results.

The next phase involved using qualitative content analysis methods and procedures for data analysis, interpretation, and validation of data emerging from quantitative and qualitative results that end in emerging new themes. This ICT-MR Internet communication research developed a more detailed understanding of the results from applying qualitative content analysis on the collected data and developing a validation process for the resulting data, as well. That results presented knowledge of the primary research purpose of endorsing moderation, reconciliation, and democracy amid conflict.

2.6.4 Mixed-Method Procedure for the ICT-MR Internet Communication Research

There are different types of mixed-method procedure; the ICT-MR Internet communication research emphasizes only one type, which was the mixed multiphase method. This refers to one conducting several mixed-method approaches, such as the mixed-method explanatory or sequential approaches; this form of research is most common because programs usually implement fields in which multiple phases of the project stretches over time. Creswell's (2014) multiphase explanatory Mixed-Method is explained in Figs. 2.8 and 2.9.

The coding procedure for Internet communication research was as follows (Creswell and Clark 2011): The researcher

- collected the data, such as text files, images, and videos;
- prepared the data for analysis;
- wrote field notes for the collected data;
- Read the data and obtained a sense of the material;
- Coded the data and assigned labels that related to the theme of the research;
- Coded the text for interpretation used in the research report; and
- Coded the text for themes that can support thc research (see Fig. 2.10).

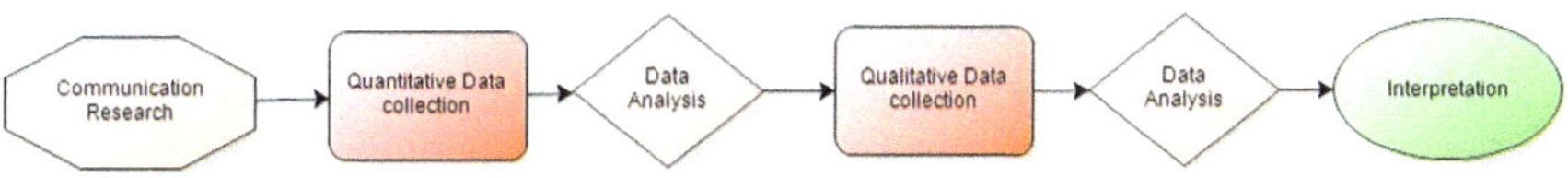

Fig. 2.8 Explanatory sequential Mixed-Method. (Creswell 2014, pp. 215–237)

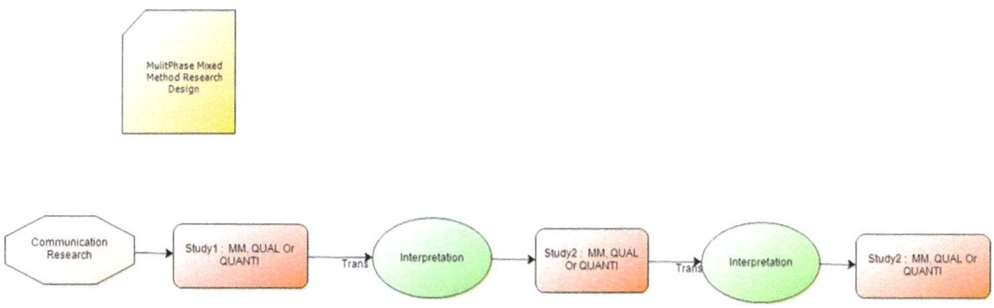

Fig. 2.9 Multiphase method. (A source from Creswell (2014, p. 221))

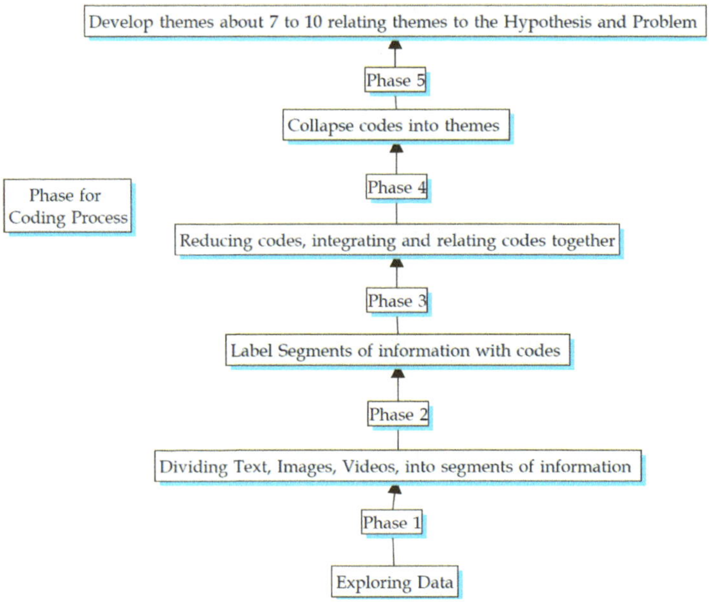

Fig. 2.10 Coding process. (Creswell and Clark 2011)

After coding that data, I started the process of analyzing the data. The overall process for analyzing the data was as follows (Creswell and Clark 2011):

1. The researcher must explore the databases of the dataset for the material.
2. The researcher starts to develop a coding procedure out of the data.
3. The researcher develops findings—a description and themes that relate to the research hypothesis and problem.
4. Presenting the description of the data and the emerging themes.
5. Interpreting the findings
6. Validating the findings

Figure 2.11 illustrates the analysis process.

2.6.5 Validation Process for the Internet Communication Research

The process for validating the accuracy of the information drives into different steps (Creswell 2014, p. 197) involved

1. Subscribing the raw data, such as field notes, images, videos, articles, etc.;
2. Organizing and preparing the data for analysis;
3. Reading all the data;

4. Coding the data using CAQCAS such as NVivo 10 for Windows;
5. Writing themes and descriptions for findings;
6. Interrelating themes and description of the primary hypothesis and research problem of the Internet communication research; and
7. Interpreting the meaning of the themes and description by challenging it with the hypothesis or the research problem.

The validation process is illustrated in Fig. 2.11.

2.6.6 *Devolving the Rational Model Design for Mixed-Method*

This approach was established to implement applied Phronetic social science using in mixed method research design. The Phronetic approach was used to develop the phases of progress to conduct the ICT-MR research on the Palestinian community. The process to carry out the ICT-MR was referred to as the mixed-method netnography approach (see Fig. 2.12).

First, I chose the community for the research, and then I applied the mixed-method research design. I deployed an online community and applied netnography in the online community. I collected data with the help of a CAQCAS, NVivo 10 for Windows to develop data sets for two Facebook pages—Wasatia. Info and Wasatia Vision—from the social network website, http://www.wasatia.info. I applied the

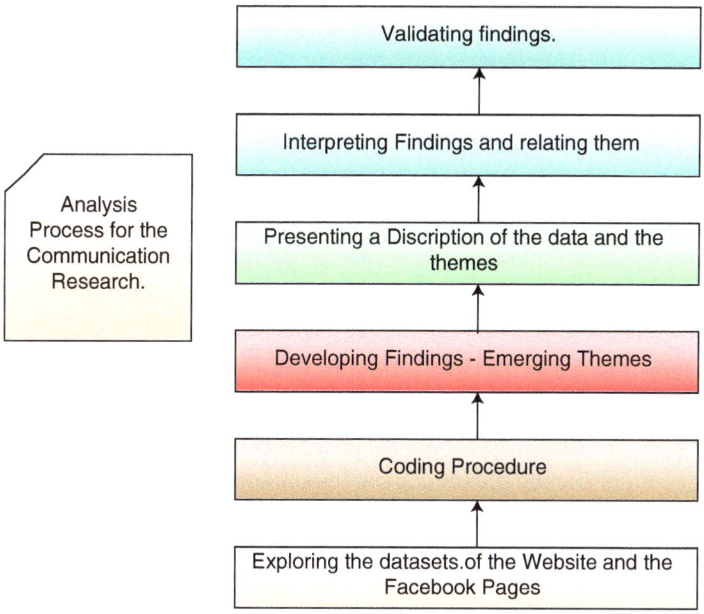

Fig. 2.11 Analysis process developed by Creswell and Clark (2004)

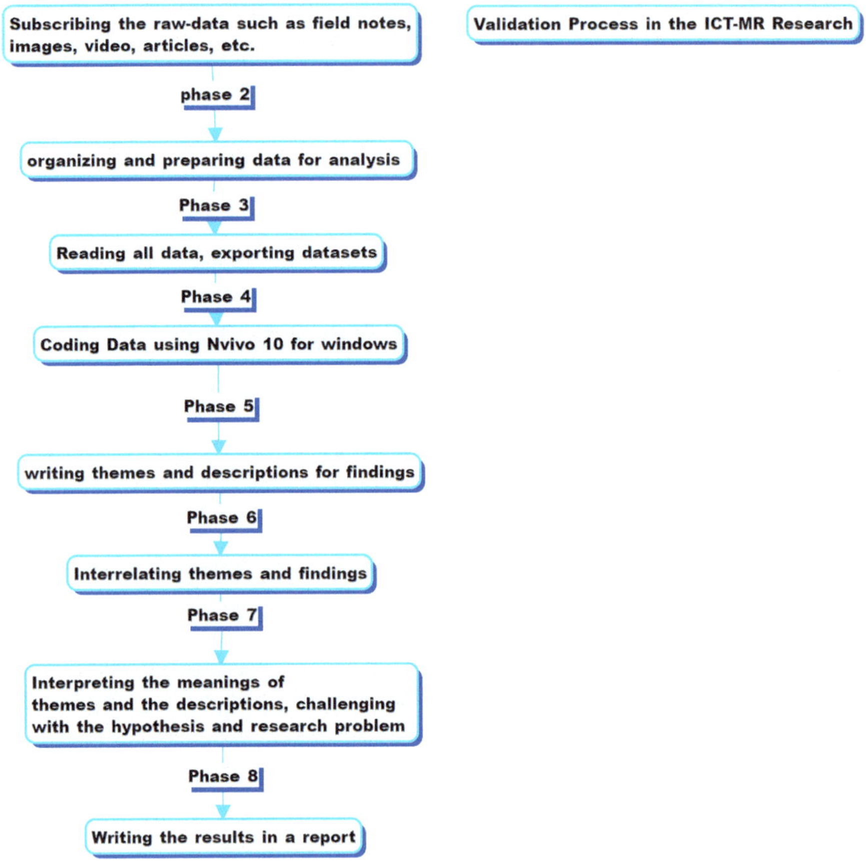

Fig. 2.12 Validation process, as adopted from Creswell (2014, p. 197) as phases for validation

qualitative content analysis method and procedures for the research using CAQCAS, NVivo 10 for Windows. I collected qualitative and quantitative data using NVivo 10. I then interpreted and validated procedures to provide a follow-up of interpretations by applying the explanatory sequential multimethod approach. Finally, I provided evaluation and concluding remarks in the final stage of the research.

2.6.7 The Complex Mixed-Method Intervention-Experimental Research Design

This method was applied to complex mixed data requiring intervention, and the intervention in this research was netnography and the influence of ICT on the community for social change. The method is "*providing reasons for the collection and timing of the qualitative data, taking steps to reduce the potential for introducing*

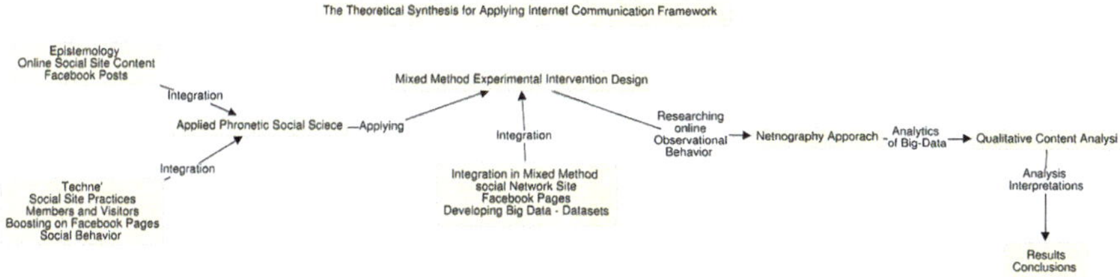

Fig. 2.13 Rational model for Phronetic Mixed-Method netnography research design

bias if the qualitative data collection occurs during the trial, and determining what collected qualitative data will best augment the experiment" (Creswell and Clark 2018, p. 20). Figure 2.13 shows the applied procedure for a Mixed-Method to develop analytics for the ICT-MR.

Figure 2.14 shows the discourse for the Mixed-Method experimental intervention procedure as defined by (Creswell and Clark 2018). The Mixed-Method is illustrated by the integration of both Facebook pages and the social network site to develop qualitative mixed data from both and then applying qualitative data analysis. This represented the first phase of the research analysis. The second phase was done by having an intervention procedure, which involved applying the Phronetic netnography approach. The CAQCAS software was used to develop the qualitative data and derive from it quantitative data and to apply quantitative data analysis; the CAQCAS is NVivo for Windows 10 was used. Finally, these data led to interpretations that examined and explored the hypothesis positively and negatively, the figure is more explored in detail in Chap. 8.

2.6.8 The Phronetic Complex Mixed-Method Netnography Integration

The Mixed-Method involved establishing a study integrating the online social network site and the Facebook pages to generate a context-specific qualitative data collection and convey this to link to a quantitative data collection through a display using CAQCAS that summarized how qualitative results informed each aspect of the quantitative results. This integration also involved interpreting data at the end of the study, making this the best method applied in the social network analysis, as illustrated in Chaps. 8 and 9. Figure 2.15 describes the Mixed-Method experimental design that was used to strengthen the contribution of using such method in the netnography approach; the intervention in the Mixed-Method was netnography.

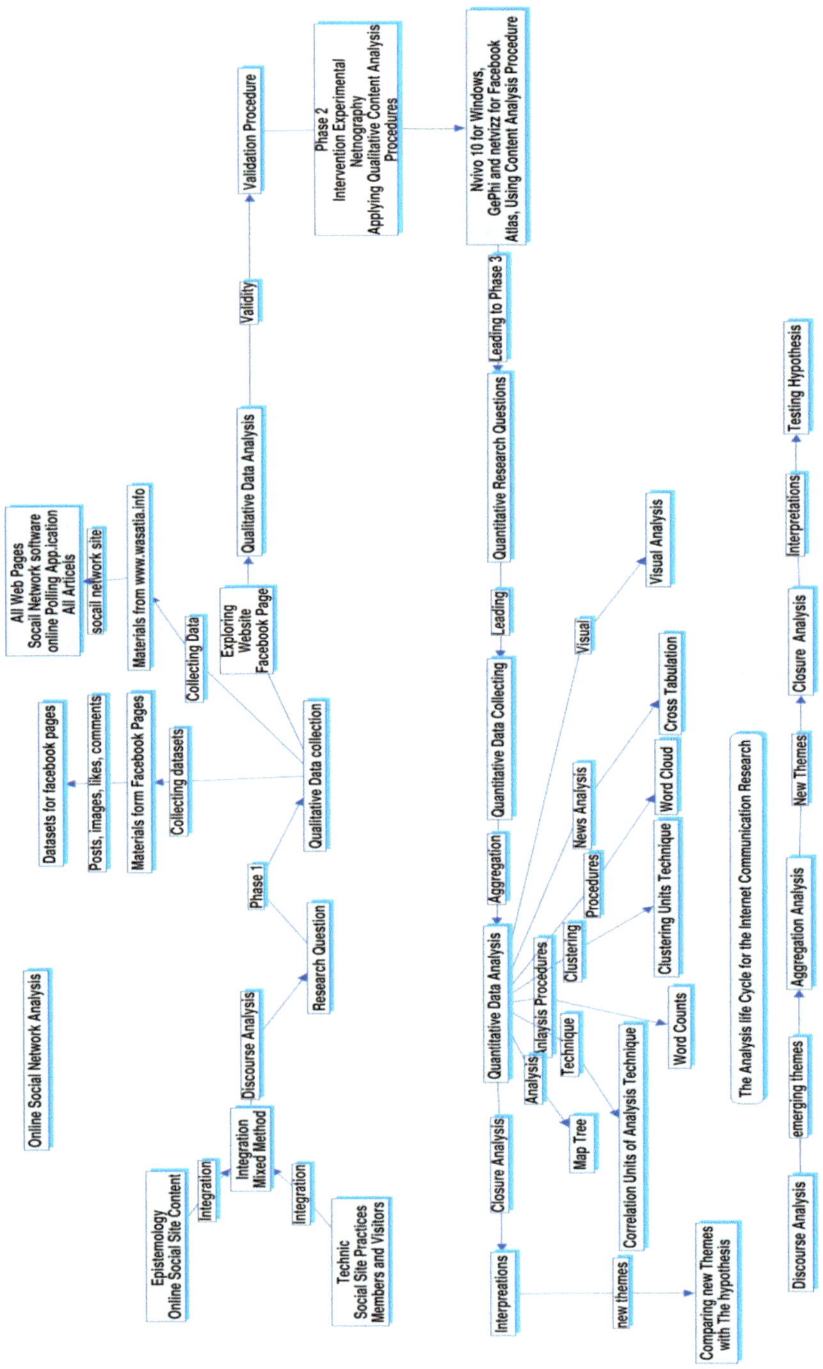

Fig. 2.14 Applied Mixed-Method experimental research design procedure

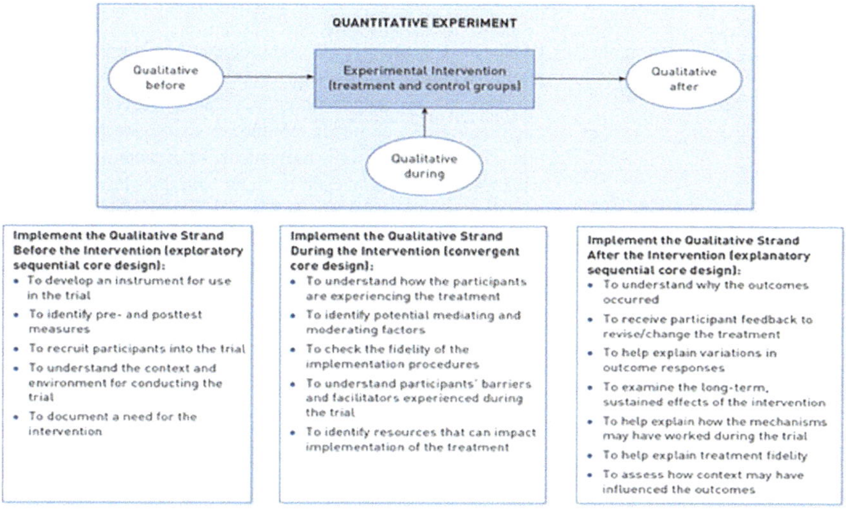

Fig. 2.15 Adding qualitative data into an experimental design to form a Mixed-Methods experimental design. (Source: Creswell, John W. Designing and Conducting Mixed-Methods Research (On the electronic version of Kindle Locations 2330–2331). SAGE Publications. Kindle Edition)

2.6.9 Applied Qualitative Content Analysis in Qualitative Research

The features and the steps for the qualitative research used to apply qualitative content analysis are defined and discussed in Table 2.1.

In the ICT-MR, I encountered the issue of the evolvement of radicalism by using ICT as a tool for the radicalization of society. The hypothesis involved researching the effectiveness of ICT to foster moderation, reconciliation, and democracy (Wasatia Circle)[5] in the middle of the conflict. The data were collected via datasets, which were exported from the website of the NGO Wasatia and from two Facebook pages (Wasaita.info – Wasatia Vision). The data were collected from the content of the website, such as articles, news, and the Auschwitz news on the trip conducted by the participants of the project. After collecting the data, analysis techniques were implemented to develop new themes that were used to examine the hypothesis. I worked on data collection and on data analysis using qualitative content analysis methods and procedures, as described in Chaps. 8 and 9. I was the Phronimos and worked on data representation using computer assistant qualitative content analysis software (CAQCAS), such as NVivo 10. In addition, I applied data validation using content analysis techniques, as defined in Chap. 9. Table 2.2 illustrates the steps, which are conducted to apply qualitative research and to examine the hypothesis in the research.

[5] Wasatia Circle discussed in Chap. 4.

Table 2.1 Characteristics of content analysis

Characteristics of qualitative research	Description of the characteristics
Empirically grounded method	Exploratory in the process and predictive or inferential in intent. Example: Words, signs, significance, symbols, and logic, examining data, printed data matter, images, or sounds and text to make an understanding of what those data would mean for people and how they enable or prevent a situation and what information conveyed by them, and does those elements or interpretations relates to the Internet communication research.
Contemporary content analysis transcends traditional notions of symbols content and intents	How does the development of the media technologies shape the attention of its users to communicate and its role in influencing the culture in assigning significance to what is being analyzed?
Contemporary content analysis has been forced to develop a methodology of its self.	It enables researchers to plan, execute, communicate, reproduce and critically evaluate results through analysis

Table 2.2 Methods for analysis: elements of qualitative research steps

The problem	The use of ICT as a tool for the evolvement of radicalism in the middle of conflict (Israeli – Palestinian).
Hypothesis	ICT can foster moderation, reconciliation, and democracy in the middle of the Israel Palestinian conflict. And apply Netnography methodology to develop qualitative data structure for the ICT-MR communication project, to investigate, examine and interpret the effect of ICT on the Palestinian society.
Sample	Taken from the website: www.wasatia.info, Facebook page: Wasatia.info – Wasatia vision.
Data collection	Datasets from Wasatia articles, Wasatia news, Wasatia Auschwitz news. Facebook datasets from Wasatia.info and Wasatia vision pages.
Analysis	Word Counts, Word Cloud Analysis, Tree Maps, Correlations units of analysis and clustering analysis. Using NVivo 10 for Windows to analyze the data using Qualitative- Quantitative Content Analysis method described in (Chap. 8) and implemented in (Chap. 9). Using Krippendorff validation methods outlined in Chap. 9 (Krippendorff 2013, p. 329).
Results and discussions	Straight description of the data organized to "*fit*" the data findings of the analysis. Reporting the results of the analysis see Chaps. 9 and 10 for the conclusion

Part III
Internet Communication Technologies (ICT) for Reconciliation Process "Episteme Approach"

Chapter 3
Reconciliation in the Middle of Conflict Exposed in Internet Communication Technology (ICT)

3.1 Introduction

> *A true and a lasting peace also requires a culture of peace, that is a comprehensive, society-wide system of values, beliefs and attitudes the interplay and influence of which in and on the civil society would lead citizens of the Middle East – Arabs, Israelis, Palestinians – in their daily lives, on the ground, to put a premium on peace, to desire peace, to seek peace and to stand for peace. – Proceedings of the United Nations Department of public information's international encounter for the European Journalists on the Questions of Palestine, June 9-11-1993, London.* (Bar-Tal and Bennink 2004, p.11)

Having introduced the research methodology, the figure below serves to build up the theoretical foundations for reconciliation in the middle of the conflict, to explore how ICT can expose the reconciliation process (Fig. 3.1).

Increasing technological advancements and innovations in communications have revolutionized how individuals think and behave. Such new technologies have provided a source of opulent information to people everywhere, whether the intentions are good or bad. With the growing number of global ICTs, such as Facebook, YouTube, Instagram, and Twitter, there is an abundance of information, which is also ambiguous or mentioned as fake data or news. "*Big data*" are the massive amount of collected data over the Internet, which are readily available, influencing the behavior of people mostly online and within the community. This behavior pattern affects people of different social levels and ethnic backgrounds. However, non-traditional media forms can be biased, where facts and truth are difficult to judge.

Cole & Crawford studied building peace through information and communications technologies. They stated, "*Information and communication technologies can be a huge aid in the effort to build lasting peace, by helping people communicate, view information, make decisions, and understand each other better*" (Cole and Crawford 2007).

I. M. AlDajani, *Internet Communication Technology (ICT) for Reconciliation*, Lecture Notes in Social Networks, https://doi.org/10.1007/978-3-030-41203-6_3

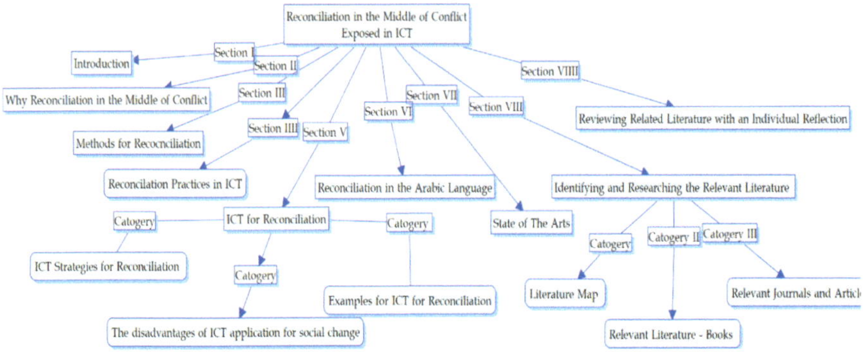

Fig. 3.1 ICT for reconciliation in the middle of conflict

People and institutions around the world influence times of peace and times of war, which are harnessing the power of ICTs in varying ways. The hypothesis for this research study was whether ICT could foster concepts of moderation, reconciliation, and democracy to influence communities towards the reconciliation process, whether it was in the virtual world or not, and whether in peace times or in war times. As Kofi Anan (1999) stated, "*The same Internet that has facilitated the spread of human rights and good governance norms have also been a conduit for propagating intolerance and has diffused information necessary for building weapons of terror*" (Bidgoli 2006, p. 28).

With the widespread modern availability of the Internet and satellite communications, public information supplied by digital social network sites plays a crucial role in forming opinions and influencing individuals and groups. For example, ICT seems to have been an essential tool for activists involved in the Arab Spring awakening. Social media utilizes platforms, such as Facebook, Twitter, and YouTube, and other conventional media, such as TV, Satellites as Al-Jazeera news broadcaster, and other news agencies, a vital tool for the Arab Spring horizon. These were used increasingly for expressing views concerning reasons, which influences people to reassert their human rights to overthrow their respective autocratic governments.[1]

Activists around the globe used online social networks and bypassing governmental restrictions and censorship, thereby making them capable of sharing an immense amount of uncensored information[2] to spread their messages to the people. As was the case in the Arab Spring, online social networking sites provided activists with the stamina and power to create the democratic movement that eventually guided ordinary citizens to rebel in overthrowing the long-entrenched authoritarian dictatorship regimes in countries, such as Tunisia, Egypt, Libya, Yemen, and the scandals of the on-going war in Syria. In such countries, the influence of online social networking sites assisted in increasing the magnitude of protest

[1]The first reported use of social networking websites by dissident groups taking part in a civil revolt was in Moldova in April 2009.

[2]For instance, the Egyptian government of ex-President Hosni Mubarak tried to censor the Internet to no avail and the state was not able to stop what is being spread over the Internet.

demonstrations. Alternatively, governments such as in Bahrain, Syria, and Iran, applied sanctions and restrictions on online social media. The online network could not be trusted; some were fake media or fake news; and the effect of this resulted in it being ineffective compared to the former regimes, which were not subject to sanctions. Those restrictions and sections illustrated some vulnerability and a disadvantage of using ICTs in those countries.

The ICTs can connect and coordinate activities by groups and individuals from different destinations and by empowering them to connect and participate in communication for peaceful coexistence, such as peace organizations and other organizations that support human rights without the censorship of the suppressive governments. Also, ICT provides rapid Internet interaction through networks, such as Twitter and Facebook, which provides information to people and groups about meetings, sit-ins, rallies, demonstrations, or peace activities. Such networking can increase the power and pressure on the government to adopt a peace agenda.

For peace activists, ICT may bring together distant individuals or groups with similar views, offering tools that took over conventional media, which is usually part of the problem rather than being part of the solution because it would be under the official control of the governments. Gilbert Achcar[3] (2013) in *The People Wants: A Radical Exploration of the Arab Uprising* addressed the question of whether the Arab Spring could have the same influence, or even if it happened at all without the use of satellite TV, as it had a significant role in inflaming the Arab uprising. He went on to list five explosive factors that he believed ignited the upheaval in the Arab World in 2011: the lack of economic growth, massive unemployment, widespread corruption, enormous social inequalities, and despotic government's void of democratic legitimacy. Achcar (2013) maintained that the mass of the people that evolved from the Arab spring came from different classes of social layers and they were united under the aspiration of having a real democracy in their countries.[4] Achcar then raised the question as to why the Arab Spring came all together at once; the response provided from this was the following: Several features of the upheaval were related directly to another revolution: the global information revolution. How one receives and shares information has transformed in the past few decades. Satellite TV, using online social networks like Al-Jazeera, has given the Arab region a common bond based on its linguistic unity. By transcending state boundaries and ignoring state censorship, this new communication technology has allowed the populations of the Arab-speaking region to follow the events in real time. In instances where repression prevented TV cameras from attending protests, as in the case of Egypt, Syria, and in Iran's presidential elections (2009), many activists used their phone cameras and YouTube to project images of struggle and repression. Some images were quickly picked up by news organizations and conveyed to the broad watching public.

---

[3] Gilbert Achcar جلبيرالأشقر a French-Lebanese political scientist and sociologist; since 2007, he has been Professor of Development Studies and International Relations at the School of Oriental and African Studies (SOAS) at the University of London, 2013.

[4] Gilbert Achcar, *The People Want: A Radical Exploration of the Arab Uprising*, translated by G. M. Goshgarian (Berkeley: University of California Press, 2013).

The Internet and online social networks have become the way for communication. Interestingly, the younger generation uses ICT more than, the older generations, as they can understand easily how the technology works and are more interactive compared to the older generation. The young, educated, and politicized find themselves amid a political stream, leading the population to revolt and rebellion against their entrenched authoritarian regimes. This new breed of activists makes intensive use of ICTs, such as Facebook, Twitter, and You-Tube, which allows them to network with ease and speed to relay their messages.

Terrorist groups used the Internet social network to strengthen their ideologies and recruit more members into their sects worldwide. For that reason, ICT should start to deploy Internet social networks in conflict areas for reconciliation to develop a grasp of what causes conflicts in different areas and to try to recruit members who want to aspire moderation, reconciliation, and democracy in the middle of the conflict. According to the UNCTED (2015), "*Terrorist groups continue to utilize the Internet and social media to facilitate their activities, including incitement to commit terrorist acts, radicalization to violence, recruitment, training, planning, a collection of information, communication, preparation, financing, and execution of attacks*" (Special meeting of the Security Council Counter-Terrorism Committee and its Executive Directorate 16–17 December, 2015).

The development of ICTs can help break down the psychological barriers, which prevent opposing groups from working together, thereby ideally contributing to new senses of trust and familiarity between the two parties (Dahlberg 2007). Reconciliation is needed, as the alternative is a frozen conflict between two rivals. For example, the Israelis and Palestinians conflict for more than 60 years, and their children and grandchildren inherited this conflict. Reconciliation is for future generations who want to thrive beyond the conflict.

According to Leiner[5], "*reconciliation is a comprehensive strategy that includes civil society, as well as political leaders. It involves grassroots activities, lawyers, psychologists, religious leaders, historians, the popular media, literature, women, and men. Reconciliation must start now, amid violence and conflict; moreover, "the process of reconciliation is long, and the alternative is to live in violence and danger and to become a victim or perpetrator of the murde*r" (Leiner 2016a, b, p. 45). Reconciliation among conflicting groups is one crucial element of and a condition for effective democracy. To summarize, how can ICT be a means to an end, as a tool for directing the community toward adopting moderation, reconciliation, and democracy in their daily lives, which adheres to peace and reconciliation via democracy, not to foster hate.

The research case study is the Palestinian community for their need to adopt moderation and reconciliation in the middle of the conflict. For that reason, the researchers measured whether any kind of democratic deliberation was present through the Internet communication research and whether it led toward an

---

[5] Professor Martin Leiner, is the director of Jena Center for Reconciliation Studies at the Friedrich Schiller University in Jena – Germany, and the lead professor for the Heart of Flesh Not Stone Project.

understanding of moderation, reconciliation, and democracy in the Palestinian community that would at the end of the research adhere to the reconciliation process.

According to Bar-Tal and Bennink (2004), "*reconciliation is needed to build a peaceful relation between members who conflict; however, reconciliation will not substitute conflict resolution, as it was never to substitute agreements signed between two parties*" (p. 12). Instead, reconciliation is an outcome and a process, and it has different significant definitions. In this research, the focus was on reconciliation in the middle of the strife; Leiner (2016a, b) referenced this notion in the "*Hölderlin Perspective.*" Reconciliation is needed in the Israeli Palestine conflict because this conflict penetrated the cultural infrastructure and the psychological barriers for both societies of the Israeli and Palestinian nations. Leiner emphasis reconciliation as a process and peace is the outcome.

3.2 Why Reconciliation in the Middle of Conflict?

Reconciliation has a broad meaning for the understanding of the word, as it conjugates to many different aspects in different domains to construct peaceful relations. Reconciliation.

> *is the restoration of the relationship between individuals, groups, states after violence, war, genocide, civil war, gross human rights violations like segregations (Apartheid), enslavement, or similar activities. Reconciliation as policy requires a long-term strategy with many practices with multiple levels.* (Leiner 2016a, b, p. 185)

Reconciliation is needed when

> *societies involved in a conflict evolve widely shared beliefs, attitudes, motivations, and emotions that support adherence to the conflictive goals, maintain in the conflict, delegitimize the opponent, and thus negate the possibility for a peaceful resolution and prevent the development of peaceful relations.* (Bar-Tal and Bennink 2004, p. 13)

Social beliefs, where fear, anger, and hatred become part of social ethos in society, are formed in conflict, adopted by societies, maintained from government institutions, and supported as part of the collective social memory, which further stagnates the conflict. The fuel for the continuation of the conflictive relations develops obstacles to the progress of peacebuilding. These kinds of conflicts do not change by day and night, nor by peace agreements; these conflicts require a social change within the society, including a new belief for coexistence and harmony. However, to reach this social change, the society must introduce a social change toward the reconciliation process (Bar-Tal 1998, 2001, 2007). Therefore, the reconciliation process requires a social change to develop peaceful relations between entities in conflict. It requires effort to overcome social, cultural, and economic obstacles that are perpetuated by conflict.

Two types of conflicts need reconciliation; (inter or intrastate), one evolves into an inner group with one state, and the other is the outside group, such as two states in conflict or presenting different social, economic, political capitals. The inner group applying reconciliation has to commit to social change, which means having a new belief in a united in one belief. The concept would promote peaceful relations as a new form of interaction to develop a stable foundation for cooperation and peaceful acts that symbolize the relation, such as empathy for the other and accepting the narrative of the other. The other type that needs reconciliation is the between states that are in conflict, which may involve only the leaders and agreements signed by them without concerning their nations. This ignoring of the nation's opinions stagnates the hope to change the negative relations into peaceful relations, which are based on recognition, respect, the right to coexist, and the right to cooperate justly. The objective of reconciliation entails a process to change the motivations of the goals and beliefs, attitudes, and emotions for most of both societies and nations in conflict. This process of reconciliation changes the type of relationship between the societies and evokes harmony and empathy as people develop respect for one another.

The development of tensions among groups is one of the properties of conflicts evolving in social life, whether between individuals or groups, because of conflicting ideas between groups. Reconciliation usually involves nations that have conflict within an ethnic, religious, or ideological concept. Those conflicts might have different goals that need to be met on both sides, which can create harm socially, culturally, and economically, such as the inner Palestinian conflict between Hamas and Fateh.

According to Bar-Tal (2004), reconciliation is the formation of the peaceful relationship between societies that have been involved in the conflict, after a previous resolution is achieved (p. 14). Additionally, the Jena Center for Reconciliation Studies in Jena, Germany adopted the notion of the "*Hölderlin Perspective,*" including the concept of reconciliation, thereby emphasizing reconciliation during conflicts. As Leiner explained in the book *Latin America between Conflict and Reconciliation, Hölderlin Perspective* derived from the German poet Friedrich Hölderlin (1770–1843), who wrote in his novel, *Hyperion:* "*Versöhnung ist mitten im Streit und alles Getrennte findet sich wieder*" (Leiner and Flämig 2012, p. 16). In English, this sentence reads as the following: "*Reconciliation is in the middle of the strife, and all that was separated finds each other again*" (p. 16). This notion shows that reconciliation can be part of conflict transformation, which transforms conflict relations to peaceful relation within the conflict.

Reconciliation is known to many scholars as requiring the support of the majority of society; however, in the aspect of the *Hölderlin Perspective*, reconciliation requires one to be in the middle of the conflict, and it can work even within the minority by changing this minority into the majority. Reconciliation has a holistic combination as an outcome and as a process, where the process is the reconciliation, and the outcome is peace; both are combined in a recurrence of action according to the permanent effect on both parties in conflict. The focus of reconciliation is to reach a stable and lasting peace. This process provides mutual recognition and acceptance for the narrative of the other by accepting their interest and goals;

providing mutual trust and recognition, developing cultural, economic, and political relations; and respecting the sensitivity and consideration of the other parties' needs and interests (Bar-Tal and Bennink 2004, p. 15).

When reconciliation becomes a process within groups in conflict, both groups become united in political, economic, cultural systems, thereby developing a single political entity. However, in conflicts between nations, it becomes a relation between two states that have the acceptance and the recognition of the other, forgiving each other for past adversaries, restoring good relations with good intentions, and developing a political restructuring of former relations. Therefore, reconciliation calls for a restructuring mechanism of social, cultural, economic, and political spheres. This restructured system can be implemented on both outer-groups, reconciliation between two states, and within the inner-groups, reconciliation within entities of a state.

The structural elements that evolve from reconciliation include opening and maintaining regular channels of communication between the government of the two states; reducing violent threats and tensions on borders, demilitarization, disbarment, and reduction of military manpower on borders; developing joint institutions and organizations; developing free and open trade; developing cooperative and economic projects; exchanging vital information; developing free joint ventures in different areas, such as free and open tourism; and mostly exchanging cultural social spheres(Bar-Tal and Bennink 2004, p. 16).

In addition to the disarmament, demobilization, and reintegration (DDR) programs, the reconciliation process contributes to security and stability by disarming combatants, removing them from military structures, and socially and economically integrating them into societies (Taies Nezam and Marc 2009).

Reconciliation is needed when societies involved in a conflict evolve widely shared beliefs, attitudes, motivations, and emotions that support adherence to the conflictive goals, maintain in the conflict, de-legitimize the opponent, and thus negate the possibility for a peaceful resolution and prevent the development of peaceful relations. (Bar-Tal and Bennink 2004, p. 13). Therefore a definition must be stated for reconciliation, according to Leiner (2018) "*Reconciliation can be defined as the overarching approach to conflict which focuses on the process of rebuilding relationships. Its goal is to create normal, trustful and if possible good and peaceful relationships" (p.179)*. He also defines the phases for reconciliation in phases; reconciliation in the phase of justice, reconciliation in the phase of truth, reconciliation in the phase of building resilience (Leiner 2018, p. 182). He also states some preferences for applying the reconciliation process such as; that all parties must be integrated in the reconciliation process, that it must be promoted publicly, that it must inquire and investigate into the exact needs of all parties concerned, and that the preparations must be considered before entering the reconciliation process (Leiner 2018, pp. 180–181). The research tries to approach those preferences as basic needs for the reconciliation process in ICT.

3.3 Methods for Reconciliation

The proper implementation of ICT requires a move toward the advancement of reconciliation and for the transformation towards democracy, endorses individuals and groups for reconciliation and democratic values. ICT as a tool for reconciliation can pave the way for a change of recognition for the other by replacing the negative image of the relationship between both entities of conflict to a renewed positive image driven by prosperity for both entities in all levels of society, such as cultural and economic capitals.

The methods for reconciliation are diverse, depending on the type of conflict, whether it is an interstate conflict or intrastate conflict. According to Bar-Tal and Bennink (2004), to influence change from conflict to reconciliation, there should be a social structure mechanism for social change and a psychological structure for change in the conflict sphere. When developing a social network site, one should emphasize those structures while developing the social network site and highlight those methods when starting empirical work, such as the analytics produced from the datasets imported from the social network sites, the analysis, and the interpretations. The concept of the research was to develop a social network site that adhered to the reconciliation process. First, the research must explore the reconciliation methods introduced by (Bar-Tal and Bennink 2004, pp. 20–38) for social change, and these are according to the following principles:

- **The societal beliefs about groups' goals.** The aim of this principle is to develop a peaceful relationship with a former enemy, thereby transforming conflict goals and social beliefs. The reconciliation process must help obtain new goals and beliefs that hold the aspiration for both parties in conflict, post-conflict, shaped by conflict resolution agreement, or in the middle of the conflict to adopt reconciliation in the middle of the conflict. It should introduce both goals using ICT to develop a social influence on society, as ICT can reach a vast spectrum of society and can build a deliberative correspondence between the stakeholders in concern. For example, online deliberation can develop a transformation for the justification of the goals and social beliefs that out breached or maintained the conflict.
- **Social beliefs about the rival groups.** The image of the adversary group has to change, asking two parties to define the adversary images and ways in which these can be transformed to develop a peaceful relationship. In other meanings, it is important to recognize its members and legitimize its new role in advancing toward reconciliation. This approach develops the desire of both parties to develop and maintain the acceptance of the other. Also, the other important aspect of this principle is to elaborate on the victimhood of the other.
- **Societal beliefs about one's own group**. This principle elaborates the social belief of someone's own group, as well as its part in the conflict and its new role in reconciliation. The previous part, where the group was glorified by its actions and no wrongdoing, including self-praising, ignoring, and censoring information that might harm or give negativity toward the group, must change. In that sense,

the group must take responsibility for its actions in the conflict. It can be understood as self-recognition of actions being taken in conflict.
- **The social beliefs about the relationship with the former enemy**. The transformation of the relation between the two groups who were in recent conflict must change. The new relations between the two groups must stress for its importance of cooperation and its friendly relations by respecting each other's needs, goals, and general wellbeing.

The Society Believes in Peace This is a significant aspect, as it can be referred to as the prosperity of reconciliation by developing guidelines to achieve it. According to Bar-Tal and Bennink (2004), the process for reconciliation must describe a new definition of social beliefs and stress a multidimensional nature for peace and reconciliation. This aspect involves outlining the cost and benefits of reconciliation and the methods to achieve it by introducing the aspiration for living in peace and considering the two sides' interests for prosperity and the pursuit of happiness.

The Jena Center for Reconciliation studies, under Leiner (2016a, b), presented the following new elements for an approach toward the process of reconciliation (Tolliday et al. 2016, p. 186):

1. **An orientation toward the past**. The past should never be neglected, and it should be part of the process of reconciliation. The past must be dealt with by learning from mistakes of the past and emphasizing that the future is the path to a reconciliation process.
2. **Truth**. Truth refers to the acknowledgment of truth by both parties, including the victims and their perpetrators. Accepting the truth in the narrative of the other and acknowledging the wrongdoing from both sides should lead to a reconciliation process.
3. **The importance of conserving the past**. The remembrance of the past is essential. Remembrance can involve memorials, museums, books written by historians, and conservation of a historical site of suffering. Therefore, the history of the past should never be forgotten.
4. **The importance of guilt**. The individual perpetrator must accept his or her guilty of doing wrong. He or she should be punished in different ways, which should be accepted by both parties.
5. **The importance of forgiveness**. Forgiveness depends on the verbal expression of guilt; forgiveness might reveal the disturbing aspects of attitudes and drive attitudes toward the reconciliation process, as well as drive the acceptance of change toward the other.
6. **Accepting empathy**. Empathy refers to understanding the narrative of the other, as well as respecting each other's narrative of history, culture, and future aspiration, which must be a part of the reconciliation process.
7. **The importance of emotions**. Emotions can allow healing a victim's feelings of atrocities done by a perpetrator. The emotions must be restored to the original sentiments, neglecting hate and animosity for the other. Emotions can lead to the acceptance of the perpetrator into the community after the acceptance of an

apology, or punishment or reparations. Therefore, the changing of the image of the perpetrator must be accepted by the society.

8. **The vision for a shared future**. Both parties have to understand the vision of a better future for both parties in conflict to work toward the reconciliation process. The relation between the identity of the victim and the perpetrator should be stable at all times, with no change, not for future perspectives or historical aspects.

Those principles stated above were the epistemology part of the research, as part of the Applied Phronetic Social Science (known as "*Phronesis*") is to adopt epistemology and "*Techné*" to approach practice wisdom, thereby embedding those parts in an online social network site integrated with social applications, such as Facebook, Twitter, and You-Tube. The "*Techné*" part of the Applied Phronetic approach that described by (Flyvbjerg et al. 2012). Which can be adopted for reconciliation process and implement methods for reconciliation in accordance to Bar-Tal's work on the methods for reconciliation (Bar-Tal and Bennink 2004) and Leiner's (2016a, b) overview of different strategic practices for the reconciliation process in (Tolliday et al. 2016, pp. 188–203).

3.4 Reconciliation Methods Applied in Practice Using ICT

1. **Apology**. Apology illustrates confrontation with the past and taking responsibilities for the actions done during the conflict by admitting to these mistakes. If this type is not dealt with, reconciliation is not possible because each party portrays themselves as the victims, harboring negative feelings; those negative feelings might result in instigating new conflict. As Bar-Tal and Bennink (2004) stated, "*Apology is a formal acceptance for responsibility for misdeeds carried out during the conflict and an appeal to the victims for forgiveness. It implies pursuing justice and truth*" (Bar-Tal and Bennink 2004, p. 29). ICT can contribute to developing apologies through online social networks such as in Forms, You-Tube and Video and other ICT applications. For example, it can be a unilateral apology using different ICT applications where they express sorrow about what happened during the conflict and in the past. ICT can be mean to express forgiveness by accepting apologies in online social media between the perpetrators and their victims.
2. **Truth and reconciliation commissions (TRC)**. The commissions are a way to deal with the atrocities done in the past by revealing the truth about what happened in the past. It is considered a mechanism to serve justice for victims of past atrocities. In this act, compensation is not possible, but it reveals acts of violence, discriminations, violations of human rights, and other deeds of racism in the conflict. Some of the aims of the truth and reconciliation commission involve establishing a comprehensive record for the cause of the brutal acts, nature, and the extent of the violation of human rights. Those TRC gave amnesty

to perpetrators, as they became part of the victimhood for their disclosure of violent acts in a political context. TRC worked on restoring the dignity of the victims by giving them the opportunity to recount their experiences and share these. Lastly, this involves preventing human rights violations and recommending measures for reparations. ICT can help find experts in (TRC) using online search engines, and find other cases including witnesses, perpetrators, and victims by locating them online and spreading information to millions of viewers of the TRC.

3. **Public trials.** The public trials are a significant part of the reconciliation process, bringing a perpetrator to justice who committed human rights violations and crimes against humanity. This method aims to acknowledge and reveal the suffering of the victim and to recognize the violent acts done by the perpetrator. When trials are carried out, and the perpetrator receives punishment, a type of retribution occurs for the victim; therefore, the justice was carried out. However, those trials are subject to only the perpetrator and not the group, which gives them the opportunity to be a part of the reconciliation process. ICT can make the trials public for all audiences and expose what happened in the conflict and develop archive to store memories and history to learn for future references.
4. **Reparations** This method is most appealing in the reconciliation process, as it requests both parties to take responsibility for wrongdoing and to compensate the victims of this atrocity committed by the two entities in the conflict. This method indicates the admission of guilt and the recognition for the suffering of the other, where the victim is willing to forgive his or her perpetrator. ICT with its online media can develop a medium for donations to support reconciliation processes, for example, buying shoes for the kids in Gaza, which happened in 2014, some Israelis have paid for it, even it is in the middle the Israeli-Palestinian conflict.
5. **Writing a common history**. The method calls for recreating the history of the past, as agreed on by the groups who conflicted; the past that can be learned from its atrocities and the idea of *"never again."* It takes historians from both sides to agree on a shared history and to negotiate an agreed past of events for both nations that conflict. It also provides a basis for a new collective memory, which leads toward reconciliation. ICT can document archives for the history and memories of nations that are in conflict. ICT can develop a common history with nations in conflict by building a blog with documents, photos and videos, as both store their memories and histories online, which develops an understanding of the narrative of the other, and closer to empathy.
6. **ICT in Education**. This method is essential, as it is the only way to change the psychological barriers of the past, thereby promoting reconciliation. Education can involve peace studies and reconciliation studies to influence the students and members of society. Education constructs the students' ways of thinking, such as their values, beliefs, attitudes, motivations, skills, and behaviors that endorse the reality for the reconciliation process, preparing them to live in an era of peace and reconciliation. ICT can expose reconciliation theories and

develop online courses to teach reconciliation process. ICT can help explore reconciliation methods, theories, and practices.

7. **#HashTags#,** ICT can bridge communication between nations in conflict, as it can help expose its views presenting it on the Internet, and help widespread it all over the globe. Today, ICT is electronic and part of daily human life, as it is part of people's mobility, social, communicative, and political habits. ICT can be used to spread information about new strategies for reconciliation that can influence nations in conflict. Most importantly, ICT serves as a channel for communication to send messages for peace and reconciliation to different rivals in conflict. For example, on **Oct 27, 2017**, A Beirut and Telaviv were chosen as the best tourism cities to be visited in the Middle East. A beirute.com, a media company, tweeted, ***"Please remove us from this Narrative."*** The Tweet went viral; the Mossad Hashtagged hashtag **(#TelAvivLovesBeirut)** went viral, and Lebanon rejected it very negatively because they did not want their city to be affiliated with Tel-Aviv. This kind of communication, which is in the open domain of the Internet, went viral to everyone around the world; this kind of message can be sent over the Internet to express the need to reconcile with the other. Also, it illustrates the strength of the Internet and that it can bring strong rivals to communicate with each other.
8. **Published meetings between presentations of the group who conflict**. When conflicted groups have joint meetings, with their leaders or their academics, it should be published. This illustrates that both groups are working for reconciliation. For example, in October 2017, Tzipi Livni, the Minister of Justice in the Israeli government requested for recognition on her official page. She wanted Israelis and Palestinians to cooperate for peace to fight crimes of terrorism and denounce that the Israeli prime minister, who keeps saying there is no one to have peace with, is working with Palestinians as a countermeasure against terrorism. The Israeli and the Palestinians are meeting together to provide security for both nations. These kinds of meetings, if announced, illustrates that both are working to better the lives of their people, which can be an aspect for reconciliation process; if published, it would be a better aspect to spread that there is a process for reconciliation in the Israeli/Palestinian conflict.
9. **Exposing non-governmental organizations (NGOs) empirical work for the reconciliation process**. Non-governmental organizations, either from NGOs that are part of the society or from an international community, can contribute to the process of reconciliation. NGOs can illustrate that peace relations can bring benefits to societies in conflict, such as by spreading messages about the importance for the reconciliation process, about the prosperity of reconciliation, and the benefits of having peaceful relations with past enemies. ICT can help connect NGOs that are against the war to develop communication within each other, such what happened between Israel and Lebanon, where some NGOs were against the war.
10. **Online Joint projects**. Those projects are significant, as these can help bridge barriers between enemies who conflict by facilitating and researching psychological reconciliation. It can connect members of two groups from different

levels of societies in the same project. This provides opportunities for different members from different nations in conflict to have personal encounters, which would foster the understanding for the narrative of the other and would teach them about peace relations, as well as why it is crucial for both nations.

11. **Culture exchange using ICT**. ICT can help nations in conflict exchange cultures, by online videos and other interactive chat applications. This exchange can help each opponent in conflict learn about the other, such as through translating books, visiting artists, and learning academics from different conferences, exhibitions, and festivals. ICT uses games and other online documentaries to help expose other cultures that are in conflict.
12. **enhancing Empathy in conflict**, using virtual reality in ICT to expose different perspectives on what is happening in the conflict. There was a presentation in Jena, on the 4th of September 2018, presented by Eran Halpine, were he used virtual reality to expose different perspectives on the same scene of a checkpoint from the different views of a soldier, a Palestinian and a neutral observer. They found that it increases more empathy with the Palestinians when observed from the Palestinian perspective.

3.5 ICT for Reconciliation

This ICT-MR Internet communication research was supposed to promote society toward a reconciliation process in the context of the Israeli-Palestinian conflict. However, since the concept of reconciliation, which is merely understood in the Palestinian society, this ICT-MR Internet communication research focused more on advancing the understanding of the concepts of moderation, reconciliation, and democracy in the Palestinian society, and the aim to advance the reconciliation process and means social change towards the reconciliation process, that would assist in defining and exploring the hypothesis of the research.

In the ICT-MR Internet communication research, the terms of moderation, reconciliation, and democracy and how these are correlated to one another are explored. This phenomenon is referred to as the Wasatia Cycle, described in detail in Chap. 5. One of the primary objectives of the ICT-MR Internet communication research was to adhere to the reconciliation process in the Palestinian community.

This Internet communication empirical work sought to use the tool of ICT toward communities in conflict—to be open for reconciliation in the midst of conflict; however, it needs support and encouragement. The reconciliation process essentially approaches for forming, the environment conducive to attaining reconciliation in the middle of the Israelis-Palestinian conflict.

According to Brantmeier (2009) ICT is utilized as a means and a tool toward understanding moderation, reconciliation, and peacebuilding which would pave the way for the reconciliation and bring diverse societies in conflict and with divergent interests together to work toward a common goal (Vrasidas et al. 2009, p. 214). ICT can be a tool for reconciliation, as it works to mend a breakdown in communication and build relationships to bind people to groups, nations, or regions (Vrasidas et al. 2009, p. 216). Moreover, reconciliation is a process involved in times of wars, conflicts, and robust peace; however, lasting peace in a protracted conflict is built over time, with many different approaches. It requires different actions by different people, in different ways, and at different times during the conflict (Cole and Crawford 2007). The concept of reconciliation has become increasingly popular in scholarly discussions on building sustainable, peaceful inter and outer group relations, as researchers have used the term in referring to group reconciliation methods (Bar-Siman-Tov 2004; Nadler 2010).

According to Bar-Tal and Bennink (2004), "*Mass media can be a very powerful tool for promoting reconciliation*" (p. 32), as it can transfer knowledge through various layers and cover every part of the globe and the population through a large sphere of publication. The media can spread the notions of new peaceful goals toward entities in conflict to help develop peaceful relations by giving people the opportunity to understand others. One of the most critical aspects of digital social networks is that it opens channels to communicate with leaders about messages of moderation, reconciliation, and democracy can be exchanged (Bar-Tal and Bennink 2004, pp. 12–32).

Reconciliation is a complicated process, long-term structural changes in social and economic strategies must be within the process in an ongoing progression to create a conflict transformation of a stigmatized conflict. It is *"the reestablishment of 'good' or at least 'normal' relationships after a violent conflict like war, after mass atrocities like genocides, after crimes against humanity, or after other massive injustices inflicted to one group by another"* (Leiner 2015a, b).

Reconciliation is a process of understanding to establish a mutual interest between conflict groups and nations, which can develop the respect for the differences of narratives, and ultimately to reach the concept of forgiveness (Vrasidas et al. 2009, p. 216). According to Andrews (2000), reconciliation has two models. The first one has negotiated forgiveness: *stating that forgiveness is inspired from the actual dialogue between the perpetrator and the victim.* The other model is "*unilateral forgiveness,*" which is *a type of process which is contained entirely within one individual; it neither engages with nor is in any way dependent upon the position of the perpetrator* (Andrews 2000, p. 75–86).

Brantmeier (2009) stated that the process of reconciliation is *"healing and transforming for all parties involved; win-win outcomes are possible when common ground is established"* (Vrasidas et al. 2009, p. 216). The attempt to the reconciliation process requires dialogue exchange and an understanding of the interrelatedness of both parties that are in conflict. ICT can promote dialogue between groups or nations in conflict; Brantmeier (2009) stated that ICT could be *"transdialogic, transdisciplinary, and transnational"* instrument (Vrasidas et al. 2009, p. 216). The reconciliation process can be introduced differently, but the process is not an excuse for exemption from wrongdoings. In addition, it is not in opposition to an alternative to the truth or impunity from justice. It is not a religious concept, perfect peace, and an excuse to forget. It is not a matter of merely forgiving someone (Bloomfield et al. 2003, p. 14).

The process of reconciliation involves finding a way to bridge a vision for a better future for both parties in conflict. It is the reestablishment of relationships between both entities to develop a closure from past acts of violence and aggression. It is the society's extensive and long-term process of creating profound change in social, economic, and political-cultural spheres. It is a process of acknowledging, remembering the suffering of the other, and learning lessons from the past (Bloomfield et al. 2003, p. 14).

ICT creates a massive dialogic exchange which leads to a massive exchange of social capital, such as the sharing of social capital influencing the cultural capital that influences the economic capital, which is usually created by the elites who are part of the circle of power and the decision-making process in the conflict (Vrasidas et al. 2009, p.183). From those conditions, one can conclude that extensive use of ICT in the process of reconciliation can create a conflict transformation for reconciliation in conflicts. As stated by Brantmeier (2009), *"It is not to be reduced to a discontinuous series of instantaneous mechanical equilibria between agents who are treated as interchangeable particles; one must reintroduce into it the nation of capital and with it, accumulation and all its effects"* (Vrasidas et al. 2009, p. 217).

The disadvantage of using the ICT for reconciliation and its ambiguity is that it concerns and influences those who have access to ICTs. Conversely, those who do not have it, do not have access to it, so that they will be neglected from the process, or do not understand how it works (Hattotuwa 2004). One of the main faults of using ICTs is that it can be used by terrorist groups to recruit more terrorists from the globe; in addition, ICTs can be used to harm the stability of a state in various fields of social, capital, and political spheres, and ICT can help terrorists in sending information cyphered to their members, and to recruit radicals into their groups.

3.5.1 Examples of ICT for Reconciliation

Desmond Tutu portrayed the importance of media during the Truth and Reconciliation Commission procedures. *In no small measure, it had high regards for the hearings because of the press playing a role through "radio, TV, [and] the newspaper, [which] gave the TRC and its victim's hearings saturation coverage"* (Gibson 2006, pp. 409–432). Gibson (2006) also maintained that public radios could effectively *"pierce the consciousness of nearly all South Africans"* (p. 416).

Cambodian Reconciliation Project About ICT for the Cambodian Reconciliation Project:

The Cambodian reconciliation committee CRC. www.crcfresno.org. It is a non-profit organization; they work on education effort for the Cambodian people. The Cambodian Genocide Project: A student-led organization researches and digitally collect Khmer Rouge survivor stories. The upfront for the Cambodian reconciliation tribunal is projected to be USD 61.6 million. (Vrasidas et al. 2009, p. 227)

ICT in Education Policy The Cambodian Ministry of Education adopted the policy entitled, *Policy, and strategies: Information and communication technology in the teaching in Cambodia.* (Vrasidas et al. 2009, p. 227). This was used to utilize ICT to increase access to primary education and improve the quality of education. Moreover, it created a stable globally competitive workforce through the Internet for its people.

The YAD-B-YAD Project Translated from the Hebrew to English as the Hand-by-Hand This project is an educational research project to foster dialogues between Jewish and Palestinian Arabs children in Israel. The project illustrates promoting a culture of mutual understanding and tolerance. The idea is to create bridges between young Palestinians; this is the sector of Arabs in Israel and young Jews in conflict; the project prompted youth from both parties to share with each other ideas that are related to cultures and social exchanges, including identity narratives for each other. The project is implemented by designing a bilingual site for Arabic and Jewish children. The site has elements in Hebrew language and in Arabic language, side-by-side, which illustrate the different activities done online by the site in both languages. The participants can write in Arabic and in Hebrew; this kind of element translates both languages, thereby making the site need technological software and intensive financial investment. The site develops inter groups with participants from both parties, and they are responsible for using the language they prefer; English seems the neutral language often used. The site promotes online activities between participants to share their social and cultural views of each other (Vrasidas et al. 2009, pp. 205–207).

Tibetan NGOs The online presence for Tibetan NGOs, such as The Tibetan Center for Human Rights and Democracy, used the Internet for consciousness-raising and awareness via information dissemination about the Chinese/Tibetan conflict. In addition, the Central Tibetan Schools Administration has a strong presence on the Internet (http://www.tcewf.org), which helped to educate the people (Vrasidas et al. 2009, pp. 229–230).

Myanmar The uprising protests of Buddhist monks on the streets of Myanmar provide a light incident of the use of ICT to globalize a political uprising against a military regime. However, the military shut down all Internet access, such as Internet cafés, phone lines, and Internet services, to disconnect the voice of the protestors from reaching the outside world (Vrasidas et al. 2009, pp. 217–218), thereby showing the negative powers of the elite to stop the spread of the reconciliation process via ICT.

Radio The Voices of Reconciliation is a radio station in Sri Lanka. The station promotes the advancement of social and cultural capitals between the people to promote social and cultural unity. This use of media reaches many people to inspire the process of reconciliation; the radio station can be heard online.

Iranian Presidential Election The Iranian presidential election which erupted an online revolution in (2009),[6] but as it continued to evolve, the international community tried to intervene to provide more evidence. Protestors used YouTube to upload all videos of the violations from the police and the *Basij* (a parliamentary group in Iran). However, the government disconnected the use of YouTube to isolate the protesters from spreading their message of the atrocities committed by the regime.

Burundi During the fighting in Burundi, online discussion groups were hosted by Burundinet and the Burundi Youth Council. These allowed Burundi of different backgrounds to discuss the situation that causes the debate, debate root causes, and figure out ways to move forward after it became difficult to meet physically (Cole and Crawford 2007). Therefore, ICT was used here to spread awareness where it had become impossible to do so before.

In conflict, there is always a power struggle between the oppressed and the oppressor; the reconciliation process can shift a role change in power between the oppressors and the oppressed to dominate, subordinate, and modify the dynamics of power to reinforce the inequality between the oppressed and the oppressor. The role of ICT in the presence of shifting powers is to moderate the inherited power difference between parties engaged in the Reconciliation process (Vrasidas et al. 2009, p. 211). These following notions from Brantmeier (2009) indicate the phenomenology of ICT in Phronetic social science for a social, economic, and political change. ICTs can facilitate the advancement of reconciliation and peace in conflicts, as it utilizes the following (Vrasidas et al. 2009, pp. 219–232):

- ICTs utilize peacemaking and peacebuilding, as it bridges both physical and temporal distance, and it connects people in conflict in a virtual community connecting to its physical one.
- ICTs provide a broadcast for a voiceless citizen, who operates with limited knowledge of social-political, economic, and environmental conditions in his or her country and to the world.
- ICTs can provide information from various arguments of perspectives or different narratives that help to counter ideological campaigns.
- ICTs can also harness and enhance social capital and build disconnected relations between nations in conflict.
- ICTs can disrupt both ideological campaigns and the control of information by groups in the circles of power in conflicts.
- ICTs provide a nonviolent means for building a culture of peace through unique and individual forms of forgiveness, which may be negotiated or unilateral.
- ICTs can promote national renewal, unity, and solidarity among the general populous who are healing from violence and oppression.

[6] "Iran To Hold Presidential Election In June 2009" (Reuters). Radio Free Europe/Radio Liberty. *7 September 2008*. Retrieved 2 December 2008.

- ICTs have the potential for fostering connective relationships that occur post-conflict and create a reconstruction process.

  Hattotuwa (2004) stated,

  The possibilities of using ICT for peacebuilding dovetails seamlessly into conflict transformation, since it can augment the efforts of peace builders by enhancing channels, avenues, and possibilities for communication, information and knowledge sharing, collaboration, empowerment and discussion in virtual space, even when physical, real-world meetings are impossible on account of geographical distance or political sensitivities. (p. 11)

The following are features of ICTs for advancing peace and reconciliation (Cole and Crawford 2007):

- ICTs can provide extensive information for the populous to expose a real understanding of the conflict.
- ICTs can provide the tools to process extensive information in a brief time.
- ICTs can improve the help in decision-making skills by giving an understanding of the specific consequences of a wrong or right decision.
- ICTs can contribute to reducing scarcity and providing information on conflicts to change relative scarcity, which can improve efforts for reconciliation and peace.
- ICTs can help support an existing relationship and create new relationships.
- ICTs can help people understand each other by creating a closer gap between the two societies and by providing an understanding of the culture of the community.

By strengthening the ties between individuals and communities and improving their ability to share, learn, and interact with one another, ICTs enhance society. More excellent information also means a better ability to understand situations and act accordingly (Cole and Crawford 2007). Table 3.1 illustrates the tools of ICT to foster reconciliation and peace.

ICTs utilize social and political consciousness awareness and change, which can contribute to the initial stages of the process of reconciliation. Table 3.2 illustrates the domains of ICT for catalyzing peaceful conflict transformation (Vrasidas et al. 2009, p. 223).

ICT can help to assist in a change of the conflict. For example, in the Palestinian and Israeli conflict that is stigmatized for a long time, ICT for reconciliation can be the tool to reach a process of reconciliation between the Palestinian and Israelis in their conflict. Brantmeier (2009) stated that ICT could be "*counter-hegemonic*" because it can take its own discourse, whether for a good or bad, as it can change the ideological campaign and control the flow of information by groups of people that have the power of control ICTs. (Vrasidas et al. 2009, p. 219). ICT is thought to develop a practice that can be used to lead a process of social reconciliation; moreover, social reconciliation is "*social forgiveness between individuals or groups requires the involvement of two or more parties in an interaction of apology and forgiveness and the willingness to embark on a new relationship based on acceptance and trust*" (Lambourne 2001, p. 319).

The term has been broadened when utilizing ICT for social reconciliation, as a process "*to heal, to bring different ethnic groups together toward a common identity*

Table 3.1 ICT for reconciliation

Features for advancing reconciliation and peace	ICT – tools
Providing information	Internet connectivity
	Mobile phones & personal data assistants.
	Geographic information systems.
	Satellite imagery
	Forums
	Radio
	Chat.
Help people process information	Websites – a social media website.
	Data visualization tools.
	Online dispute resolution tools.
	Virtual command center
Improve decision making	Games and simulation.
	Online dispute resolution.
To help reduce scarcity	Mobile phones.
	Handheld portable devices
Support existing relationships	Social network tools
	Online collaboration tools.
	Virtual realities.
	Internet Café
	Tele-centers
Support pupils understand each other	Translation software
	Blogs
	Social network tools
	Multimedia

Note: From Ronald Cole and Teresa Crawford (2007)

Table 3.2 Catalyzing peaceful conflict throw cross-media combining offline and online and pure online activities by Brantmeier (Vrasidas et al. 2009, p. 223)

The Domains of ICT	Actions to process a peaceful conflict transformation
Consciousness awareness	Distribution of information
Political mobilization	Organizing political protest, e-campaigns, marches, and political activism.
Dialogue	Truth recovery, knowledge creation of different narratives and cultures, the establishment of perspectives, change conditions.
Reconciliation	The establishment of common ground and assert for forgiveness.
Renewal	The transformation of social, political, cultural, and economic capitals.

Table 3.3 Strategies for implementing ICT for social reconciliation (Perera 2015)

Good governance using ICT	E-Governance is a distinct value-adding system representing good governance characteristics.
Social inclusion	Promote equal opportunity for everyone irrespective of his or her age, gender, ethnicity, or any type of discrimination, in having an equal chance to prosper in society.
Reconciliation fundamental elements	ICT has a vital role to play in improving communication, facilitating negotiations, strengthening transparency, and more importantly building trust among different stakeholders in society.

for sustainable development" (Perera 2015). The social reconciliation, applied in ICTs within three strategies, as described in Table 3.3. Social reconciliation is defined as reconciliation engaging on social change towards the reconciliation process. The empirical work of this communication research focused on social change and on practices on social reconciliation. ICT can implement the pillars for social reconciliation: good governance, social inclusion, fundamentals of social reconciliation including forgiveness, and restoring relationships and building trust between communities (Perera 2015).

The empirical work also includes online polls, discussion, forms, online interactivities and some offline workshops, such as the Interactive reconciliation workshop which is the Auschwitz Trip, more discused in details in the next chapters.

ICT has the power and potential, extensive coverage, and acceptance to affect the social, cultural, and economic capital for transformation toward a reconciled society; people must use ICT as a strategic power toward the advancement of reconciliation. In the Israeli-Palestinian conflict, society could use ICT to empower the reconciliation process in the societies.

3.5.2 The Disadvantages of ICT Applications for Social Change

The disadvantages of the use of ICT for reconciliation can be exemplified in the following aspects:

- In the use of ICT in developing countries, where wealth and economic stability is the primary concern of the populace of the community, the unfortunate part of the community would be deprived of ICTs and its applications because of its high value for the end-user.
- Illiterate people who might not know how to use the ICT applications might not be capable of using ICT and exposed for information that helps expose virtues of reconciliation.
- The utilization of ICT for false information in promoting fake videos and spreading them over the Internet can also pose a risk.

- The use of ICT by terrorists to endorse their extremist ideologies can close the gap between different extremists that indoctrinates the same values to cooperate and work together and enlarge their circles; therefore, this use would spread their ideologies of hate and animosity, with ICT assisting them in recruiting new members. For example, one can consider the use of Facebook pages and YouTube videos by the Islamic state in Iraq to promote their way of thinking to recruit new members into their terrorist organization.
- The government might control and have access to organizations that oppose them, which might make users feel at risk if they use ICT applications.
- The control of information by the government that is using ICT to promote its agendas for political motivations and gains is also an issue.
- Technical vulnerability, hackers, and crackers can hack and attack the online applications and also destroy social network website, and all online activities and increase more in attacking participants who are working on reconciliation, using cyber wars.
- ICTs, whether through social media, blogs, wikis, webinar, and other ICTs, could provide a standard tool for the advancement of humanity, or radicals could use it as a negative tool, a weapon used for the promotion of alienation, hatred, and animosity among communities. For Example; Godwin's (1999) Law[7] stated, *"As an online discussion grows longer, the probability of a comparison involving Nazis or Hitler approaches"* (Godwin 2016) . The scale of an argument over the Internet that reaches animosity approaches Godwin's (1999) Law; the discussion becomes irrelevant and meaningless, so one should be aware of the excess of Internet communication that can become biased and shallow. The ambivalence in this example was that communication was used to establish a regular narrative sphere of communication, but it ended in hatred and animosity between both parties that initiated the online conversation. This type of conversation may happen in debates between Israelis and Palestinians when they start comparing occupation to Nazism. At that point, Godwin's Law was applicable, as it was meaningless to continue the conversation, but in the end, it all depended on the central issue of the conversation.

3.5.3 ICT Strategies for Reconciliation

Strategies for the reconciliation process through the aspects of work done by Leiner (2016a, b) are as follows:

1. Political and legal provisions.
2. The creation of collective security, with disbarment and crisis management.

[7] Godwin's Law is a social law of Usenet named after Mike Godwin (godwin@eff.org) concerning Usenet "discussions". It reads, according to the Jargon File: As a Usenet discussion grows longer, the probability of a comparison involving Nazis or Hitler approaches one.

3. Apologies and symbolic acts, which spread empathy and understanding.
4. Reparations and other attempts at justice restoration.
5. Cooperation regarding aspects of economic, legal, and international issues, such as help in cases of disasters, including fires in Haifa Israel; on November 15, 2016, Palestinian firefighters went to Haifa as part of a mission to help Israelis firefighters to control the fires that went all around the Haifa district, burning nature and houses around.
6. Co-operation in civil society, such as programs that can help the understanding of reconciliation and the benefit of reconciliation, such as hands for peace or seeds for peace.
7. The acceptance of history, which means the confrontation of each historical narrative of the other.
8. The confrontation of the historical narrative of the individuals between victims and perpetrators. The right for the victims to know, rituals, purification, encounters between victims and perpetrators, truth, and reconciliation commissions.
9. The integration of both groups on reaching a prosperous collective future in the cultural, economic, and political sphere.
10. Intentional strategies to humanize the image of the other, such as overcoming negative stereotypes, commissioning schoolbooks, and changing the image of the other in media by changing educational materials about the other in a respected sphere toward reconciliation.
11. To change the discourse of religious leaders from the aspect of war to the aspect of reconciliation and respect, understanding the narrative of the other.

3.6 Reconciliation in the Arabic Language

In the Arabic language, reconciliation has a different word; some scholars have referred to reconciliation as *Musalaha*, which infers that reconciliation may also lead to coexistence. *Ta'ayoush* refers to invoking ideas of conciliation and appeasement in light of the divisions and tensions that have arisen from the conflict. In other words, terms for reconciliation in the Arab world may vary with different concepts that may not appeal to both sides of the conflict. Such terms of reconciliation may lead to *moujalassa*, which means the act of sitting together; in this concept, people sit to discuss issues with their enemies to reach consensus on matters evolving with negativity on both sides. The other term that reconciliation may be understood is *moujayara*, which means a person will respect another as a neighbor. Other terms include *moussamaha*, meaning forgiveness, which may please some in conflict and others may understand it as a weakness; *moussafat*, which means a resolution with the other; and *tardiay*, which means to please the other and accept his or her terms as part of reconciling. Alternatively, *taswiya* means a settlement and an agreement, and lastly, *tatbi* means normalization, which is in harm's way within the Israel Palestinian conflict. The abbreviations of this study for the definition of reconciliation in the Arabic World is taken from the international center for a transitional justice research report, *Not Without Dignity*, by Rim El Gantri and Karim El Mufti (2017).

These terms in Arabic are often misunderstood because these always depend on the situation that is evoked. For that reason, it is a vital cause to achieve an understanding in the Arab world about the interpretation of reconciliation and to focus on achieving the prosperity of reconciliation in the Middle East. The word reconciliation in the Arabic word as it must be understood is "Musallah" developing empathy through understanding the other; this term is accepted in this research as it also accepted by (Bar-Siman-Tov 2004), which can also mean accepting to reconcile in the Arabic language.

3.7 Conclusion

ICT is acknowledged for being a means for developing social reconstruction in the interest of social justice. This aspect introduces that one can use ICT to renew social relations in different contexts in conflict, criticizing the deterministic social model affecting the conflict, and transforming ICT into an instrument of social transformation for social justice (Vrasidas et al. 2009, p. 20). ICT can be a means for the reconciliation process (Cole and Crawford 2007) by introducing ways of using ICTs for peace and reconciliation (Vrasidas et al. 2009, p. 221). Those tools also can be applied to implementing methods for the reconciliation process. Table 3.4 illustrates applied ICTs for the methods for reconciliation to implement social change.

Table 3.4 ICT for reconciliation process

Methods for the reconciliation process	Social change	Examples of ICTs
Apology	Provide information	Internet connectivity
Education		Mobile phones and personal data assistant.
Published meetings		
Truth and reconciliation commissions		Geographic information systems.
Public trails		
Reparations payments		Forums
The work of NGOs		Radio
Joint projects		Chat
		Videos
Writing a common history	Help people process information	Social network sites and portals.
Published meetings		
Truth and reconciliation commissions		Data visualization tools.
Public trails		Online dispute resolution tools.
Reparations payments		
The work of NGOs		
Joint projects		
Education	Improve decision making	Virtual command center.
Public trails		Games and simulations
The work of NGOs		Online dispute resolution tools.
Joint projects		
Apology	Reduce scarcity	Mobile phones.
Education		Hand held portable device, such as iPad, Tablets.
The work of NGOs		
Joint projects		
Apology	Support relationship	Social network tools
Published meetings.		Online collaboration tools
Truth and reconciliation commissions		Mobile phones
Public trails		Virtual reality
Reparations payments		Telecenters.
The work of NGOs		
Joint projects		
Apology	Help people understand each other	Translation software
Truth and reconciliation commissions		Blogs.
Public trails		Social network tools.
Reparations payments		Multimedia
Joint projects		
The work of NGOs		

Chapter 4
Internet Communication Technology (ICT) in Social Transformation

This chapter focuses on the question why and how ICT supports social change towards reconciliation process by introducing several customs of ICT, especially amid conflict. The chapter elaborates on several theories and concepts. The chapter also explores and examines ways in which ICTs are applied social, cultural, and economic capitals in different areas of society. The chapter includes the basis for the Internet communication research hypothesis—ICT can foster precise concept such as moderation, reconciliation, and democracy thereby promoting these in the middle of the conflict and adheres the ability of ICT to develop social change towards the reconciliation process than for animosity and hate.

Having introduced the research why reconciliation in the middle of the conflict and its vital contributions to developing a social change in a protracted conflict, This chapter explores ICTs as a mean of communication, the ambivalence of using ICT, the ICT-MR communication Research, ICT as an average of intrusion the figure below shows the structure of the chapter. ICTs for peacebuilding, ICTs in digital social networks. The figure below illustrates this chapter (Fig. 4.1).

4.1 Introduction

ICTs have evolved dramatically over the last two decades; these have transformed the lives of the people and shaped them differently in developing a means of communication. Today, the Internet is part of society's everyday habits, which derive from a heterogenic world, where habits of using media transformed into posting family photos, commenting on friends' daily lives, being a part of an online community, shaping the interaction of individuals in their lives, watching movies online, or being active in a group for different purposes (Couldry 2012, p. 17) (Fig. 4.2).

According to Aldous Huxley (1958), one finds what he is looking for

I. M. AlDajani, *Internet Communication Technology (ICT) for Reconciliation*, Lecture Notes in Social Networks, https://doi.org/10.1007/978-3-030-41203-6_4

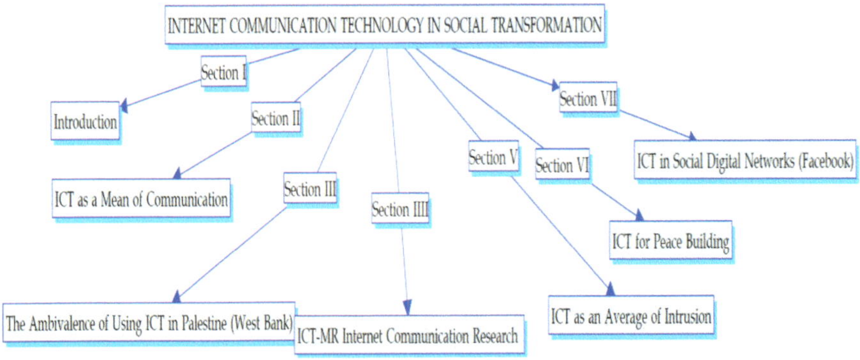

Fig. 4.1 Internet research in ICT in the social, cultural and economic transformation

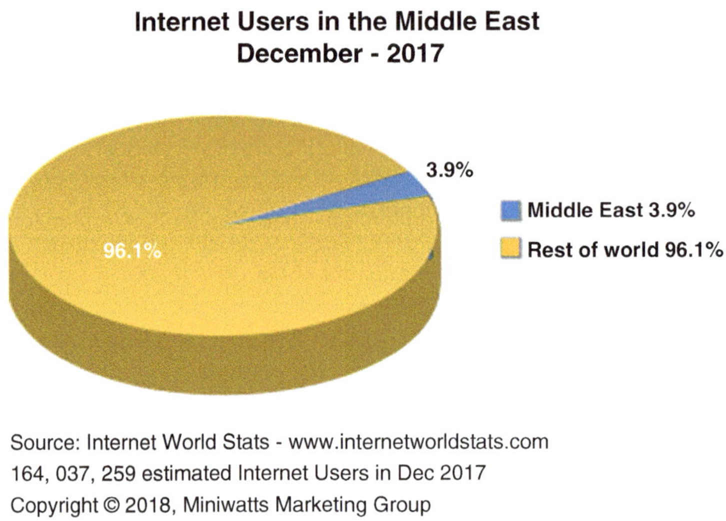

Fig. 4.2 Internet users in the world by regions 2018. (Source Miniwatts Marketing Group)

> through *mass communication, [which] in a word, is neither good nor bad; it is merely a force nor, like any other force, it can be used either well or ill. Used in one way, the press, the radio, and the cinema are indispensable to the survival of democracy. Utilized in another way, they are among the most potent weapons in the dictator's armory.* (Huxley 1958)

ICT cannot be removed from the modern world; however, it can evolve into either something better or something worse. Using ICT developed several open-ended structures, which advanced a massive stimulation of cultural production. According to Yochai Benkler (2006), the cultural production occurred for its low cost and the ease of use, which created massive use through the public domain. People used ICT to develop an infrastructure of communication, which became the public sphere in the social and political domains. Moreover, ICT became a critical process of centralization in the social, economic, political, and cultural sphere and evolved to become a significant part of the social, cultural, and economic sphere of the globe (Couldry 2012, pp. 20–24).

Social online networks develop an online practice such interactivities in Facebook, Blogs, Chats, Polls, Forms and other online application such as peace games, and online surveys, which is massive, as these give one the right to show, be shown, or post and be displayed on with unauthorized control. All users hold social networks accountable when using online social networks for different purposes. For example, YouTube provides a domain for actors and individuals to post and discuss videos, as it is cheap and relatively unregulated, where individuals can broadcast, and big enterprises can use the same robust infrastructure to advertise. Several governments have prohibited the regulation of today's online social networks, as it has become a tool for mass communication for individuals and groups around the world (Couldry 2012, pp. 43–50). Blogs can help develop online journals that can have massive information about reconciliation, blogging is used to expose visitors for information and interact by commenting in a live manner, and the last comments appear in front of the subject. Such reconciliation blogging is the National Catholic Blogging series "*Reconciliation in Chicago*." Chat is another method of using ICT application for reconciliation, for example, "*conflict chat*," which introduce various online visitors to chat online and debate about reconciliation. Another example for a chat for reconciliation is the live chat done by CBC indigenous chat, where visitors on a live feed debated reconciliation and how it works and how it looks like. Polls were used in a massive manner to help advance reconciliation such as Polls for peace and reconciliation; the poll introduced a survey of peace and conflict resolution process in any context. Forms were based on introducing various topics about reconciliation and educating users on how to apply reconciliation, the methods, and strategies. Such as reconciliation forms in conflict resolution. Today mostly is used Facebook, Twitter, and You-Tube for such methods, because of its massive audience.

ICT has a significant role in the Israeli/Palestinian conflict. The following represent some posts that were run over the discourse of the search on 09 May 2014 using You-Tube:

1. *A Video on Crash Course on World History* was about the conflict between Israel and Palestine. The video posted on YouTube was watched and viewed by 2,478,714 liked by 36,121 and received 3208 dislikes John Green in 2015 provided a crash course on different topics on conflict resolution. His channel had over 4,354,432 subscribers on a YouTube Channel.
2. *The Israel/Palestine Conflict: A Brief History* was watched by 600,073, liked by 10,373, and disliked by 1997.
3. *The Truth of the Israelis-Palestinian Conflict* was watched by 1,754,162, liked by 7422, and disliked by 3900 (Channel 2012).

Table 4.1 illustrates Internet usage and world population statistics as of 31, December 2017.

The table shows the use of the internet in the Middle East regions, as compared to populations and the Internet usage of data. The Middle East would include the countries of Bahrain, Iran, Iraq, Israel, Jordan, Kuwait, Lebanon, Oman, Palestine (West Bank), Qatar, Saudi Arabia, Syria, United Arab Emirates, and Yemen.

Table 4.1 Middle East usage compared to the rest of the world statistics 2018

Middle East region	Population (2018 Est.)	Pop. %	Internet users 31-Dec-2017	% Pop of region	Facebook users
Total Middle East	254,438,981	3.3%	164,037,259	64.5%	116,482,942
Rest of the World	7,380,319,981	96.7%	3,995,403,425	54.1%	2,002,577,210
World total	7,634,758,428	100.0%	4,159,440,684	54.5%	2,119,060,152

Usage of Internet Users updates:
(1) Internet Usage and Population (Resident population)Statistics for the Middle East were updated on December 31, 2017.
(2) Facebook subscribers were also updated on Dec 31, 2017.
(3) Population estimates are based on data from the United Nations Population Division.
(4) The most recent Internet stats come mainly from data published by Nielsen Online, ITU, Facebook, and trustworthy local sources.
(5) Data on this site may be cited, giving the due credit and establishing an active link back to www.internetworldstats.com.

Source: https://www.internetworldstats.com/stats5.htm

The Middle East users according to their countries, the source is from Miniwattas Marketing Group the source: https://www.internetworldstats.com/stats5.htm.

Middle East Internet users, population and Facebook statistics 2018

Middle East	Population (2018 Est.)	Users, in Dec/2000	Internet Usage 31-Dec-2017	% Population (Penetration)	Internet % users	Facebook 31-Dec-2017
Bahrain	1,566,993	40,000	**1,535,653**	98.0%	1.0%	1,100,000
Iran	82,011,735	250,000	**56,700,000**	69.1%	34.6%	40,000,000
Iraq	39,339,753	12,500	**19,000,000**	48.3%	11.6%	17,000,000
Israel	8,452,841	1,270,000	**6,740,287**	79.7%	4.1%	5,800,000
Jordan	9,903,802	127,300	**8,700,000**	87.8%	5.3%	5,300,000
Kuwait	4,197,128	150,000	**4,104,347**	97.8%	2.5%	3,100,000
Lebanon	6,093,509	300,000	**5,546,494**	91.0%	3.4%	3,600,000
Oman	4,829,946	90,000	**3,310,260**	68.5%	2.0%	2,630,000
Palestine (State of)	5,052,776	35,000	**3,055,088**	60.5%	1.9%	1,700,000
Qatar	2,694,849	30,000	**2,644,580**	98.1%	1.6%	2,300,000
Saudi Arabia	33,554,343	200,000	**30,257,715**	90.2%	18.4%	18,000,000
Syria	18,284,407	30,000	**6,025,631**	33.0%	3.7%	4,900,000
United Arab Emirates	9,541,615	735,000	**9,385,420**	98.4%	5.7%	8,700,000
Yemen	28,915,284	15,000	**7,031,784**	24.3%	4.3%	2,352,942
Total Middle East	**254,438,981**	**3,284,800**	**164,037,259**	**64.5%**	**100.0%**	**116,482,942**

Source: https://www.internetworldstats.com/stats5.htm

According to several studies, ICT is a tool for dealing with poverty, as well as a social and cultural misunderstanding between communities and nations. People can use ICT to empower individuals and groups in the community to analyze and overcome obstacles of social exclusion in stigmatized conflicts. However, can one use ICT to focus on empowering precise concepts within the community, such as moderation, reconciliation, and democracy? This Internet communication research would intricate this theory within the discourse of the study.

4.2 ICT as a Mean of Communication

One of the theories included in this study was Bourdieu's (1977) theory of culture and social reproduction which negotiates between physical structures and practices. Researchers have used this theory to study the class-based analysis of social satisfactory; to endorse the cultural capital that possesses the influence on the economic capital, which is necessary to access the circle of powers that are decision makers; and to influence decisions concerning public policies that affect the community (Vrasidas et al. 2009, p. 218). The utilization of ICT could influence Bourdieu's (1977) theory of the relationships between social, cultural capitals and provide a link to economic capital for the reproduction of the existing social order that enables the transformation within the community in the midst of conflict. This theory provided a base for the hypothesis of this Internet communication research, which was to study the influence of a change in the community from the intentionality of hatred and animosity to reconciliation, moderation, and democracy by utilizing ICT for that purpose.

ICT is an instrument that exposes and connects individuals or groups together without boundaries or restrictions unless opposed to the application software. People can use ICTs to empower persons or groups to be part of the global conversation that can change the world. By using ICTs, communities can become networks and infrastructures through which they converse with one another. ICT can evolve from private versus public, trusting versus cautious, sharing versus withholding, and collaborative versus individualistic, which falls in the sphere of action of the individuals or groups in the community who have used online social networks (Nussbaum 2012, p. 68). One can utilize ICT as a tool to communicate between cultures and societies, as based on respect for cultural differences and narratives and with the aim of achieving the participation of individuals and groups who have barriers and obstacles in conflict to communicate (Vrasidas et al. 2009, pp. 3–15). The case study in this communication research is the frozen conflict between Israelis and Palestinians who wanted peace, and Israelis and Palestinians who wanted a continuum a process of war and occupation. I thought ICT could focus on directing the two entities, who wanted reconciliation and moderation together, thereby empowering them to move the stagnated conflict toward reconciliation or social change and not a frozen peace.

ICT has changed the way people live, learn, and perform. People have used ICTs to develop newly structured media, such as the move from a limited number of terrestrial television channels to millions of online channels spreading information around the globe, creating fast and continuous access to the Internet. These include the transformation of all radio and news channels to a digitalized online newspaper websites. ICT has industrialized a massive growth in online content delivery networks for exchanging material and information, including photos, film, television, and music. ICT has contributed to the advancement of social media networks, such as Facebook, Twitter, and YouTube, which are most used today. These may be utilized to ease communication and exchange of knowledge and information. ICT has also contributed to the development of media applications for phones, such as iPhones, Android phones, Blackberries, and other mobile devices, which are called *smartphones* (Couldry 2012).

As the use of ICT has become more massive, ICT has contributed to education, development, social inclusion, and cultural understanding. ICT has influenced issues, such as human rights, the eradication of poverty, freedom of speech, and equality among nations, and it has preserved the importance of international organization, scholars, governments, and activists. ICT has developed interactions and integrations of ICT members in everyday activities, as well as virtual communications on all levels, whether the individual level, global scale, governmental level, or social scale. Individuals have used ICT to create the opportunity for unlimited communications of levels using something called *virtual reality*. Virtual reality adopts the usage of ICT in different domains over the Internet utilized by individuals and groups from all over the globe (Vrasidas et al. 2009, pp. 3–15).

Such societies are defined as *information societies*: *"A rhetoric focuses on a wide scale of implementation of ICTs without questioning fundamental assumptions that might unjustly exclude the disadvantages"* (Vrasidas et al. 2009, p. 8). One of the most critical aspects of ICT in the development of the information society that can help in widening the development of the community in all aspects of capitals, social, cultural, and economic capitals. Information societies can also evolve the society for a social change in a definite sphere. ICT provides an alternative vision and practice for the advancement of education, development, social justice, and cultural understanding. According to Castells in this book "The Rise of Network Society", he referred to the social capital change affect by ICTs "*The Information Technology Revolution*", in which ICTs develops what here refers to as the "*Network Enterprise*", the controls all aspects of social, cultural, economic capitals (Castells 2010, p. 163).

ICT can be used as a mean to advance the development for education, contribute to enhancing educational knowledge of students in school, provide an authentic and engaging learning experience, and prepare students for better interaction in life (Cuban 2003, pp. 41–45). Researchers have considered ICT a tool to improve traditional learning and teaching, such as online courses and other e-learning courses (Vrasidas and Glass 2005, p. 10). People can use ICT for education to promote ease of access to educational information for the underprivileged communities using online educational programs that can reach poor communities or individuals who have no access to such information. The use of blogs, social networks, and e-newsletters can raise awareness of technological development for all ICT users. ICT can advance social justice in promoting learning environments for individuals and communities to take actions and to participate in social activities and workshops by influencing the community for social justice (Vrasidas et al. 2009, pp. 8–9).

People can use ICT as a sphere to express, share, and enrich cultural values in communities. In addition, it can influence the understanding of cultural, political, economic, and educational challenges, thereby eliminating the distance of international education. There are no boundaries for accessing educational information through ICT. It can harness the three characteristics to empower work for peace, justice, and equity, which are critical for emotional literacy, collective witness, and collective intelligence (Zembylas et al. 2004). Today, the use of ICT to access information and network learning is embedded in complex socio-political, cultural, and economic contexts (Vrasidas et al. 2009, p. 8). ICT has developed social inclusion for individuals, groups, and communities to societies, which will add to the communities' social and economic capitals. According to Warschauer (2003) Social inclusion refers "*to the extent to which individuals, families, and communities are able to participate fully in society in terms of economic, resources, employment, health, education, housing, recreation, culture, and civic activities*" (cited in Vrasidas et al. 2009, p. 11).

Warschauer (2003) also stated that ICT for social inclusion is interpreted *as "e-inclusion"* or "*the use of ICT to overcome social exclusion and improve economic performance, employment opportunity, quality of life, social participation, and social cohesion*" (cited in Vrasidas et al. 2009, p. 12). ICT can lead to improving the skills of its users in different social and cultural aspects, as in improving the skills of its users and employability, as well as to better health awareness from accessing online health services. People can use ICT to develop better engagement with the public issue and increase democratic participation (Vrasidas et al. 2009, pp. 11–14). It is a powerful asset for social capitals in the society and an essential element in the advancement of the cultural capital in the society.

ICT enhances the *cultural embeddedness* in national and global societies, which refers to "*a powerful tool in the process to facilitating intercultural education, reconciliation, and peace, as ICT has a potential to surpass limits imposed national border, facilitating cultural understanding among individuals and communities, including those in groups that are in conflict*" (Vrasidas et al. 2009, p. 12). ICT is an instrument for exploring peaceful ways to interact in dialogues and support conflict transformation; in sharing the realities and narratives of people's recent conflicts, ICT is a useful tool to deal with poverty, social inclusion, and cultural understanding to try to build a more secure future for the communities in conflict (Vrasidas et al. 2009, p. 13). Therefore, people can use ICTs to collaborate with different people and communities, whether these are local or international communities. ICTs provide an alternative vision for peace and reconciliation.

According to Bekerman and Horenzsyk (2009), ICT is a valuable educational tool for reconciliation, especially in cases of two groups who cannot meet face-to-face because of a psychological barrier or real obstacles (Vrasidas et al. 2009, p. 205). People can use ICT to contribute to the reduction of cultural misunderstanding and conflict (Hart 1998). In addition, people can use ICT as a means to combat the culture of war and violence, which is equally useful with rich resources provided for raising awareness and suggesting appropriate actions (Dangwal and Srivastava 2015).

4.3 The Ambivalence of Using ICT in Palestine (West Bank)

Some of the ambivalence of using ICT in conflict transformation should be considered, such as the lack of accessibility for poor societies. The Internet in Palestine has a minimum speed, which may not be adequate to use ICT as social media for the interaction of members. In Palestine, people also have a lack of expertise and experience in IT knowledge. Youth generally have good knowledge in using some social networks, such as Facebook, but not the other social networks, such as Twitter, blogs, and Wikis. ICT hardware and software are costly, and usually, Palestinians may not have the financial resources to purchase such products.

In conflict situations, communities may lack the incentive to share or connect with others; this implies that ICT in times of war is not sufficient; consequently, ICT can be used to demonize others or blame others. People may use ICT to expose the community and its culture within the conflict, which adversely influences conflict transformation. In addition, security monitoring for social networks, such as Facebook accounts, denouncing the freedom of expression and speak which are one of the fundamentals of freedom. Such as for example, the monitoring of Palestinian Facebook accounts which can result in arresting Palestinian Facebook users in accordance to what they share in their accounts. This issue develops a lack of trust in the Palestinian communities to share ideas and narratives within the conflict transformation, which will decrease the usage of ICT. This was done by the right-wing government in Israel to East Jerusalem, as convicting them of incitement, instead most East Jerusalemites feel sorrow for their brothers in the West Bank and in Gaza, even if the East Jerusalemite morns for sorrows of Palestinian whom conducted crimes against the state of Israel, even if it was on Facebook or other online social media, some got convicted and sent to jail for some months. Palestinian Authority does the same; if you speak about politics against the Palestinian Authority and its cooperation with Israel on security, the Palestinian might get convicted and sentenced to jail.

ICTs can also be accountable for distributing fake news, which makes ICTs unreliable in context and conflict. The exposure to fake news results in the spread of misinformation within a community. For example, Facebook is combating fake news by adding tools, which are used to illustrate disputed issues and subjects; these will disappear from users' feeds, and this option was being utilized in the German national election held on October 2017 (ICT – Facebook To Fight Fake News In Germany, 2016). There are more other features for the disadvantages of using the ICT in conflict transformation, but the concerns are on the higher average of coverage for promoting reconciliation, moderation, and democracy from the influence of ICTs.

4.4 ICT-MR Internet Communication Research

This ICT-MR Internet communication research developed a social network site for the Wasatia movement in Palestine to establish a social network environment as a base for an academic library with articles, as well as publish workshop activities for endorsing moderation, reconciliation, and democracy within the Palestinian society. The social network platform of the ICT-MR involved embedding theories of online deliberative democracy for developing a virtual reality where they develop their own online profiles that influenced an online community. In addition, it involved integrating other approaches through articles from expert thinkers, writers, and scholars about moderation and reconciliation, impelling democratic values for the visitors and members of the social network site. This ICT-MR also attempted to strengthen the cause of Wasatia NGO as a driving force for moderation, reconciliation, and democracy in the Palestinian society.

This ICT-MR examined and studied the driving force for ICTs to spread notions of moderation, reconciliation, and democratic values in the Palestinian society. The primary objective was to steer members toward moderation and reconciliation, rather than to hate and animosity in the community of the Israeli/Palestinian conflict. The expectation from such conduct develops an ego-centered network that had, at its core, the values of moderation leading to reconciliation and advancing democratic concepts within the society. Therefore, this ICT-MR was used to develop a forum in the social network site, which would host the realities and publish articles to expose the historical realities that the community faced. Visitors and members shared their thoughts in a deliberative manner to influence social change towards reconciliation process in the Palestinian society online. Moreover, this ICT-MR social network site had a set of tools for members to receive the knowledge to recognize the logical and cultural consequences of the practical choices that they made. In addition, the site was used to develop a compelling socialization experience that reminds participants how much their thoughts were important in the development of their communities.

4.5 ICT as an Average of Intrusion

ICT does not have the capacity to liberate people or to elevate poverty by narrowing the gap between digitally knowledgeable people or non-digitally illiterate people who are connected (Hattotuwa 2004); however, it can be a tool for exchanging ideas and knowledge that can approach conflict transformation and understanding. ICT interventions can transform conflicts by producing social networks and channels of communications that are speedy and secure, thereby allowing more participation, accountability, and transparency between stakeholders who are in conflict (Hattotuwa 2004). Therefore, ICT develops a process of engaging and transforming relationships, interests, and discourses to change the society that supports the continuations of violent conflict against aggressions (Miall 2003).

In social development, the success of the ICT platform is often undermined in serving the actual conflict and mistaking it for making it worse, thus widening the gap between the two communities in conflict. Conversely, ICT platforms can be a primary tool for the development of the society, as a phase of conflict transformation and influence in the community to enhance the process and apply peace-maintaining and peace-building initiatives through repeated efforts for a change of intentionalities in the socio-political fabric of the community. This aspect results in ethnopolitical relations between the two communities to empower a transformation toward social change on the positive side (Hattotuwa 2004).

At the United Nations "The World Conference against Racism, Racial Discrimination, Xenophobia, and Related Intolerance" was held August 31 through 8 September 2001; members from 143 states expressed their conviction that ideas of racial superiority could no longer exist among conflicts and in between communities. In addition, they demanded ICT to resolve this material progression of racism and racial discrimination from using the influence of ICT in the conflict areas (World Conference Against Racism 2001). The ICT platform influences the conflict in different aspects; it can shift conflict resolution and management as making a change in the characteristics of the conflict itself. This definition focuses on the"*change[s] in the characteristics of a conflict*"(Botes 2001, p. 1). The ICT platform can stimulate a shift in the unique aspect of the conflict by changing the features that evolved the conflict. Therefore, ICT helped in communicating a suitable, sustainable, and transformative manner between communities in conflict. The objective was to generate conflict transformation for the central issue of the conflict. In addition, the objective involved reviving stagnant dialogues to benefit the conflict transformation process by providing the space for sustained dialogue and interactions between two entities in conflict (Hattotuwa 2004).

Vayrynen's (as cited in Miall 2003) approach characterized the effect of ICT platforms by influencing the conflict in four characteristics: *context transformation, structure transformation, issue transformation, and personal or group transformation*. A full view of the explanation of the effect that ICT has on the characteristics of the conflict transformation is shown in Table 4.2 (Miall 2003).

Table 4.2 Description of the intrusion of ICT platforms

Content transformation	ICT platform can transform the conflicting ideologies of both communities that are in conflict, thus influencing the participants that are using the ICT platform.
Structural transformation	ICT will change the structure of the conflict by impelling a change in the essential conflict features.
Issue transformation	ICT will affect the transformation of conflict statements, held by the two entities that are in conflict and vice versa.
Personal or group transformation	ICT will create a change in the profile of the entities that are in conflict also distressing the groups that are interacting with it.

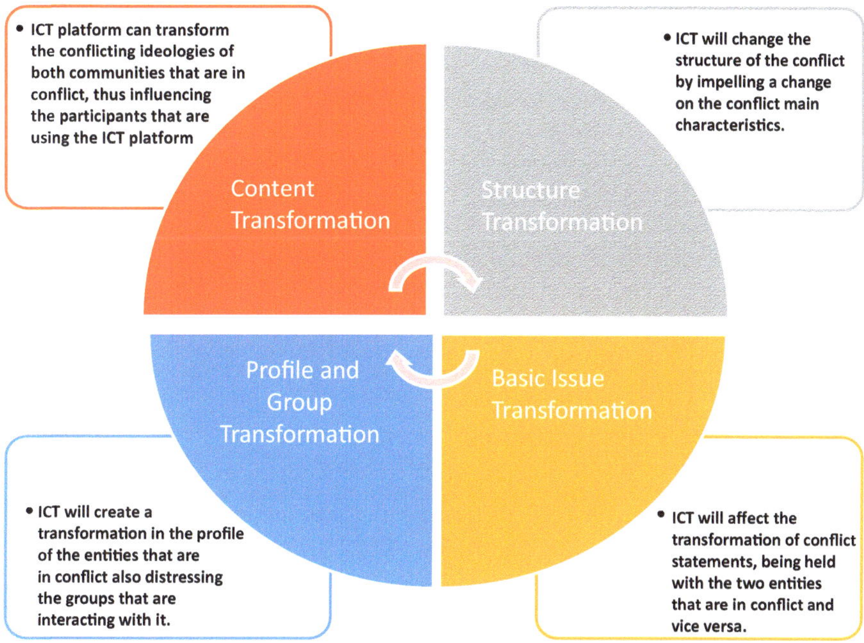

Fig. 4.3 ICT platform from Hattotuwa (2004, pp. 39–68)

Figure 4.1 illustrates the intervention of ICT on the characteristics of a conflict (Hattotuwa 2004, pp. 39–68). The figure illustrates the cycle from necessary issue information to profile group transformation, where groups begin to develop their own profiles in accordance with their necessary information. After the intervention of ICTs, it directs the content of the communication to content transformation between users, which develops a structural transformation that influences the substance of the conflict (Fig. 4.3).

The figure demonstrates the effect of ICT platforms on the characteristics of conflict transformation. According to Pool (1983), ICT Platforms "*is a technology of freedom since it enhances the freedom of choice for individuals and intensifies bottom-up relations in networks of organizations and individuals*" (cited in

Hattotuwa 2004, p. 51). Pool (1983) further stated the ways in which the above definition enhanced interaction between the two entities in the conflict in creating new characteristics for the content of the conflict.

> *influence when used for peacebuilding after a ceasefire or a peace agreement, when the dynamics on the ground are relatively more receptive to the need for sharing information through dialogue and exchanging knowledge, collaborating, appropriating technology and developing mechanisms (both physical and virtual) for communities to deal with conflict creatively and non-violently.* (p. 13)

ICT platform interventions are at the core of the conflict transformation in all its aspects because these enable information to flow not only radically with subverting existing patterns, but also through circulating knowledge and power information through differing organizations, groups, and individuals to produce and share information between one another. For example, Facebook, Tweeter, YouTube, and other social networks can influence the aspects of conflict transformation from spreading infinite or censored knowledge.

In new technologies, I tried to uncover the appropriate usage of the role of ICT in this ICT-MR Internet communication research. The aim was to develop an online community through the influence of ICTs on the ICT-MR platform to build awareness for moderation, reconciliation, and democracy in the online community. The goal was to move members of certain intentionalities toward the understanding of moderation, reconciliation, and democracy.

4.6 ICT for Peace Building

Peacebuilding in the Israeli-Palestinian conflict was not sustainable until today. With ICTs working around the globe, hatred and violence can spread using fast-growing ICT platforms worldwide. Where hatred and violence reside, there is reconciliation (*Hölderlin Perspective).* This ICT-MR sought to use the tool of ICTs to influence communities in conflict to be open to evolve to moderation and lead these to reconciliation already present in the midst of conflict.

The concept, illustrated in the *Hölderlin Perspective,* indicated how reconciliation was present already in the middle of a conflict (Leiner and Flämig 2012, p. 16). In the context of reconciliation, this *ICT*-MR sought to uncover the role of ICTs in the reconciliation process to influence reconciliation in the Palestinian/Israeli conflict. This ICT-MR shared an initiative in Palestine that encouraged moderation, reconciliation, and democracy within support from the Wasatia NGO in Palestine. ICTs can influence the values of moderation, reconciliation, and shared knowledge in the midst of conflicts to influence the peacebuilding process.

This ICT-MR focused on using ICT in conflict countries, such as Sri Lanka (Hattotuwa 2004), where ICT was used in the peace process, as an on-going initiative for the building of an on-going peace process. ICT usage influences the mobilization of the local and the grassroots, which may influence peacebuilding and the balances of powers (Richmond and Tellidis 2014, pp. 1–5). Another historical usage of ICT in conflict and post-conflict environments includes Kenya's preventive violence network (Jorgic 2013, pp. 2–5). Moreover, other examples include Uganda's and Kyrgyzstan's election monitoring (Karefe 2012, pp. 249–258), Sudan's low-tech adaptations for the community (Larrauri 2013, pp. 71–86), the communications of Cyprus' civil society empowerment (UNDP 2008), and Sri Lanka's civic journalism (Hattotuwa 2004, pp. 39–68).

4.7 ICT in Social Digital Networks (Facebook)

Social media became the arms that connect a community with its individuals. Nussbaum (2012) viewed the social media as a *Global Brain*, which is *"an emerging, collectively intelligent network that is formed by the people of this planet together with computers, knowledge bases, and communication links that connect them together*" (p. 69). Social digital networks evolved through the developing of new technologies, such as blogs wikis, Facebook, Twitter, Google+, and YouTube, making it easy for individuals and groups to communicate within the community as well nationally and globally. Social Digital Network became part of the newly developed communication technologies, and now it is referred to in the context of the ICT.

Social Digital Networks (SDN) created virtual communities, developing communication between local individuals and groups in a local community, and then shifting communication from local to global communities. Those virtual communities enabled individuals across the world to express their ideas to countless numbers of others within a virtual community. Moreover, these helped promote relationships among diverse people and groups to close the gap in understanding different narratives between the various communities. In addition, these offered a new venue for setting new goals and resolving various issues (Rehm 2000, pp. 153–172). SDN helped provide the space for a variety of activities from information gathering to conducting international cooperation between individuals, groups on the local level, and in the international arena.

Researchers stated, "*Power goes to those who can control the flow of information. But when all information, all music, all art, all images, all ideas are digitized, then everyone can access, alter, create, and transmit anything anywhere anytime*" (During and Sherif 2003, pp. 203). Social digital networks have its disadvantages; it allows irresponsibility toward sensitivity for the others in a virtual community. Moreover, allows hostility with more than verbal comebacks as consequences. ICT offers relaxation of social consciousness; that results in its inability to operate as space where diverse opinions and ideas are actually heard (O'Connor and Skulnick 2009, pp. 235–253). Online virtual communities that are using the online application for networking, such as Facebook, MySpace, and social networking websites, create social networking among members of the same network. The Facebook page is an important virtual community; its users are not faceless or anonymous, as nothing can be posted on the Facebook page anonymously. Even if the profile created on Facebook is anonymous, there is always a physical person that contributes to that profile (O'Connor and Skulnick 2009, pp. 235–253). Researchers stated, "*Each virtual community exists in a unique relationship with physical communities within imagined space*" (Fox 2004, p. 55).

Facebook evolved into one of the most used sites for the online social networking in the world, the reason is for its ability, to have public profiles on its pages and cooperate unique features within those public profiles. EDUCAUSE (2006) stated,

> *The concerns about Facebook center on its being public even though it feels like a private forum, although some students understand how and when to separate private from public content, many lack the discretion to present themselves and others appropriately online.*(p. 2)

This ICT-MR Internet communication research worked on developing the social network site and utilizes the use of the Facebook pages as an online social network to develop a virtual society that adapted the Reconciliation process in the Palestinian society.

4.8 Conclusion

ICT can be used to manage resources whether these are human, financial, or technological. Controlled all over the globe, ICT affects the intrastate and regional ethnopolitical conflicts, such as the Israeli/Palestinian conflict. Therefore, peacemakers can continue to engage with the complexities of peace process without having to be in the same region, with repositories of information that are accessible securely and confidentially (Hattotuwa 2004, pp. 39–68).

ICT can develop the Network Enterprise according to Castells (2010), that enhances and develops social, cultural and economic capital change. ICT works as a communication and information switchboard connecting individuals, networks, and institution (Kozinets 2015, p. 63). In that sense, ICT can be a haven for endorsing knowledge and empowering communities for a better future perspective. From the many theories and examples in this chapter, ICT was endorsed as a tool, and a means to prove and implement those theories. This chapter introduced how can be a tool in different fields of study, therefore it proves that the notion of ICT can foster values of moderation, reconciliation, and democracy in the middle of conflict can be investigated, this is the hypothesis of the communication research, the empirical work starts in the empirical study part of this communication research.

In the context of that historical approach, ICT usage can be an effective tool to approach conflicts and endorse social change towards the reconciliation process, enhancing values of moderation, reconciliation, and democracy in the middle of the conflict, but still, can ICT be a tool towards reconciliation process in the Israeli Palestinian conflict. In the next chapter definitions and a more thorough understanding of theories and concepts that are endorsed in the communication research. Such as "*Wasatia*" concepts and moderation in Islam, the Wasatia as a Movement, The Founder of Wasatia in Palestine, ICT in Applied Phronetic Approach towards the Wasatia Cycle.

Chapter 5
Internet Communication TEchnology (ICT) for Moderation in Islam "*Wasatia*," in Palestine

5.1 Introduction

This ICT-MR Internet communication research endorsed some central concepts. The following treatment of these concepts is not intended to be comprehensive but to explain ways in which the concepts are being used in the present Internet communication research. This chapter introduces Wasatia as a concept and Wasatia as a movement in Palestine, the founder of Wasatia in Palestine, ICT in applied Phronetic social science towards the Wasatia Cycle.

Having explored the transformation concepts from using ICT, the Fig. 5.1 below serves to build up the theoretical foundations for the research design in researching "Wasatia" as a concept and movement in Palestine, and a biography of its founder. The movement and its organization are the sample study in the research.

5.2 Wasatia as a Concept

This section discusses Wasatia as a concept and as the central defining element of the movement in Palestine. This conceptual outline helps focus the recent case study using the ICT-MR Internet communication research. The section explores the definition of moderation in Islam, the scope, and the objectives of Wasatia as an Islamic concept from the perspective of Mohammed Dajani Daoudi who created the movement in Palestine in January 2007.

Dajani (2014) explained that Wasatia was a term derived from the Arabic linguistic origin "wasat" meaning middle, as in the middle of the road and middle ground, or center, as in the center of the field and centrism. In a religious context, it means balance, moderation, temperance, and justice, as a middle objective between adversaries. The English equivalent for the word Wasatia is moderation. The concept of

I. M. AlDajani, *Internet Communication Technology (ICT) for Reconciliation*, Lecture Notes in Social Networks, https://doi.org/10.1007/978-3-030-41203-6_5

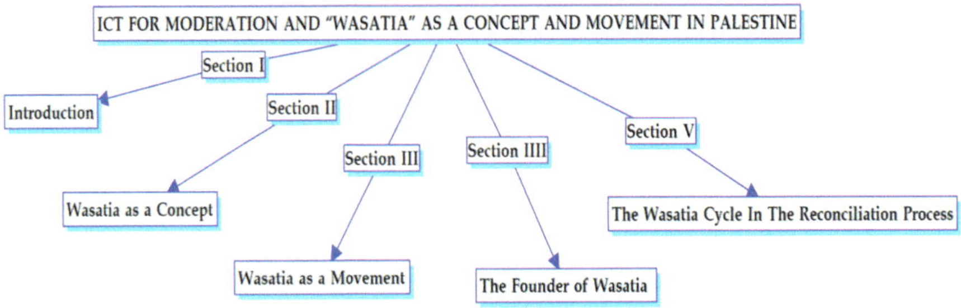

Fig. 5.1 ICT for moderation

Wasatia emphasizes avoidance of extremism and the rejection of radicalism (Dajani 2014).

Dajani (2014) quoted the concept from the following Quranic verse: "***And thus We have created you a mid-ground nation… Thus, have We made of you an Ummatan Wasatan (justly balanced***)" (p. 1).[1] "*Wasat*" appeared in verse number 143 in the second Surah in the Quran titled *Surat al BaQnCArah*, which was composed of 286 verses, thus implying its denotation. This verse was preceded by another verse, which stated that God guided whom He willed to a straight path, making the believers infer the Wasatia path. To confirm this meaning, verse 143 continued to say, "*And thus we have made you a just community that you will be witnesses over the people and the Messenger will be a witness over you*" (Quran, *BaQnCArah*, V143). Being a witness in the presence of God requires a believer to be moral, devout, honest, truthful, sincere, and pious in order to tell the truth in that he or she is bound to testify.

Wasatia reflects the peaceful spirit of Islam. (Dajani 2009) named his book in English, *Wasatia: The Spirit of Islam.* Wasatia adopts liberal values of equity, peace, tolerance, pluralism, freedom of expression, the rule of law, respect for human rights, brotherhood, free choice, tolerance, soft dialogue with the other, freedom of the faith and its establishment on a foundation of conviction, and free will, as well as peaceful coexistence with the other.

Wasatia provides a total guarantee to the faithful of the liberty to believe, to think, to discuss, and to observe. All these are values advocated in Islamic Holy Scripture and traditions (Dajani 2007).

These themes are reflected in the following verses:

- "*Call men to the path of your Lord with wisdom and kindly exhortation. Reason with them in the most courteous exhortation*" (Bee Surah; verse 125).
- *"The believers are a band of brothers. Make peace with your brothers and fear God, so that you may be shown mercy"* (Chambers Surah; verse 10).

---

[1] "How do you Say Moderate in Palestinian? Wasatia" by Gill Troy, *The Jerusalem Post*, Feb 07, 2012.

- *"Had your Lord pleased, He would have united all mankind. But only those whom He has shown mercy will cease to differ. To this end, He has created them"* (Houd Surah: verse 118)
- *"Had your Lord pleased, all the people of the earth would have believed in Him. Would you then force faith upon men?"* (Jonah Surah; verse 99).
- *"There shall be no compulsion in religion" (Al-Bakara Surah; verse 256)*
- *"Call men to the path of your Lord with wisdom and kindly exhortation. Reason with them in the most courteous exhortation"* (Bee Surah: verse 125). (as cited in Dajani 2014, p. 20)

The Holy Quran also confirms religious freedom in a number of verses, such as the following:

- *"Say: This is the truth from your Lord. Let him who will, believe in it, and him who will deny it"* (Cave Surah; verse 29).
- *"You have your own religion, and I have mine"* (Unbelievers Surah; verse 6)
- *"Among His other signs are the creation of heaven and earth and the diversity of your tongues and colors. Surely there are signs in this for all mankind"* (Greeks Surah; verse 22). (as cited in Dajani 2014, p. 20)

Prophet Mohammed (PUB; 570–632 A.D.) endorsed the concept of Wasatia as an essential moderate value in his saying, *"Do good deeds properly, sincerely and moderately. Always adopt a middle, moderate, regular course, whereby you will reach your target (of Paradise)"* (Al-Bukhari, Volume 8, Hadith 470). The WASATIA values that are exemplified by Dajani (2014) and accentuated in the NGO Wasatia bring a broader and more rational understanding of Islam to Muslims, as well as to non-Muslims. Wasatia values, as Dajani (2013) wrote, can be simplified as follows:

- To clarify the distortions to which Islam has been subjected to by different countries in the West;
- To educate Palestinians on taboo topics, such as Holocaust history;
- To seek comprehensive answers for the deeply religious, political, social, and economic crises inflicting the Palestinian society;
- To work for ending the Israeli military occupation through negotiations and peaceful means; and
- To spread and promote Islamic tolerant concepts, values, and principles within the Palestinian community.

One of the core values is to encourage the practice of moderation among Palestinians to moderate religious radicalism and bigotry by reducing political extremism. In addition, it can bring a message of peace, moderation, justice, coexistence, tolerance, and reconciliation to the Palestinian community through vocal civic leaders. It can also teach creative and critical thinking and open-mindedness to empower a potential for leadership in society. In addition, Wasatia works to deconstruct mythologies, distortions, and misinterpretations to promote knowledge and empathy for the other.

Wasatia also encourages moderation as a value shared by the various thinkers and philosophers, as well as all faiths; therefore, it could become a fruitful foundation for dialogue to achieve peace and reconciliation (Dajani 2013, p. 7). People can use Wasatia to organize workshops to facilitate dialogue and outreaching to the press and social media. In addition, it is used to conduct presentations to high school students; endorse meetings for discussions on religious and political topics; and conduct workshops, lectures, television interviews, and study tours. In addition, the doctrines and characteristics, which reflect Wasatia as a moderate concept in Islam, are adopted by Dajani (2013) as those basic principles for the Wasatia movement, mainly the obedience to the word of God, setting good examples for others to follow, and forbidding violence against people. It was stated, "*Fight for the sake of God those that fight against you but do not attack them first. God does not love the aggressors*" (Al-Baqarah Surah; verse 190).

Wasatia confirms Islamic doctrine that maintains a Moslem is only responsible for himself or herself. The believer has an open and direct channel to God. He or she prays directly to his or her creator to seek guidance, forgiveness, and blessings. For example, "*Should my worshipers question you about Me, tell them I am near. I answer the prayer of the suppliant when he calls to Me; therefore let them answer My call and put their trust in Me, that they may be rightly guided*" (Al-Baqarah Surah; verse 186).

Wasatia calls for equality among all believers, without any consideration of race, color, social status, or financial ability. In Islam, there is but one scale to judge the faithful: a person's belief in God and the deeds He portrays. Islam requests that the affairs of society be conducted in council. The Quran affirms "*and their affairs are consultative among them*" (Surah Surah; verse 38). Therefore, Dajani (2013) took the initiative to form the Wasatia movement, the Wasatia Democratic Party, and the Wasatia Graduate Academic Institute for doctorate research and studies. The goal was to refute extremists of Islam to promote peace, justice, moderation, coexistence, and tolerance within the Palestinian community.

5.3 Wasatia as a Movement

The Wasatia movement was established in January 2007 as a charitable, non-profit, and non-political initiative to change the extremism trend. It declared its goals as the following (Dajani 2009):

- To bring a broader and more rational understanding of Islam to Muslims as well as to non-Muslims; to clarify the distortions to which Islam has been subjected to by different countries in the West;
- To educate Palestinians on taboo topics, such as Holocaust history;
- To seek broad answers for the deeply religious, political, social, and economic crises inflicting the Palestinian society;

- To strive and work for ending the Israeli military occupation through negotiations and peaceful means;
- To spread and promote Islamic tolerant concepts, values, and principles within the Palestinian community;
- To encourage the practice of moderation among Palestinians to moderate religious radicalism and bigotry and reduce political extremism;
- To bring a message of peace, moderation, justice, coexistence, tolerance, and reconciliation to the Palestinian community through vocal civic leaders;
- To teach creative and critical thinking and open-mindedness; and
- To empower potential for leadership in society.

Wasatia encourages one to lead the road for reconciliation as a value shared by the various thinkers and philosophers, as well as all faiths; therefore, it could become a foundation for dialogue to achieve peace, democracy, and prosperity. Wasatia structures its programs on thematic pillars, such as peace, state building, governance reform, education, women's empowerment, religious and political centrism, and civil society development. Moreover, in fostering a culture of moderation, Wasatia works to prevent Palestinian children from being exposed to the education of incitement, hate, and violence, where they can co-exist in peace, prosperity, and harmony with the Israelis. Toward this end, Wasatia aims to bridge the gap between two civilizations – between the civilization based on the Christian Bible and the Torah and the civilization based on the Quran (Dajani 2008a, b).

This ICT-MR communication research considered Wasatia as a model for the research, as a case study to encourage its principles in the community. Wasatia was an Islamic moderate religious movement, calling for a negotiated peace and reconciliation with Israel that would help in bringing creative solutions to the acute economic, social, and political crises plaguing the Palestinian society. As such, Wasatia advocated the establishment of a tolerant, democratic, tolerant state in Palestine through fostering a culture of moderation by attracting Palestinians who were well known as temperate in their Muslim religious beliefs and proud of their cultural heritage. Wasatia also required that Muslims respected the religious identity of other religions, such as Judaism and Christianity, which were part of Israel and Palestine (Dajani 2010, p. 10). In addition, Wasatia aimed to give birth to a political vision that would stand on the foundation pillars of peace, state-building, good governance, democracy, education, women's empowerment, religious and political centrism, and civil society development (Dajani 2007).

To achieve its goals, Wasatia disseminated information by publishing and distributing books and articles freely to the public, as well as organizing workshops on moderation, conflict resolution, peace, and reconciliation topics. These were useful tools for the promotion of moderation and reconciliation through the practice of empathy and advocating Holocaust education by supporting Palestinian trips to the Nazi death concentration camp of Auschwitz in Poland. Those trips helped Palestinian scholars and students to witness the historical memorial of the Jewish suffering at Auschwitz. By witnessing means of emphasizing "*to see*," as the Aristotelian way of knowing the consequences of actions, helped one to understand

and acknowledge empathy for the suffering of the other or the suffering of their enemy (Knoepffler and O'Malley 2016, p. 26). In addition, Wasatia strove to build bridges of understanding with other religions through interfaith dialogue.

5.4 The Auschwitz Trip (2014)

The Wasatia Ngo, with the direction of Dajani Daoudi, associated with a project called the Heart of Flesh Not Stone Project, the Project was constructed and affiliated by Friedrich Schiller University, associating the Jena Center for Reconciliation Studies, led by Professor Martin Leiner, and the Ethics Center led by Professor Knoepffler and Dr. O'Malley and Ben-Gurion University of the Negev, and Tel-Aviv University. The project amid to teach empathy and the suffering of the other by encountering your enemy. The trip took place in March 2014, was Palestinian students visited Auschwitz-Birkenau State Museum in Poland, and Israel Students visit Palestinian camps in outside Bethlehem. After the trip, an Israel newspaper wrote about the trip "Haaretz," and it was misinterpreted in Arabic news article, this caused an uproar at the project by the Palestinians society. They marched into Dajani Daoudis office at Alquds and hanged a Sign that he is a traitor working for Israelis and that he has adopted normalization, the concept is wrongly understood by Palestinians and Israelis, which aim to adopt the Zionist ideology of the land is only for the Jews given to them by God. Wasatia promoting Moderation, reconciliation, democracy was overlooked as normalization which destroyed the Ngo's reputation and was very hard to conduct other trips that have the same properties. Dajani Daoudi's car was torched by chemicals; no one claimed the incident and the police.[2]

5.5 The Founder of Wasatia

Professor Mohammad Dajani Daoudi was born to one of the famous Jerusalemite Arab families "Al-Dajani" which was honored by Sultan Sulieman The Magnificent whom he bestowed the name "Daoudi" to the family, for building a magnificent shrine above Profit David Tomb in Jerusalem in Zion Square at the Mount of Zion, outside the gates of Zion Gate in the old city of Jerusalem, the shrine was completed in 1524.[3] He is a Professor in Political Science, was the founder of Wasatia in Palestine (West Bank); he advocated openness, understanding, and tolerance to a religious-blind society that would transcend the sad experience of religious enmity. It was a similar dream as Rev. Martin Luther King Jr. dreamt of a color-blind

---

[2] Epstein, Nadine (July–August 2014). "Mohammed Dajani Daoudi: Evolution of a Moderate". Moment. Retrieved July 27, 2014.

[3] Kalman, Matthew (April 20, 2014). "*Palestinian Teaches Tolerance via Holocaust*". The New York Times. Retrieved July 27, 2014.

American society in his famous speech at the Lincoln Memorial in 1963: "*I have a dream that my four little children will one day live in a nation where they will not be judged by the color of their skin but by the content of their character*" (Sirimarco 2005). To paraphrase this quote, Dajani (2007) believed, "*I have a dream that our grandchildren will one day live in a nation where they will not be judged by the faith they hold, but by the content of their moral character,*" the opening ceremony for the Wasatia Movement – 2007. Nevertheless, Dajani (2014) did realize that religious tolerance by itself would not end the Israeli occupation of the Palestinian territories and establish the State of Palestine. As his Wasatia work progressed, Dajani became particularly disheartened by the spread of violent extremism among the Palestinian youth of his society, as well as in Israel. However, there were many discrepancies between both groups: "*Extremism among Palestinians is labeled as 'Muslim terrorism,' while extremism among Israelis is labeled 'Jewish ultraorthodox radicals'*" (Dajani 2014, p. 20).

Dajani (2014) linked religious extremism to the occupation, which created a state of repression, closure, and despair under which the Palestinians lived. In his view, only through the ending of the occupation and the establishment of a Palestinian state could one counteract this spread of violent extremism bolstered by the overlapping forces of poverty, closure, and loss of identity. Dajani (2014) advocated for peace in many journals, where he advocated for peace, understanding, and tolerance between Arabs and Jews. He viewed the spread of moderation within the Palestinian and Israeli societies as the basis for reconciliation, whether on the human level (between an individual and another), the national level (between Fatah and Hamas), the regional basis (between Israel and Palestine), and the international arena (Islam, Christianity, and Judaism).

While all other nations have gained independence, the Palestinian nation remains among the few still under occupation and far from exercising a right to self-determination. In his speeches, interviews, and presentations, Dajani (2008a, b) made an impassioned plea for having the Palestinian/Israeli conflict resolved based on a just and comprehensive, though painful, compromise for both parties as a prerequisite for a stable democratic secular Palestinian state (pp. 191–219). Fatah and Hamas are cast as opposing forces, with Wasatia in the midst of these. These clashing political visions resulted in targeting Dajani's (2014) moderate and peaceful approach by labeling it as a Western tool to undermine Palestinian national cause. According (Barakat 2017), Dajani had evolved from having extremist ideologies but to believing in moderation and moral values as the right path needed to be taken. To achieve peace with Israel, he strongly opposed both Fatah and Hamas and saw in their approach a lack of real leadership, which was urgently needed. He felt both parties lied to the people to enhance popularity, instead of being honest with the people by curbing their unrealistic dreams. He opposed Hamas adopting a violent approach to liberate Palestine and its insistence not to acknowledge Israel and to reject any negotiations with Israel to reach compromises with Israel. In addition, Dajani (2015) opposed Fatah for not being honest with the people and its double standard of claiming to be a peace partner to get funding from the United States, the European Union, and Japan. However, he urged its members to oppose normalization, incite against Jews, advocate the boycott of Israel, and even use violence in resisting the occupation.

Wasatia remains the only association among opposing forces in the midst of an acute crisis. It remained faithful to its principles of moderation and reconciliation and its advocacy for diplomacy and dialogue by condemning the use of violence and terrorism against Jewish civilians. Dajani's (2015) stance came at a high personal cost: He faced public animosity for his pacifist views that climaxed in outrage when Dajani, in March 2014, accompanied a group of 27 Palestinian students to the Nazi death concentration camps in Krakow and Auschwitz in Poland. This visit was to educate them about the Holocaust as part of Heart of Flesh, not a Stone project. Palestinians denounced him as a *traitor, king of normalization*, and a *defeatist*, forcing him to submit his resignation from his academic posts as founding director of the American Studies Graduate Program and rector of libraries at al-Quds University. He was threatened in his work on reconciliation, and his car was torched, thereby causing him to seek refuge in the United States for an indefinite period.

Recommending an ICT platform for the promotion of moderation, reconciliation, and democracy required oversight, guidance, and a commitment to these ideals for influencing people in Palestine and Israel. While there were not many organizations that fulfill these requirements, Wasatia is a different NGO that promoted moderation, reconciliation, and democracy in Palestine. To reach the community as a whole, the initiative for a future perspective was to form the Wasatia Democratic Party and the Wasatia Academic Graduate Institute to train the teachers and clerics on topics of peace, justice, moderation, coexistence, empathy, comparative religion, and tolerance within the Muslim community (Dajani 2014). The goal of these initiatives was to hold the center away from the heart of radicalism, and to de-radicalize, those youth lured to radical ideas.

Wasatia sought conflict resolution in dialogue, not in boycotts, and in negotiations, not confrontation, which held the promise of guiding the Muslim community in Palestine in the search for the right path for peace and justice (Dajani Dajani 2014). Wasatia endorsed the Arab Peace Initiative, as launched by the Arab Summit Conference held in Beirut on March 28, 2002, which called for a comprehensive peace with Israel, recognition of Israel, and normal relations with all 22 Arab countries, in return for Israeli withdrawal from Arab territories occupied since June 1967.

5.6 The Wasatia Cycle in the Reconciliation Process

This ICT-MR communication research boosted Wasatia NGO in launching an ICT platform to encourage toward moderation, reconciliation, peace, and democracy in Palestine. Wasatia was fiercely committed to and had already been struggling to achieve these objectives. Consequently, it could efficiently guide and facilitate online conversations and debates on extremism and moderation to large segments of audiences. Creating awareness, via the Wasatia ICT platform, would ultimately increase audience numbers' Likes on articles, posts, shares, and visits to the Facebook page (Fig. 5.2).

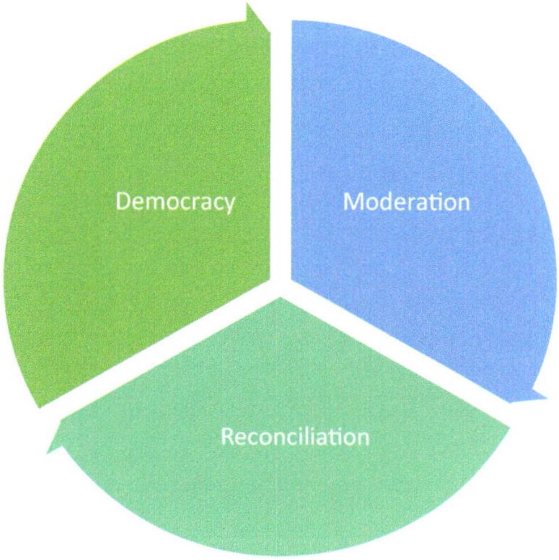

Fig. 5.2 Wasatia circle developed by Dajani, showing the relationship between moderation leading to reconciliation to empower democracy

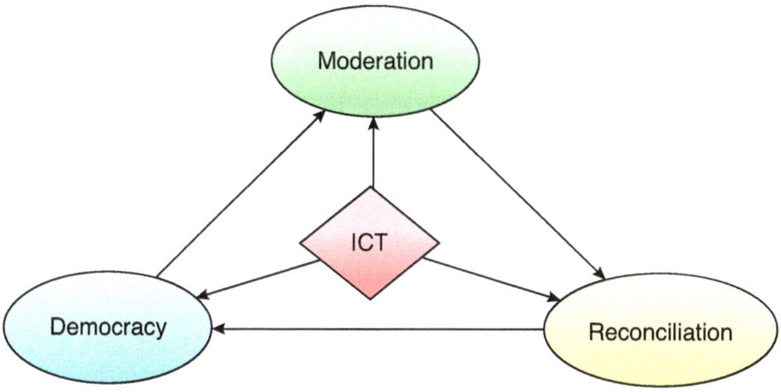

Fig. 5.3 The Wasatia Cycle is a theory of change

The ICT platform was used to develop a social network site; present slide shows; and publish online articles, workshops, and activities. The goal involved transforming the audience who wanted to be part of Wasatia's vision for moderation to members of the Wasatia NGO by interacting with polls, social network forums, and articles, posts posted in the Wasatia Facebook page that is integrated with the Wasatia's social network website www.wasatia.info. Moreover, integration between Facebook pages activities and the social network site was developed through ICT-MR communication research, which led to the spread of democratic, moderate, and reconciliation values. The ICT-MR platform illustrated the concept of Wasatia circle,[4] which demonstrated the relationship between moderation (Wasatia), reconciliation, and democracy (see Fig. 5.3).

[4] Kneopffler, N., and O'Malley, M. (2016). Wasatia: A Way to Enduring Peace in Palestine. In: Teaching empathy and reconciliation in the midst of conflict. Edited by: Dajani Daoudi, M., Dajani Daoudi, M., Leiner, M. and Barakat, Z. Jerusalem: Wasatia Press, Jena Center for Reconciliation Studies, Ethik-Zentrum Jena. Pp. 19-29.

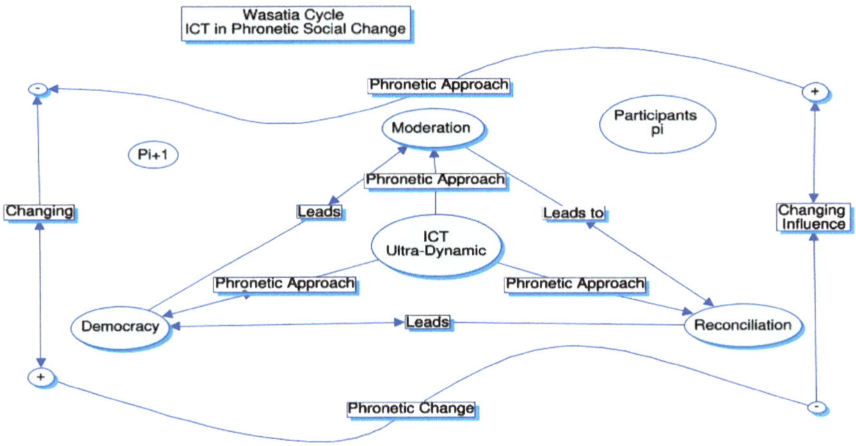

Fig. 5.4 ICT in applied Phronetic social science, ***"The Wasatia Cycle"***

The Wasatia circle elaborated that moderation led to reconciliation, empowering democratic values, thus creating a political system characterized by political stability, economic development, and social prosperity. It allowed respect for human rights and the rule of law and put an end to authoritarianism, corruption, and prejudice within the community, thereby leading to equality in the pursuit of happiness. Therefore, it resulted in a democratic society and moderation (Wasatia), which created an environment of trust, empathy, tolerance, acceptance, temperance, and the desire to conduct dialogues while respecting the narrative of the other, thereby creating an atmosphere for open-minded dialogue with the other. Reconciliation, by allowing the willingness to seek innovative solutions, was enhanced by dialogue with others to understand the narrative of the other. Empowering democratic values guided negotiations in good faith with respect and adhered the inclination to make painful sacrifices to achieve sustainable, just peace (see Fig. 5.4).

As detailed more in Chap. 2, Fig. 5.3 illustrates that ICT is the element of power that initiates the practice theory for Bourdieu (1977) and develops a power relation according to Flyvbjerg (2001). In addition, the epistemology used was the Wasatia Circle embedded in the ICT platforms, and the "*techne*" refer to the ICT applications used by participants, which were the Phronimos, according to Flyvbjerg (2001). The phonetic approach led to social change towards the reconciliation process, as illustrated as the Wasatia Cycle.

5.7 Conclusion

Knoepffler and O'Malley (2016) described Wasatia as having remarkable analogous principles and dynamists with an ancient Greek. As such Wasatia has to promote moderation and to transform actions successfully to steer conflicts for moderation

(Knoepffler and O'Malley 2016, pp. 19–29). Knoepffler and O'Malley (2016) compared Wasatia to Aristotle's ethical philosophy as a *"means between the extremes"* and which is helpful and relevant to steer the Israel/Palestinian conflict for moderation as a mean between two enemies (p. 29). Wasatia offered a vision of an ethical action with Muslim teachings, which benefited not just Palestinian, but also all Muslims, to steer them away from radicalism and enmity to moderation and reconciliation. Knoffler and O'Malley also emphasized that Wasatia was a moderate way to alter the stagnated conflict and was an alternative road for both Israelis and Palestinians to alter hearts of stone to hearts of flesh.

The 1933 Peace Nobel Prize *winner, Norman Angell of the United Kingdom,* stated in Oslo at the award ceremony:

> *The obstacles to peace are not obstacles in the matter, in inanimate nature, in the mountains, which we pierce, in the seas across which we fly. The obstacles to peace are in the minds and hearts of men. Only by intellectual rectitude and in that field, shall we be saved. There is no refuge but in truth, in human intelligence, in the unconquerable mind of man* (Haberman 1991, p. 172).

State of the art in applying the Wasatia Cycle concept is the implementation of ICT applications to improve virtues in the community, attitudes, emotions, trust, forgiveness, acknowledgment, recognition, communication among conflicted parties in a protracted conflict. Wasatia Cycle concept would facilitate better collaboration and a more integrative approach within dealing with both communities or nations that are in conflicts.

Chapter 6
Internet Communication Technology (ICT) for Deliberative Democracy

Having researched moderation in Palestine and Wasatia as a movement, the democracy described in the Wasatia cycle are more elaborated to evolve into deliberative democracy, the figure below serves to build up the theoretical foundations for the research design in researching Online Deliberative Democracy exploring its impact on social, cultural, economic, political transformation for a democratic society.

6.1 Online Deliberative Democracy

6.1.1 Introduction

This section concentrates on deliberative democracy for a better understanding of different approaches for implementing the discourse of deliberative democracy in digital social media, mostly details the online social networks, which is considered part of the online deliberation in a deliberate process. However, the chapter summarizes different models of democracies considered within the deliberation process or part of deliberation or competitive to deliberation. The chapter also illustrates ways in which the models of democracy may be in effect in deliberative democracy and vice versa. Figure 6.1 demonstrates the different types of popular models; however, this research focused on online deliberative democracy to study in what ways deliberative democracy differed from other democratic models. This chapter addresses definitions of the main types of democracies to grasp a better emphasis on the differences of each (Fig. 6.2).

Types of five democracy model, as developed by Martin Liner and Ayad Dajani to show the illustration of relations between different types of democracies leading to deliberative democracy, this directs the chapter to explore online deliberative democracy using Internet communication technologies.

I. M. AlDajani, *Internet Communication Technology (ICT) for Reconciliation*, Lecture Notes in Social Networks, https://doi.org/10.1007/978-3-030-41203-6_6

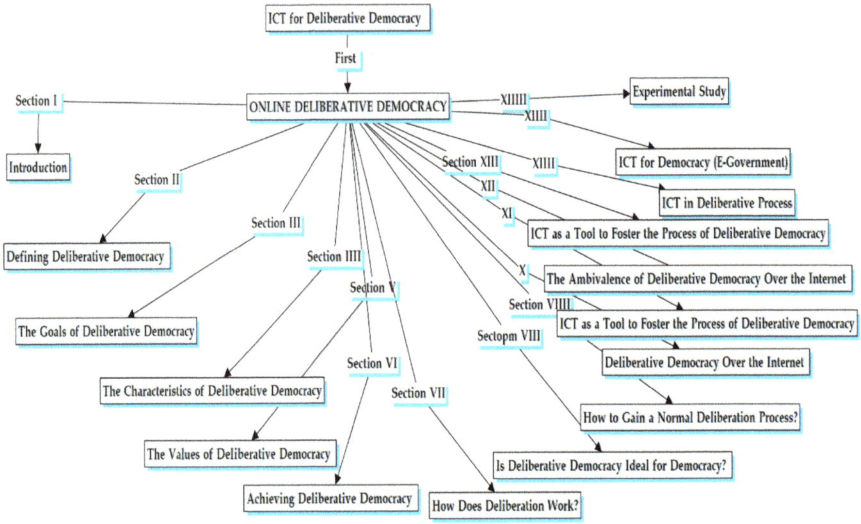

Fig. 6.1 ICT for deliberative democracy

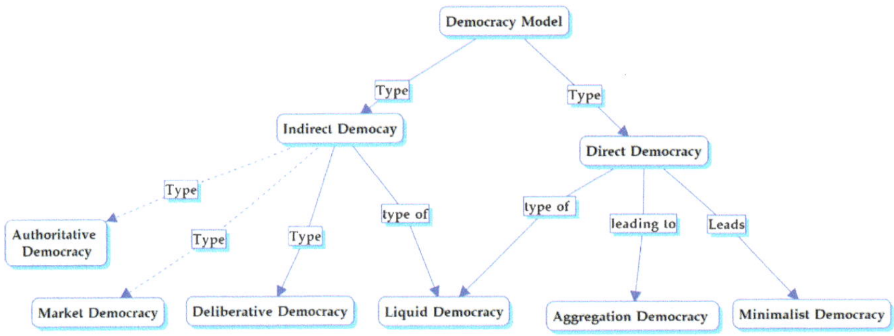

Fig. 6.2 Types of democracy

There are two necessary forms of democratic models: direct democracy and indirect democracy. Direct democracy depends on votes accumulated from all participants taking part in a democratic process, with all participants involved in the process. From the word direct, one can understand the concept of direct democracy, viewed as a *pure democracy*. Indirect democracy depends on representatives gaining votes of participants in a democratic process, where the ideal role is to decide on initiatives that are of concerns to the voters themselves

Figure 6.3 demonstrates models, which have a different kind of process in a democratic system. For liquid democracy is between direct and indirect democracies combining them both. The liquid democracy protects the rights of the participants to exclude their votes from the representative and interact in the democratic

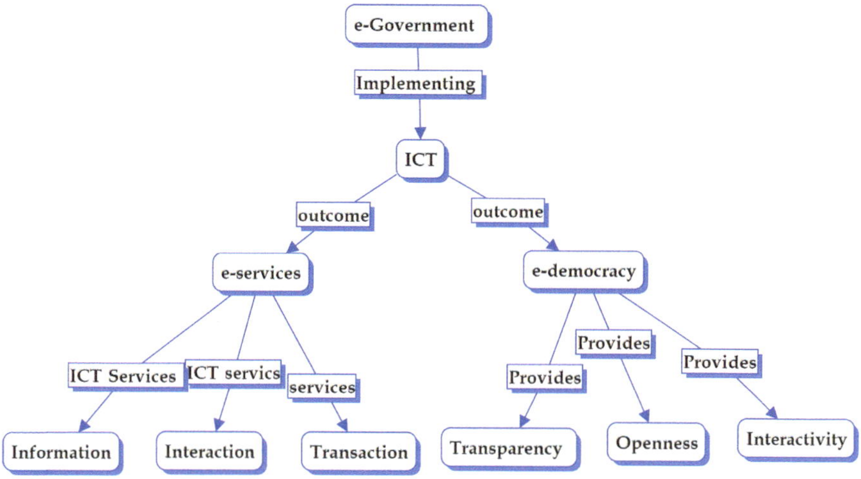

Fig. 6.3 Development of governments utilizing ICT, e-Services, and e-democracy

process. However, liquid democracy promotes using electronic media for the democratic process and the use of virtual community in the process of voting.

As for deliberative democracy, Democrats contend that laws and policies are based on reasons that all citizens can accept. The political arena should be one in which leaders and citizens make arguments, listen, and change their minds.

With indirect democracy, voters depend on representatives they elect in a free and democratic election. This model has two types of process: one is an aggregative democracy and the other minimalistic democracy. Aggregative democracy depends on representatives being elected in a democratic process to engage in issues of concern to the voters, where voters may interact with their representatives in any matters that are of concern. Minimalistic democracy is when people give their representatives the right to run the government, where they are not allowed to interact with the representatives on decisions each representative is making according to his or her knowledge and judgment alone. Voters are assumed to have no clear view or information of agendas being discussed in the political arena; from that perspective, one can assume the presence of a minimalistic democracy.

The deliberative democracy process, as presented by Cronin (2008), should be reasonable and sufficiently inclusive in the deliberation of opinions and participants for the debate to be well-formed (pp. 1–5). In his groundbreaking study, Dryzek (2002) argued that democratic theory was now dominated by a deliberative approach. Moreover, deliberative democracy has become a key focus of research in contemporary democratic theory. It is the most widely debated conception of democracy in modern times. Many scholars believe that it provides a stronger model of democracy. It is growing as a critical approach to the study of democracy and political behavior.

For the present study, deliberative democracy was used as a theory that was especially appropriate for understanding and evaluating democratic discourse of ICTs that applied to online social networks and the political conditions of the people engaging in social networks. Because this ICT-MR involved studying the phenomena for the influence of ICT on societies that were online, participants in the research were participating within the democratic principles of deliberative democracy through the social network site. Moreover, deliberative democracy requires the following three conceptions: public deliberation, political culture, and the definition of citizenship. The deliberative democracy theory raises five fundamental questions:

1. Why deliberate: Is the idea of deliberative democracy coherent?
2. How to deliberate democratically?
3. Who Deliberates?
4. Where to deliberate?
5. With what consequences people interact with those whose political views differ from their own.

Other questions arose as to what was related to deliberative democracy include, for example:

1. How can democratic ideals and institutions best be implemented?
2. What is the nature and value of deliberation?
3. Who was involved in political decision-making?
4. What is the relationship between deliberative democracy and political representation?
5. How can equal citizens and their representatives make conscious decisions for their society in the face of the fundamental disagreements that are inevitable in diverse societies?
6. What is the relationship between the representatives and their constituents?
7. To what extent do representatives can modify their positions through parliamentary debate?
8. What are the institutional aspects of deliberative democracy theory?
9. How will deliberative quality in political decision-making be achieved?
10. What are the role of communication and social networks in influencing public opinion?
11. Is deliberative democracy the ideal goal of free speech?
12. How does social movement organization use the media to frame the organization discourse?
13. What responsibilities do the media have in maintaining or promoting democracy?

6.1.2 *Defining Deliberative Democracy*

One definition of deliberative democracy is the participation of the population for political adaptation of a decision-making process, which advances a political argument. It refers to the idea that citizens and their representatives owe each other a mutually acceptable reason(s) for the laws they enact. It is the assembly of people to affirm the need to justify decisions made by citizens and their representatives (Gutmann and Thompson 2004). Citizens and representatives are both at the heart of democracy and a mandatory part of much public decision processes.

Deliberative democracy is the process by which individuals sincerely hold arguments with the merits of competing with other arguments in discussions that also hold merit. Moreover, political equality is the equal consideration of policy preferences (Fishkin 2009). Political participation is the behavior of a member of the mass public directed at influencing the political agendas, whether they are for the public interest or for political governance. Most citizens participate in events, should genuinely engage citizens and consequently, affect decisions, and become part of the decision-making process. The basis of deliberation is on the principle that a legitimate democracy grows out of public deliberation among citizens. Its compelling proposition is that "*in politics, it is not only power that counts, but good discussions and arguments too*[1]" (Steiner 2012). A deliberation is an approach to decision-making in which citizens consider relevant facts from multiple points of view to converse with one another (Bonnemann 2009).

Deliberative democracy strengthens citizen's voices in governance by including people of all races, classes, ages, and geographies in deliberations, which directly affect public decisions. As a result, citizens are influenced and can see the result of their influence on the policy and resource decisions that affect their daily lives (Bonnemann 2009). Dryzek argued that democracy was a form of self-government and that deliberation was the fragment that tested the quality of the discourse that went on in the sequence of making a decision, whether it was a governmental decision or a public one (Dryzek 2002).

Competitive democracies are based on elections, votes, and representatives. Deliberative democracy is a particular type of democracy that always focuses on the various elite of people, with power that will interact with those elections in which voters vote, and it is intended to be a peaceful way for those elites to take office and compete, regardless of what the people want in the first place. Therefore, there is a need to distinguish between aggregative democracy (representative democracy) or deliberative democracy, while in deliberation, a preferable viewpoint is considered. Conversely, an aggregative method is a consensus that all decisions are right, and people take the one that has more effect on the quantitative and qualitative method regardless of its merits.

Deliberative democracy is known as a better way to have preferable viewpoints and combine these into one viewpoint that is preferred by all. As for aggregative methods, these are more based on quantitative methods, regardless of the qualitative outcome of that method. In other sense, aggregative is the majority rule, regardless of the consensus within the community or views held by others. Using deliberation, one can ensure that the consideration of the merits drives the results by competing

[1] Jurg Steiner is a professor emeritus in the department of political science at both the university of Bern and the University of North Carolina Chapel Hill.

for arguments and not distorted by some pattern of domination of elite groups who have more power in the society. In other words, deliberative democracy considers political equality because it will bind people and treat collective decisions equally in the practice of deliberation. Alternatively, in aggregative democracy, decisions are collective in case these arise from arrangements of binding collective choices that include the interest of each person bound by that decision. However, for deliberation, a decision is a collective in case it emerges from an arrangement of binding collective choices that establish conditions of free public reasoning among equals, who are governed by those decisions. In those conditions, citizens are treated equally by not equally considering interest. This means that some interests may be discounted by arrangements of binding collective choices, by offering justifications for the exercise of collective power framed regarding considerations that can acknowledge with respect for all reasons (Cohen 1998). In this case, one could recognize the fault received from aggregative democracy and the ways in which deliberative democracy interacted with collective decisions in accumulating all decisions by considering these equally.

Deliberative democracy, as a deliberative view, is a framework of social and institutional arrangements, which facilitate clear reasoning and consideration among equal citizens. The main reason for a theory of deliberative democracy is to tie the exercise of the power to the conditions of social reasoning and to empower citizens, thereby allowing them to appreciate their power to participate in public discussions, which are relevant to the political processes that influence their lives. As Cohen highlighted, deliberation sets the community for political equality and substantively comprises egalitarian political values of the common interest. Deliberating the egalitarian concept of political equality, freedom of expression, and moral liberties provide the transition to democracy (Cohen 1998).

Deliberative democracy involves a process of deliberation, which is the process of rational consideration of issues for which decisions are part of the process. In deliberation, it is expected to show rational arguments for a systematic problem, and finally to arrive at a common conclusion for a fixed issue or a debated matter (Mančić 2012). Through this, it can be originated that this model of democracy has several advantages over the aggregative model most famous concept because it develops the mechanism to reach an acceptable decision, where participants or citizen have the same consensus on the initiated proposal.

The aggregative model has a setback that the decision comes down to collecting the votes followed by the option of the number of votes. In addition, the decision is taken by the highest scores in the votes, disregarding the concept of the decision decided for how the votes were voted. This model sounds attractive but is vulnerable to various forms of manipulations. One of the kinds of manipulations that occur is the concept of Condorcet paradox (Pavličić 2010), which is the social choice theory, where the decisions are taken by votes made by voters on an initiated proposal. In addition, the decision would be to encounter the number of voting, regardless of its proposed solution. For example, a small society has a situation where citizens will have to choose between three options. A committee must ensure that the elimination between the three options was justified; in such a situation, a Condorcet paradox might occur because of the misrepresentation of the representatives in a committee, such as chairperson of a committee who might determine the sequence in which proposed issues introduced the agenda set for discussion (Gehrlein 2006).

Another situation where Condorcet paradox might occur is when the election is made, and voters have voted for three candidates, but they have the same majority vote. The outcome of the election might be manipulated by the committee through another chairperson of that committee who might change the sequences of elimination. This will result in a majority rule, which disqualified one of its primary winners and gave it to another winner, through eliminating one before the other because voting was manipulated on the agendas conducted by one of the committees in a chairperson (Gehrlein 2006).

Aggregative deliberative democracy does not involve the addition of aggregative votes meaning votes are being disregarded where most votes are considered. All participants in the debate will reach the so-called unimodality preferences (Fishkin and Luskin 2006), which means that the final votes present a selection of average voters who are selected to vote on a solution to the fixed problem, where the outcome will result in disregarding the rest of the votes. However, it remains unclear precisely what unimodality preference means. In sum, unimodality is a moderate point combination whose preferences about which there is less or greater substantive agreement (Mančić 2012).

6.1.3 The Goals of Deliberative Democracy

Deliberative democrats agree that a primary aim of deliberation is to justify political decisions, including laws made by citizens collectively and their representatives. It also aims at fostering the common good for the society. Fishkin (2009) asserted that it was the "*deliberation by the people themselves with political equality [...] One strategy for achieving political equality would realize it through mass participation – in theory; everyone participates, and his or her views are somehow counted equally*" (p. 80). Gutmann and Thompson (2004) added the following:

> *deliberative democracy seeks to justify only decisions that collectively bind people, decisions in truly voluntary associations should be less subject to its demands. Politicians, professionals, and corporate officials who have power over other people should be held to publicly defensible standards. In more personal and more voluntary relationships, people should be freer to follow their own distinctive calling* (p. 34).

One of the critical aspects that make us need deliberative democracy is the moral disagreement in political arenas between different factions. That promotes the legitimacy of collective decisions made by either one of the political parties. Deliberation can help those who do not agree or did not get what they need to come and accept the legitimacy of a collective decision. Another aspect of the need for deliberative democracy is that it will encourage the participation of the public in the democratic process participating in public issues. Deliberative democracy can give a broader perspective on a standard issue at hand and provide checks and balances on any decision debated by both parties that are in question.

6.1.4 The Characteristics of Deliberative Democracy

One of the essential features of deliberative democracy is its pure democratic historical nature because it is considered a dominant part of direct democracy. Aristotle was one of the first theorists to defend the value of a process in which citizens publicly discuss and justify their laws to one another. He also argued that debate with participants on a related subject would reach a better decision than an expert working alone (Gutmann and Thompson 2004).

The interaction between democracy and deliberation is that the process of deliberation is purely procedural, and it is inclusive in its practice with democratic values. *Habermas*;[2] who had much credit in exploring deliberative democracy and its practice in our time. He argued that what made deliberative democracy was it included, in its process, all democratized values that needed to be practiced by the citizens in a democratic state (Habermas 1996).

The second characteristic of deliberative democracy was its ability to justify the decisions of the policymakers to their fellow citizen whom they represented. For instance, a leader should give reasons for his decisions and responsibility for the reasons citizens would give in return. Giving reasons and being responsive in return is a common conception characteristic of democracy, where reasons are meant both to produce a reasonable decision and to express the value of mutual respect, and where these are the aims of gaining a democratic value. The third characteristic of deliberative democracy is accessibility (i.e., the need for the information to be accessible to all citizens to whom it addresses). It means that information regarding citizens of the community should not be confidential, and the accessibility for it should be to all the people without any discrimination. Reciprocity is the fourth characteristic of democracy that deliberative democracy has viewed as mutual benefit from exchanging information between the various competing parties.

The fifth characteristic of deliberative democracy is a rationalization of partisan politics. Deliberative democracy promotes the respectful, interactive process of decision making, as it promotes the moral disagreement rule in a debate. It can support other practices of mutual respect such as the economy of moral disagreement, which makes the two parties comply with the views of each other rather than only respect their opinions. It is a tool to justify decisions and laws made for citizens, where citizens can check those decisions and find out whether these were made or if those decisions had the mutual benefits that were between the citizen and their representatives. Moreover, he argued that he was the decision maker for his own

[2] Jürgen Habermas is a German sociologist and philosopher in the tradition of critical theory and pragmatism. His most famous theories are on communicative rationality and the public sphere. One of his famous books are The Theory of communicative Action. 1981.

people. In the face of disagreement, deliberative democracy told citizens and their representatives to continue to reason together as a continuation of a moral conflict. The sixth characteristic of deliberative democracy is that any democratic decision must be binding to all citizens. In a deliberative democracy debate, the participant argues with one another, not for the sake of arguing, but in the interests of reaching a decision that is binding for all those concerned, thus gaining mutual respect by all parties.

Democracy is always evolving and continuing, as it is being developed from one party to another. Therefore, deliberative democracy aims for the continuation of the debate between participants and illustrates for us the seventh characteristic of deliberative democracy. It indicates that the process of deliberative democracy is dynamic, meaning decisions can change from one party to another where participants can criticize decisions made by other parties. Deliberative democracy reaches provisional decisions, which may be possible to have them, change later in the future. For example, a decision adopted will be valid for a period; however, it is feasible to revise this decision by other parties later in the future. Therefore, in a deliberative democracy, participants care as much about what happened after a decision is made and about what happened before it was made. This eighth characteristic gives an opportunity for other parties to alter a decision that had a disagreement if they believe that they have a chance to reverse it in the future, using what we call the principle of the moral economy disagreement.

A ninth characteristic is the checks and balances on decisions taken in the deliberation process, where deliberative democracy can make checks on decisions made and balance it by giving the information to the mass public. With this procedure, one can try to correct errors that occurred with previous decisions and correct errors of what one calls common misapprehension with deliberation. With deliberative democracy, the moral conflict will be sustainable and continual in the process of democracy.

6.1.5 The Values of Deliberative Democracy

The values correspond to deliberative democracy varies to the extent of its capability in enabling citizens to make the most justifiable decisions, whether as political decisions or decisions that will benefit the community. The types of values that may be gained from deliberative democracy are as follow (Gutmann and Thompson 2004):

- *Expressive values*: What we gain is mutual respect by both parties for decisions made by both sides, which is a manifestation of mutual respect. Reflecting the willingness to exchange opposing views before reaching a binding decision between two parties.
- *Instrumental values*: Reaching instrumental values is through achieving the right policies and the right decisions using standard tools to gain those values. To

enable citizens to make reasonable decisions that both parties would accept and respect.
- *Criticism value*: It is the ability to criticize wrong decisions and to reconcile its own expressive and instrumental values.

6.1.6 Achieving Deliberative Democracy

The discourse for achieving deliberative democracy in a political system, or in any community within its citizen always varies due to its repetitive and changing nature. In the following subsections, I review some of the principles that hold to reach deliberative democracy, whether for decisions made for the benefit of the government or the public; those principles were specified by Gutmann and Thompson (2004).

Procedural Principle and Substantive Principle The procedural principle holds that the procedures should apply only to the process of making political decisions in governments, or civil society and that the decision-making is subject to a process gained by deliberative democratic instruments. Substantive principle refers to the principle that whether the procedural outcomes were consistent with human rights, such as freedom of speech and expression, individual liberty, equal opportunity, freedom of religion, non-discrimination, and primary health care. Sometimes the outcome of procedure principle in deliberative democracy is not as accepted by the citizens and does not hold the substantive principle level. In this case, the procedure principle will not hold equal rights to all its citizens. Therefore, the procedural principle will fail.

The procedural and substantive principle of deliberative democracy attains liberalism to the citizens or participants, which adheres to the freedom of self-determination, and equality. The idea of a free and equal citizen would itself provide substantive moral content for principles that would reject an unfair decision even if it could be reached by procedurally just means. Withholding to this type of principle, one can achieve deliberative democracy by getting the right political decisions that will be accepted by citizens engaged in the same subject (Gutmann and Thompson 2004).

Deliberative democracy will offer reasons that can be accepted by free and equal persons seeking fair terms of cooperation. Such reasons can rarely justify unjust outcomes. The procedural principle has substantive content; to hold the substantive content, one must have democratic deliberation by free and equal citizens. One can reach this objective by having a provisional status for all procedural and substantive principles that will ensure free and equal status. The two principals will unite giving strength to the deliberative theory. Both principles should be systematically open to revision in an on-going process of moral and political deliberation to determine that the substantive justice of a particular decision depends on establishing a just process, which depends on determining the substantive justice of the process (Gutmann and Thompson 2004).

Deliberative democracy will always evolve and change from time to time as the decisions are emerging to have other decisions because there of unchangeable principles; the principle should be deliberated on in a deliberation process. Such principles are the freedom of expression or the freedom of speech, which are known as the principles of democracy, such as the rule of law, respect for human rights, equality in the act of the political process, free access to information, and free press. Bohman[3] (1998) argued that deliberative democracy had the attraction of many governments because it recaptures a robust democratic ideal "*the will of the people*" formed through the public reasoning of its citizens, as deliberation would be a public between free and equal citizens (pp. 400–425).

Consensus Principle This refers to the mutual consent as aiming at achieving consensus through realizing a common good for society or for gaining a political decision that fulfills the most profound real promise of deliberative democracy, which all citizens will accept despite their profound differences of identity. The consensus is for the common good of all citizens, such that citizen's view the proposal made from the deliberation process. In addition, it is for the benefit of all the people in the society. A Sandel (2009), a philosopher of Harvard University defines common good "*to do what is best for the community as a whole*" (Sandel 2009, p. 199) from that accept we realize the consensus gained from deliberation would be the common good of the entire society.

Pluralism Principle This principle argues that it is not always desirable to seek a general common good, rather than to try to live respectfully with moral disagreements. It also reserves the rights of the minority respecting their moral interest. It involves striving to agree on all decisions without eliminating any disagreement. Disagreements should be protected under the economy of moral disagreement.

Participatory in Decision-Making Principle This principle refers to one securing the moral values in deliberative democracy; it is a more participatory form of government, meaning that citizens will have a share in policy-making decisions. Citizens will have the chance to enjoy the benefits of taking part in deliberation to help develop virtues of citizenship, sharing issues, such as a political problem in a more public way. Decision-making by the next assembly of all citizens may not yield either the best laws and public policies or the best deliberative justifications for those laws and public policies (Gutmann and Thompson 2004).

Representative Principle This principle refers to citizens relying on their representatives to do their deliberative work and to make decisions regarding their fellow citizens that they represent, but it gives the citizen the ability to hold them accountable for decisions made on his or her behalf. The advantage is having decisions made by a leader who has more experience than their representative does. In

---

[3] James Bohman is Danforth Professor of Philosophy at Saint Louis University. He is the author, editor, or translator of many books.

addition, more models of direct democracy include liquid democracy, which is characterized by the possibility that every citizen has the right to interact in a political discussion and deliberate and take his or her vote from the representative to interact in the process of deliberation on the level of decision-making.

Fishkin[4] and Habermas's (1995) proposal for deliberative polling showed a random sampling of citizens to discuss policy positions of competing candidates, which offered a way partially reconciling the deliberative value of direct participation with the necessities of representative democracy in modern society. Deliberative democracy's fundamental aim of justifying laws by principles that citizens who are trying to find fair terms of cooperation can reasonably accept. Researchers have argued that decisions and legislation have all the authority they need if they are right or just (Gutmann and Thompson 2004).

Information availability refers to people having access to information that is accurate, adequate, truthful, and reasonable information, which they believe in, is relevancy to the central issues in consideration. *Substantive balance* refers to respect for opposing parties in that every party would take into account and respect other parties review of the main topics discussed and is considered by both sides. *Diversity* refers to considering the differences of each argument, meaning the review of each argument is diverse from the other by both sides dealing with the same topic at hand; in another word, the topic being revised has different alternative meanings. *Equal consideration* refers to equality of considering all discussed issues by parties in an argument regarding who offers them, and their meaning they should have the same merits regardless of which participants offer them. Moreover, *consciousness* refers to respecting others; each participant in a deliberative democracy should have his or her thoughts on what is right to do and what is right for the other by considering the rights of the other. In other words, participants must regard all viewpoints on the merits regardless of who offers them. Whether they are physically challenged, they should be considered in the deliberation process.

6.1.7 How Does Deliberation Work?

First, voters formulate issues to vote. Here, one should note that there is no third party to the aggregative model. Voters then present a problem, as in the first step of deliberation; a formula for this problem is processed, then one has a formulation problem. Unlike the aggregative model of democracy where no pre-given list of options considered, participants gather in one place, and the discussion begins. When the problem is formulated, then the next step is to list the possible solutions;

[4] James S. Fishkin holds the Janet M. Peck Chair in International Communication in the Department of Communication at Stanford University where he is Professor of Communication and Professor of Political Science.

rational discussion follows, where each participant has the right to state his or her reasons for or against one or more alternative solution.

Every opinion must be recognized and considered with all other views in the deliberation process. The process generates formulated problems and alternative solutions; the solutions will be discussed, and those problems that are not well formulated will be more precisely expressed. Participants should recognize all considerations of solutions and issues, as well as alternate solutions. Once this stage is completed, participants turn to the process of making a final decision that has targeted all those involved with formulated problems and solutions (Mančić 2012).

Another approach to the deliberation process is to employ the scientific methodology, formulated by Fishkin (2009), which he called Deliberative Day. Deliberative Day depends on deliberative polling. It explores what public opinion will be like if the audience is more efficiently motivated to behave like model citizens. Following this moment, a random sample is taken to fill in a survey of a typical sort. After that, the citizens are invited to come to a single place, at the expense of the polling process to engage in small group discussions and then to have more extensive plenary sessions, which provides participants time to exchange as much information as they can regarding the benefit of the common good of the citizens to reach a right decision. At the end of the polling process, the same questionnaire in the polling is given as on the first contact. The resulting changes of opinion are often dramatic, where participants discuss some democratic possibilities, and good results are having a consensus of any judgment. The views of the participants will have enhanced more should they be motivated efficiently to pay attention, to get enough information, and discuss the issue together (Fishkin and Laslett 2006). The benefit of the deliberative polling, to get experimental samples of random participants where they have incentives in effect to overcome all rational ignorance that they have before entering the procedure itself.

Participants find themselves in a deliberative poll, where they are randomly assigned to a small group. Resulting in having one voice in 15 or so, rather than one voice in a million, that makes them thrust into an interactive discussion where they offer reason and listen to the grounds of the other respectively. Replacing anonymous votes taken from millions of voters, which mean they have real votes in a small group. In addition, they vote in a secret ballot where the study of different voices can be more efficient and observing the change in opinion at the individual level without worrying about social pressures of false consensus of others (Fishkin and Laslett 2006).

The objective of having deliberative polling is to achieve the best decision making possible from a more informed and engaged electorate citizens perspective. It can be a push for policymakers interested in what the public would have to say if they were more informed by using social media and the polling systems. According to Max Weber, Social Action classified polling to an end of the rational action, where it should have valued action, practical action and joint action (Gutmann and Thompson 2004). Deliberative polling can also have a modest effect on public opinion through broadcasts and print coverage. The problem with this procedure is that one may wonder what happens if the information gained from such a procedure is shared throughout society (Fishkin and Laslett 2006).

6.1.8 Is Deliberative Democracy Ideal for Democracy?

Is deliberative democracy an ideal solution?

Cohen, discussed idealism in deliberative democracy, asserting that to reach an ideal deliberative democracy, one must seek to mirror a system of ideal deliberation in social and political institutions. As for deliberation to be ideal, it should have the following five main features (Cohen 1998):

- *Time to live*, which means that the deliberation process should be an on-going process, whose participants or members should expect it to continue to an indefinite future.
- *Sharing information by the participants*, which means all participants should share the common interest or knowledge of any information regarding the subject in the deliberative process to gain legitimacy, which is open deliberation among equals as the basis of legitimacy (Cohen 1998).
- *Deliberation process should be a pluralistic association*, which means the participants should have various preferences, convictions, and ideas, which they consider as part of their lives. In other words, participants should have narrative views, which are the outcome of their way of life, and narrative ideas concerning their ideologies and culture for their deliberations.

All those involved in a deliberation process should consider the outcome of the deliberation process in a deliberative association as a legitimate outcome of the deliberation process. Participants should also consider each one of them as equal in a deliberation process, which would result in a common interest for all participants, with mutual respect.

The aim of having such a deliberative association is to provide a model for institutions in which everyday choices are made, and social outcomes publicly justified which in that case it would legitimize the deliberation process in which the result of the deliberation process would be consistent for all participants. As for the deliberation process to be democratically legitimate, the resolutions, and outcomes should be free of any pressure that may indicate any influence from outsiders, and the reasoned agreements should be among regular participants in the deliberation process (Cohen 1998).

6.1.9 How to Gain a Normal Deliberation Process?

Participants in a deliberation process should be bound by the results of the deliberation and by the preconditions of that deliberation, and their deliberation should not be objected to any preconditions or constraints by any authority, prior norms, or requirements (Cohen 1998). The results of the deliberation or the decision of the deliberation should be sufficient reasoning to obtain the agreement on it for all participants; meaning, the participants in a deliberation process, should reason about deliberation to advance to a reasoned decision. Participants will give reasons, and those reasons will work to unit participants on a decision that fits the common good of all participants. Proposals for deliberation may not be accepted if these are not defended, even if these fit the participant preferences or their benefits. Reasons are always offered to bring all participants to agree on a decision and to be committed to that decision. To obtain that ideal deliberation, participants should remain equal. Each participant has an equal voice in the decision of the deliberation process, and the distribution of power and resources that are gained by the participants should not shape or affect the deliberation process. In other words, substantively equal deliberation means that no participants are influenced by contributions to the deliberation process. The most important aspect of a deliberation process is that participants reach consensus on a decision, where all participants consider and respect that decision. Moreover, they must be committed to that decision reached through the deliberation process.

6.1.10 Deliberative Democracy over the Internet

The Internet is the most comprehensive medium for deliberation, as it can include an entire city, country, or continent around the globe. It is part of the technology that most use, making it an essential component of the political arena. Fishkin (2009) stated the following:

> *Technology would serve to eliminate a separate bias in local face-to-face dialogues in that it can erase geography and create a genuinely national deliberation for each subgroup. In virtual space, I can as quickly be in a dialogue with someone on the other side of the country as with someone on the other side of town. Undoubtedly, this increase in geographical diversity enriches the dialogue. However, when deliberation is inserted into a federal system with state and local governments, it is not clear that this is a distinct advantage since the relevant demos for the decisions in question may well be smaller.* (Fishkin 2009, p. 81).

The Internet can assist the deliberation process due to its ease of access and use of the information, whether political or not. However, many scholars of deliberative democracy have believed that it is possible to connect deliberative democracy with indirect or direct democracy, and thus may reach the best possible system of political decision-making. Although this idea has many disadvantages over the standard theory of aggregative democracy (competitive democracy), the idea of online deliberative democracy should not be dismissed.

Many governments have asserted that the best way to rule a country is through a democratic system. In some countries, individuals are the ones who decide whether to choose who rule citizens in one way or another, such as in the Middle East and North Africa. The best way to know is through the voting system and polls that are considered an essential part of the element of democracy. However, with this kind of widespread appeal, the deficiency lies in the result of the vote that is not constrained by the individual's interest. The question here is whether individuals remain committed to most of the ballot.

The deliberative democracy model is more suitable for political decisions when it is over the Internet, as it can be deliberated and all elements of aggregative democracy are also present. It also adheres to the process of democracy by interacting with individuals and their needs to accept the decision taken by their representatives (Jankovic 2012). In the process of deliberative democracy through the Internet and for making this process successfully, some essential conditions need to be met: Firstly, the equality of all participants in the deliberation is critical. It is crucial that citizens who participate in the deliberation must treat each other as an equal and rational person. Everyone must respect the views of all other participants without any form of discrimination (Mančić 2012). The Wisegeek website features common question and answers in regards to Internet technology, when exploring Internet Forums, the article *"What is an internet forum e"* labeled the Internet forum as a means to establish such a condition that may be achievable, because Internet forums can make no restriction on any participant and do not discriminate between one with the other (Wisegeek 2018). More than that, it is expected everyone who is participating in the formation of the deliberative democratic process, to be honest when giving their reasons for the debate, and not to seek their interest, but to look for the benefit of all participants and the community to which they belong. Using forums of the Internet to foster the need for a deliberation process assure that the participants are honest in their views, and their opinions are taken into consideration for the common interest. Since moderators would moderate the forum, then all participants are constrained, to be frank with their views concerning the common good of his society. In defining a group of participants in the deliberation, it includes only those persons directly concerned with the problem, and that is what happens on Internet forums. It is relatively easy to provide a requirement that participants respect each other's opinions to guard the interest of the community in the Internet forum and not just focus on personal interest.

Secondly, the deliberation process should be made public. It must guarantee the right for everyone to participate in it, with every participant being able to interact with each other; meaning, it is focused on public opinions, where the public can give his or her opinion, and all participants can agree, disagree, or make a possible objection to other views. In that sense, the Internet can enhance views and discussions publicly viewed by all participants.

Thirdly, in deliberative democracy, procedures and problems are formulated and discussed in an Internet forum, which should be discussed only by participants who have interest or share common interests, meaning the topic should be relevant to all participants without any interference from a third party why may interrupt the discussion or have no interest in the issue of concern. Using Internet technologies and Internet forums can assure that all participants interact with the debates and share the same interest; from that sense, it is impossible to impose a topic for discussion for some participants who have no concerns and interest in the issue discussed.

Fourthly, deliberative democracy process should have a control button (i.e., a moderator or a kind of moderation); in that sense, the deliberative process will have records that are kept within proposals and discussions that do not turn to a wrong direction. However, a problem occurs regarding who the moderator of that forum is: Is he or she one of the participants or an outsider? Can a moderator participate in the discussion or is he or she an observer or a supervisor who has nothing to do with the discussion? In the deliberative process, there should be a moderator regarding large groups because a moderator will encourage the group discussion, but that moderator should be replaced after a short time, as he or she should never be involved in the deliberation process.

Fifthly, the freedom of opinions and expressions is one of the essential conditions for the success of the process of deliberative democracy, where each participant should feel free to express any opinion they want, without fear, and they will not suffer because of their views. Internet technologies can make participants free to be open to express their opinions without any authoritarian regimes oppressing these views. The disadvantages of deliberative democracy are that it should include all participants.

Direct deliberative democracy faces a significant drawback because it was not realistic; however, using the Internet for deliberative democracy makes it achievable. In using little technical knowledge, participants in the deliberative democracy process and its principles can be constrained. The general impression is that the Internet can achieve total equality among people. Some experts have suggested the Internet is perhaps the best tool for the realization of deliberative democracy and its decision-making process.

6.1.11 The Ambivalence of Deliberative Democracy over the Internet

There are many setbacks of using Internet technologies for the deliberation process, but those setbacks are most challenged and are more marginalized. One of the setbacks some authors have considered is that technology varies from one participant to the other. Moreover, no one can admit that ICTs can be the same for all those involved in a deliberation process. However, since the broad use of the Internet, one can omit this setback for a discussion because participants can send and accept messages. This method is used whether he or she has used Webmail or any ICT platforms.

Another setback of the deliberation over ICT's platforms is that the Internet is a huge market for large corporations trying to control the Internet, which may have provided them with influence on the debate that occurs in the deliberative process (Buchstein 1997). However, in a deliberation process that is online, the participants will be free to interact, discuss, argue, and write comments in any way that they feel it necessary for the discussion. It helps their discussion to form a decision-making process. In that sense, extensive cooperation must decide on which argument to support, but may not consider the participants. In this case, massive corporations may not have this influence; they may try to support the best argument that holds real common good for the public. The deliberation through the Internet has managed to bring considerable corporations to consider the common good instead of its interest.

There are other democratic values that can be implemented from using ICTs, which are introduced by (Johnson 2000, pp. 181–200), and they are as follows:

1. ICT can develop many to many communications.
2. ICT has the ability to give power to the stakeholder for its massive diversity of knowledge.
3. ICTs can empower the disempowered and disempowering the powerful .
4. ICTs can provide freedom of peach in a wide spectrum, no any kind of boundaries.
5. Full participation without any banning of any participants.
6. Privacy is not an issue when using ICTs; communication can be directed to private participants or shared with only private participants.

6.1.12 ICT as a Tool to Foster the Process of Deliberative Democracy

After understanding the process of the deliberative democracy, the conjecture is that the best technologies used in the process of deliberative democracy are ICTs. Such technologies are online social networks, such as Internet forums, interactive social Internet sites, and social Internet applications. For example, most social Internet networks include Facebook, Twitter, YouTube, LinkedIn, and Blogs, which have more capabilities for user interactions. People can register on the website, partake in discussions with their opinions, and share in topics on the main forum.

ICTs are used tremendously in daily life, and now governments are using ICTs to place arguments contextually by asking participants to contribute to topics of interest. Because forums usually devote to one central theme, the participants can choose topics within the social networks of their interest. These types of networks may also be called a message board, discussion group, bulletin board, or web forum, but these differ significantly from a blog. One user usually writes blogs and only allows for the responses of others to the blog material. Website forums usually allow members to create posts and lead to new topics of discussion. An Internet social network forum is also dissimilar from chat rooms; members in chat rooms usually chat or communicate at the same time. Alternatively, members in discussion groups post messages to be read by others whenever they happen to log on; forums also tend to be more topic-focused compared to chat rooms (Wisegeek 2018).

Blogs, Wikipedia, open news sites, and community-based open source journalism are forms of social software that are challenging and already established news media. With new protocols for content selection and analysis, authorship, and diversity of sources, these sites are promoting a more open, participatory culture, blurring the lines between news providers and their audiences (Flew and Young 2005); these methods of technologies enhance using deliberative democracy process in this age of technologies.

During the present Internet communication technology breakthrough, the Internet functions as a massive library where information can be both stored and accessed on a significant level. Gaining massive access to information for nearly everyone in the world, one of the principles of deliberative democracy is the access to free information on any topics of interest. ICT made the ease of data access possible. This condition is one of the fundamental principles of deliberative democracy that is acknowledged, where a participant does not take long to find out everything he or she needs about the issue that is being deliberated. From the influence of ICT on deliberative democracy, one can apprehend that ICT has achieved one of the essential deliberations principle conditions. Deliberative democracy constitutes "*An open discovery process... [which] assumes that the public sphere can generate opportunities for forming, refining, and revising preferences through discourse... toward mutual understanding and joint action*" (Sirianni and Friedland 2003, p. 74).

Janssen and Kies (2005) suggested some tentative hypotheses that could explain the influence of the principles of deliberative democracy on an online deliberation. The study influences the variables of Internet communication technologies to a successful deliberative democracy process. Identification in Internet social network sites would also refer to it as ICT platforms, where it is essential to identify the participants as if there were some discussion spaces requested from participants to be named. Because they are asked to register into the platform by entering all personal information to be part of the debates taking place in the online social network site. Some participants would not put their personal data, but also would like to participate and are giving the right to participate for the sake of the topic discussed. It will create an atmosphere for the common good of society itself and will have a positive influence on deliberation because participants will feel free to express their views freely without any constraint. However, another study of online deliberation argues that participants should be identified to give responsibilities of views discussed in the forum on the topic itself as well as on an online deliberation process.

Openness and freedom of speech mean that within the Internet forum, there is a possibility to limit the access and number of participant interventions to be able to have more influence on the plurality of ideas and opinions appearing in the debate of online deliberation. *Moderators* mean that Internet forums can be moderated or non-moderated. The moderator can play different roles. He can be a "*censor*" – for example by removing opinions that are at odds with the main ideology of the discussion space. Or he can be a "*promoter of deliberation*," such as in implementing a system of synthesis for debate, by giving more visibility to minority opinions, or offering background information related to the topics (Janssen and Kies 2005). In addition, a moderator can act as a filter, which is contributing to the interactivity and openness of discussions and the accessibility of public administration and institutional politics. In many cases described, there has been independent, *third-party* moderation. Similarly, as governments gain more experience with the Internet, moderation may shift to governmental agencies. In this respect, it is vital to stress transparency and care for negotiability of moderators (Edwards 2002). Edwards believed that moderators were necessary for the deliberation process and that moderators could possess the following functions, which were essential for the deliberation process to be a success (Edwards 2002):

- The moderator must determine the purpose of discussion and direct the discussion, so as not to steer away from the main topic or agenda.
- The moderator should obtain the support of political and other organizations, which will be agreed later, to consider the conclusions of deliberative processes and to implement them in the best possible way.
- The moderator must control the boundaries of the discussions in the deliberative process, in making sure that the participants hold the threads of the debates, and that all participants are invested in the debates, and that they are active and to power the discussion to take further steps to reach a concise decision.
- The moderator ensures that the results of the discussions and decisions into action, meaning that all participants will respect conclusions and decisions.
- A moderator will control the actual quality of the debate that means indicating a new line of argument in the debate, summarizing previous discussions, etc.
- The moderator has the right to exclude any participant, or to impose a ban on discussing a given period and to control participant behavior.
- The moderator can direct the discussion to be a fruitful debate in meaning he urges all participants to consider the views and arguments of other participants and to hold them accountable for their arguments.
- Agenda setting for an ICT forum as it can have an exclusive debate and can be decentralized (defined by participants), centralized (defined by organizers) or partly centralized (identified by both).

Such an example of used ICT platforms is online social network sites for governments who worked on having citizens to participate in debates that are relevant to topics concerning government. In the United States in the State of Virginia, when it launched an online social network site, the citizens could discuss the current problems that the state was facing to have an active debate about the topics in concern (Mančić 2012).

6.2 ICT in Deliberative Process

ICT-MR is a tool used for advancing moderation and reconciliation and a tool for the practice of deliberative democracy; however, there are many aspects to consider when wanting to utilize ICT in the online deliberation process (Gastil and Levine 2005, pp. 1–27).

6.2.1 Legitimacy

Legitimacy refers to all specific procedures and preconditions of deliberative democracy generate specific outcomes (Gastil and Levine 2005). ICT platforms must depend on whether there is a legitimate outcome of using these in the deliberation process, such as the freedom of expression and freedom of sharing information. The result of a deliberative process should possess qualities that are inclusive, voluntary, reasoned, and equal between participants. This aspect ensures the outcome of the deliberative process is a success due to the influence of ICT on the process.

6.2.2 Location

ICTs can work as a haven for networking and mobilize functions of many members of the online communities. These communities can work with one another despite their locations, whether they are domestic or international. Similarly, ICT is a tool that influences the technocratic elites and government leaders, running local traditional or international organizations, to interact with individuals and groups alike.

6.2.3 Laws that Promote Democracy as Human Right Laws

ICTs can have a useful role in enforcing human rights laws in conflict countries, such as providing a platform for opposition movements and challenging autocratic regimes and military dictatorships, despite government attempts to restrict access to the Internet and the use of the ICT in areas of conflict (Selian 2002). Examples include what happened in Egypt when denying their citizens Internet access to use in broadcasting the provocations of the governments against its people, and what occurred in the Islamic Republic of Iran during their presidential elections and the restriction for using the YouTube platform to hide the wrongdoing by the republic guard with the opposition.

6.2.4 The ICT Platform Influencing Conflict Transformation

Using ICT would ensure that many voices are potentially heard. ICT also motivated individuals to participate in the process. ICT platforms were a means to an end and not an end, which was the problem that many stakeholders using ICT found out about too late.

6.2.4.1 ICT for Democracy (E-Government)

This section discusses the influence of ICT on democracy and the importance of ICT as a mean for the process of democracy toward the Wasatia Cycle. From the reconciliation perspective, one of the essential features of using ICT as a tool was to prepare a structural change toward reconciliation process through the advancement of democracy. The advancement of democracy was implemented through political reconciliation, which was "*widely recognized to be one of the most important challenges for societies attempting to democratize following periods of repressive rule or civil conflict characterized by widespread and systematic human rights abuses*" (Murphy 2010, p. 1).

ICT encouraged the advancement of social reconciliation, through consciousness-raising in the society that would influence and impose political transformation toward reconciliation through promoting democratic values (Brantmeier 2009, p. 228). ICT influenced the advancement of political reconciliation that advanced toward the peaceful relationship between citizens and their governments. ICT improved the dialogs for a structural mechanism for political change between citizens and their governments, which led to an increase in the democratic participation of citizens in the governmental process; this type of process was introduced as e-Government. E-Government is

> *the major initiatives of management and delivery of information and public services taken by all levels of government [...] on behalf of citizens, business, involving using multi-ways of Internet, website, system integration, and interoperability, to enhance the services (information, communication, policy-making), quality and security, and as a new key (main, important) strategy or approach.*[5] (van der Meer et al. 2014, p. 489).

ICT was a tool used to develop a mechanism for political change that improved the democratic government-citizen relationship. It also created healthy relationships by inspiring two-way communication between governments and its citizens, regarding political agendas for both entities. ICT was a means of transparency, openness, and engagement between the citizens and their governments. The process of utilizing ICT as a tool for the advancement of democracy was referred to as e-Democracy (van der Meer et al. 2014).

---

[5] Guanwei Hu, Wenwen Pan, Mingxin Lu, and Jie Wang, "The Widely Shared Definition of E-Government. An Exploratory Study," The Electronic Library 27 (2009): 979.

ICT platforms have the ability to provide free information access for citizens regarding legal work, decision-making process, and procedures. ICT is opened to the public's participation, responsiveness, and immediate feedback. It also encouraged engagement of citizen's involvement concerning the online government process for contributing to public decisions. In that sense, ICT could expose citizens to e-government process in different aspects (van der Meer et al. 2014, p. 493). The utilization of ICT to implement transparency, openness, and engagement were within the concept of e-democracy.

Working with e-democracy could develop transparency to generate accountability; *"transparency relates to the notion of power"* (van der Meer et al. 2014, p. 494). ICT could expose the government by holding governmental institutions accountable to the public, thereby enhancing accountability to reduce forms of corruption (van der Meer et al. 2014, p. 494). E-democracy proved strength for the power of the openness concepts, which was sometimes defined as a function of transparency and interactivity. Openness *is "the measure of a government's interactive responsiveness to demands for information and services, and the immediacy of the feedback provided for citizen demands"* (van der Meer et al. 2014, p. 494).

Another reason for using e-democracy was for engagement, referring it to the citizen's interaction with the government via the Internet; this provided a new form of a dialogue between citizens and government to exchange information in the deliberation process. People using the online governmental online social sites have an essential role in reaching desired levels of e-democracy; it created a process used by individuals for contributing to public decisions, regarding government-citizen interaction (van der Meer et al. 2014, p. 494).

In e-democracy, ICT provided strength to the government to acquire the process of utilizing the concept of e-government as a notion for democratizing the community. As in parallel with the phases that contributed to e-democracy, ICT implemented phases of services, known as e-services. The stages of e-services involved implementing the stages of information, interaction, transaction, and integration between the government and its citizens. ICT could expose citizens to information, as it provides a free source of information about the government's actions and works to its citizens. ICT implements interaction between citizens and their government, through the web-based social networks, which includes online facilities, which enable transparency for the citizens. ICT integrates citizen's transactions within the governments, using the online governmental social websites (van der Meer et al. 2014, p. 494), such as the online governmental website for the renewal of passports in Sir Lanka (Perera 2015). ICT provides critical concepts of information openness, interaction, transaction, or integration within the idea of e-services that the e-government should provide. Figure 6.3 illustrates the development of governments utilizing ICT, e-services, and e-democracy that is parallel to the development of e-government.

For an example of applying the e-government, the Royal Government of Cambodia wanted to develop a national growth to establish a functional government that would focus on anti-corruption and more on legal, judicial reform and public administration reform. The government established the National Information

Technology Development Agency (NIDA). The aim was to bridge relations between the government and the citizens utilizing ICT (Vrasidas et al. 2009, pp. 224–228). E-government created open, honest, and free access to official information and system, which built confidence in the government that would increase the state behavior of conduct on the local and the international level.

Brantmeier (2009) *stated, "To reconcile the past and to forgive the government, people must be able to voice their decision and discuss their concerns all the while of building confidence in their newly developed government structure"* (Vrasidas et al. 2009, p. 225). E-governments create a balance between the national information against other perspectives to a procedure, such as accountability, which will produce a genuine knowledge for its citizens. The disadvantages of using e-government include the attempts of the government to control the use and the dissemination of new technologies and information. This issue will decrease the social capital influence, which will also reduce the use of technologies among citizens, and it will be counterproductive for the advancement of reconciliation between the governments and citizens, thereby creating a lack of trust and obeisance among citizens and the government. Another disadvantage was the distance between the citizens and the government, in the public engagement process, in which participants who have the knowledge of using ICT were dominant in participating in political issues compared to citizens who have no knowledge of using ICT. Moreover, the use of ICT provided the government with access to control information access to citizens and monitor citizens' behavior; this position influenced the confidentiality of the citizens who opposed the government at risk. Therefore, ICT security must be of high concern when implementing the ICT process in e-government applications (van der Meer et al. 2014, p. 490).

6.2.4.2 Experimental Study: Designing a Deliberative Democracy Process in an Online Social Network

This section discusses the theories of ways in which to build a social network site in regards to theories of deliberative democracy. In addition, it discusses the integrated interactions of participants with each other concerning the aspects of the deliberation process. The online social network site will have many participants integrating into a series of the deliberation process to decide on proposed issues and reasons by discussing the issues for the common good of all participants.

In the beginning, the government will have questions or proposed ideas or issues to present to the citizens. The social network site has a question or an issue presented to the participants and experts in that field of the proposed issue. All of them will participate together in discussing and reasoning the proposed issue. The question itself is not going to be a closed question or a multiple-choice question; it must be an open-ended question in a sense that it will create debate among all participants. One example of a discussed question is the following: Do you think that the Palestinian-Israeli negotiation is of benefit for both parties and should they continue? An open-ended question that a government will propose for its people can also be the following: How would the law control the way police interact with its citizens?

The participants in the online social network sites should have some functions that will let them interact with the proposed issue or question in hand. The participants, in this case, should be able to propose, vote, comment, and modify the same issue. The process of using ICT platforms to foster deliberation should be challenged in concern to the principles of the deliberation process.

Firstly, the debate must be fair on the online social network site; meaning, the deliberating community considers all proposals accepted in the deliberation process, regardless of the author's personality, friend support, or any other aspects. Secondly, scalability, where the website should be able to host thousands of contributors, reaching as many participants who are more involved in the proposed issue. Thirdly, the social network site should give the possibility for participants to give their opinions but that opinion should be regarded as accurate, reliable information. Here, the information is critical, called as an informed opinion; expert participants in the site itself should challenge this information but not participate in the debate, and their role is to give accurate information about the proposed issue only and not an opinion. Fourthly, the discussion in the social network site should have a little effort of interaction tools, meaning that the effort to interact in the debate on the proposed issue should be open to everyone and the way of interacting should take as much little effort possible. Giving an opportunity for all participants to be included in the debate. For example, to write one comment on what a participant thinks a decision should be on the debated issue. In addition, leaving more participants debating on the same question that would get many comments that would be seen and considered by all those involved in the discussion. Fifthly, the site should have no lobbying or participants interacting with one another, which would produce the apparent opinion of the participants, and there would be no gathering of support for a friend or any influence of any out norm outside the deliberation process. Sixthly, moderators of many online social network sites help moderators, but what is the best way for the deliberation process, with moderation or without moderation? Some aspects seek moderators in the process of deliberation but does it benefit an online deliberation process? Moderators in an online deliberation process might be of benefit, where they will help the discussion to be more efficient, and it would not go astray. However, some moderators might influence a group over another, and influence their agenda on a discussion and move the debate in which direction they prefer and moderate the discussion the way they want.

One of the real solutions for mass deliberation process is self-moderation, where every participant is self-accounted for his action and his way of participation in a deliberation process. However, there should be a control on self-moderation because the discussion might go in different directions, so the best is that there should be rules to be conducted on the social network site, which would give self-constrained for the participants who are also moderators.

There are disadvantages to self-moderation, such as the endless ramifications of discussion in which the discussion keeps on going with no end. Another disadvantage is rolling, this is an open topic, with no moderation and only self-moderation,

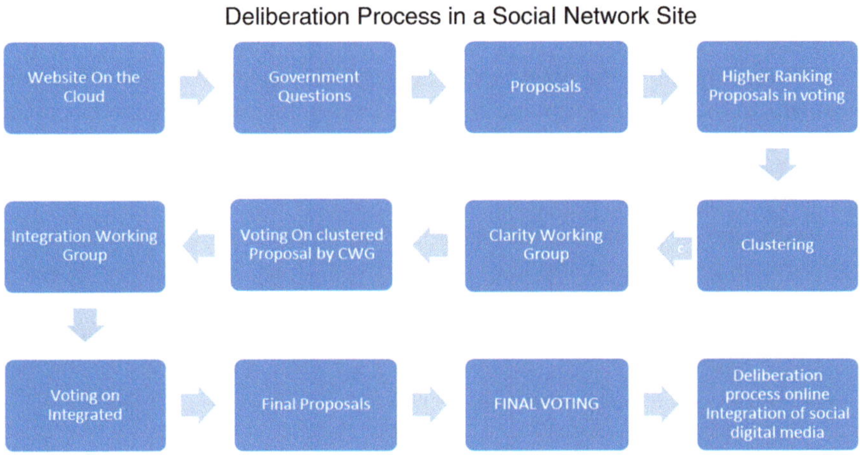

Fig. 6.4 Online deliberation process

the discussion might have much disruption from participants that are with or against the main issue, and that would harm the deliberation process. In self-moderation, the participants would have friends that favor his/her discussion and might be voted on it, that we the process of deliberation is harmed. To bypass such disadvantages in the social network site, participants will not have their own names for the account or users. They will have ID's that will only be connected to their names on an intimate level. Participants will not know who is creating the discussion and interacting with it; in that way, the deliberation process is intact and in the process, a single registration under a user-selected pseudonym with a digital signature (Fenizio 2011). As the deliberation process is being used by mass numbers of participants of thousands, how would one accommodate all comments and agreements, issue, and get a result in the process? The design of such a system is discussed at a convention about online mass deliberation process, and Fig. 6.2 shows ways in which the system or site should work (Fenizio 2011) (Fig. 6.4).

This study was conducted in a convention studying procedural and algorithmic aspects of mass online deliberation. The study was carried out by Pietro Speroni di Fenizio (Coimbra University of Portugal), Alois Paulin (Faculty of IS, Novo Mesto, Slovenia), and Cyril Velikanov (Foundation Euractiv/Politech, Brussels; Finizio 2011). This design was discussed in the convention and attempts have been made to clarify how this would work. A brief explanation of the online deliberation that was conducted at the convention was explained by Professor Pietro Speroni di Finizio as the following:

1. The social network site should be in the cloud, which means its base would be the Internet to make sure that it has an ease of access to minimal requirements.
2. There is a proposal submitted by the participants; now, those proposals are being filtered by votes, which is agreed on and which is not agreed on; the proposals

that get more votes is the proposal that will be discussed in the deliberation process.
3. Clustering includes votes for proposals that are being agreed on, should cluster together, and that is not being agreed on or like each other should also be clustered together; the way of filtering would be through online voting.
4. All proposals should be seen but sorted by the higher number of votes to the lower; proposals that have high votes mean these are very well understood and preferred by the participants.
5. After having to rank all proposals, the excellent ranking proposal in votes will cluster, and then a group will be set as a clarity working group. This group will be formed from participants who voted on the same proposal to discuss the higher ranked proposals that in the clusters. The result from the clarity working group is that one will have a unified proposal, which is best considered by all clarity groups.
6. After that the clarity groups, one will introduce what is called an integrations working group, where all participants will again form another integration working groups to engage in discussion of proposals that have a higher ranking of votes and have been clustered and clarified by groups. Thus, one will have few proposals that all participants have engaged in, leading one to submit those motions for the final vote for all participants to vote for these.
7. A final vote is proposed by the system depending on the results of the highest-ranking vote of the proposals.
8. The result will be a proposal, as suggested by all participants discussed and agreed on, and has the highest-ranking vote on a motion for a solution.
1. Because the website is online, all users can participate online and interact with it through social digital networks programs, software, or devices.
2. A proposed solution is presented.

ICT-MR communication research illustrates this principle of democratic deliberation for endorsing moderation and reconciliation in the Israeli/Palestinian conflict. Moreover, ICT-MR empirical work demonstrates online deliberation in practice toward the Wasatia Cycle, as described in Chap. 5.

6.2.4.3 Conclusion

No doubt, the ICT represents a vast arena for promoting democracy, and an online deliberative democracy should be considered. In addition, this involves developing an ideal online deliberation process among citizens or participants who are equal in shaping the decision through reasoning and having a consensus of the result for the common good of the society. The aim is for the transformation of a virtual community who has debated in an online sphere to a real community that adheres to the ideals of democracy. Therefore, this ICT-MR Internet communication research implemented the deliberation process on both quantitative, qualitative levels of deliberation by adhering interpretation and qualitative understanding and empowering the idealism toward social change towards the reconciliation process.

For example, the ICT-MR communication research designs a social network site has online applications, which adheres to the deliberation process in overall principles. Those online applications are Internet forums, polls, news web links, and members' areas for interaction, and forums for debating issues, as well as publishing articles. The ICT-MR introduces the integration of social networks sites, such as Facebook and YouTube, with other social networks sites that adhere the Wasatia Cycle discussed in Chap. 5, embedding principle of online deliberation, and benefiting from applying the Mixed-Method experimental research design, discussed in Chap. 2.

The next chapter also explores the tools of ICT platform that are used to progress in the ICT-MR communication research as part of the empirical work in the research. The ICT Platform helps to discover, collect, analyze, interpret, and examine the hypothesis of the research. In addition, illustrates the practical use of Facebook pages, which are integrated with each other and connected with a social network site, to help advance social change towards the reconciliation process.

Part IV
The Empirical Communication Research in Practice “The Techne Approach”

Chapter 7
Phronetic Netnography Research Design

7.1 Introduction

The Palestinian societies are using social media networks in broad masses, according to a poll done by the Jerusalem Media Communication Center. A research and polling center, which states that about half of the Palestinians in the West Bank and Gaza use the Internet; among these are a broad range of young people. The percentages are growing higher every day, as it is estimated that 1.5 million Palestinian use Facebook, Electronic Media, Twitter, YouTube, and LinkedIn (ESTRIN 2015).

Having introduced the epistemological research concepts, the figure below serves to build up the theoretical foundations for the research design in researching Netnography approach as the guide for the empirical work in the internet communication research ICT-MR (Fig. 7.1).

The knowledge of developing an online community within the Palestinian society is already established and is being done by different parties in the Internet industry in Palestine (West Bank). Therefore, the main idea of this Internet communication research was to develop an online community that can pave the way toward the advancement of social change towards the reconciliation process in the society. This chapter illustrates the empirical work done to develop the analytics for the research to help examine the hypothesis of the research.

The chapter explores a netnography approach to explore the empirical work done to promote the Reconciliation process via ICTs. The empirical work consists of an online social network website integrated with a Facebook page and another Facebook page, which is not integrated, to explore the integration and the ambivalence of using a social network website integrated with a Facebook Page and not integrated. The intent of integrating the social network site and the Facebook page is to develop a deliberation area for discussion among members and visitors to post new information from the website into the Facebook page; this gives the opportunity to recruit new members and spread the Reconciliation process in the social

I. M. AlDajani, *Internet Communication Technology (ICT) for Reconciliation*, Lecture Notes in Social Networks, https://doi.org/10.1007/978-3-030-41203-6_7

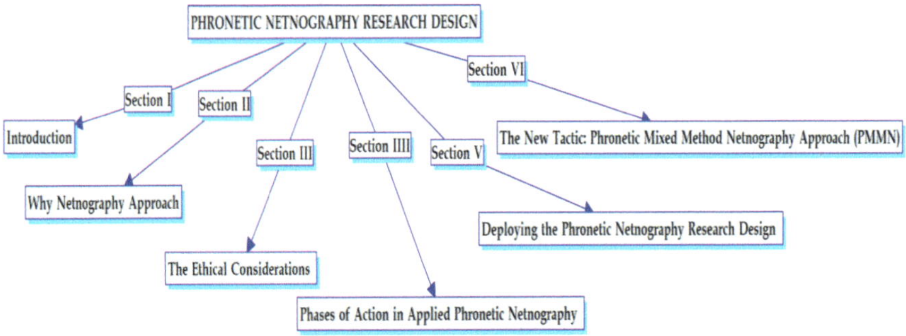

Fig. 7.1 Phronetic netnography research design

network according to notions driven by the Wasatia Ngo in Palestine meaning by spreading their content and published articles, workshops, etc. The intent of the other Facebook page is to study the social network experience without integrating it with an online social network website to view the different approaches and compare the impact they have on the society. The netnography approach was applied to the social network website and the two Facebook pages, as mentioned in earlier chapters. The chapter explores a broader aspect of the netnography approach, as defined by (Kozinets 2015). In addition, the objective of the chapter is to clarify how to conduct netnography research to implement any kind of communication research, in our case, it is the ICT-MR communication research.

7.2 Why Netnography Approach

The study was interdisciplinary research to study the phenomena for the discourse of ICTs to foster and promote the Reconciliation process. According to Kozinets (2015), netnography is acknowledged for its orientation to explore and examine the online communities including its interactions. For that purpose, I used it in this ICT-MR communication research. Kozinets (2015) stated that netnography was a

> *name given to a specific set of related data collection, analysis, ethical and representational research practices, where a significant amount of data collected and participant-observational research conducted originates in and manifested through that data shared freely on the Internet, including mobile applications.* (p. 79)

Netnography approach is utilized to pave the way toward gaining in-depth qualitative and quantitative insight into social network analysis, which would include big data and social network analytics. It aims for exploring the practice of a participative engagement, self-examination, alternative representation, a new perspective for humanist and hermeneutic studies. Netnography approach focuses on communicative actions and interactions, that are a loss over the Internet, these acts can be textual, graphics, photographic, audiovisual, and musical (Kozinets 2015, p. 80).

Kozinets (2015) offered eight advantages for applying the netnography approach, which was interpreted and summarized as follows (p. 88): The netnography approach can reveal a discourse about hidden and stigmatic behavior, which is not really present when studying natural person's behavior. It is tough to comprehend. When conducting a netnography, the researcher can distance himself from the participants or apply a degree of distance between him and the participants in the research as he is also a part of the communication interaction. That will give the researcher the ability to distance his own judgment, and it would not affect the research.

Netnography approach can be utilized for a small amount of data, where the researcher can focus on the phenomena of the study. In addition, Netnography can be applied to big data, as for large datasets and integrate computational analysis and visualization tools as conducted in Chap. 9. A netnography approach tends to focus not on the participants or the group, or the site. Nevertheless, tends to concentrate on the topic that relates to the participants, the group, and the social network site focusing on the hypothesis of the research.

There are four types of netnography approaches: symbolic netnography, digital netnography, auto-netnographer, and humanist netnography (Kozinets 2015, pp. 244–277). The symbolic netnography is a qualitative research approach that represents the online social experiences and interactions of a particular group, nations, language, cultures, and identity formations. In addition, to the traditional textual form of articles, news articles, even communication researchs, or website material to develop narratives of sharing, exploration, cooperation, conflict, exchange ideas, etc. This type of netnography relates to qualitative research techniques described as Phronetic Mixed-Method Experimental intervention Research Design detailed in Chap. 2.

The digital netnography is defined as utilizing computer-assisted software for methods of research, to collect online data, collect participant observations, and conduct qualitative data interpretations. Moreover, organizing digital tools for textual analysis and visualization analysis. This type is illustrated in Chap. 9. The auto-netnographer is a study of quantitative and qualitative behavioral study for the conducting researcher. The humanist netnography includes all data, such as visual interactions of online visualizations, video visualizations, and interpretations, such as using YouTube channels to conduct humanistic netnography research.

The ICT-MR Internet communication research develops a social structure analysis by deploying the three netnography types (i.e., the symbolic netnography, the digital netnography, and the humanist netnography) with the assistance of the CAQCAS NVivo 10 for Windows. Kozinets (2015) defined 12 principles to conduct a netnography research; these are summarized and applied in the discourse of implementing the ICT-MR communication research (Kozinets 2015, p. 98):

1. *An introspection phase*: This principle state is what the researcher or the netnographer should relate to as his or her role in the research, and the relation of the netnographer to research according to his or her current life project or life themes. I developed a social network site and the Facebook pages to conduct

the research on the emerging online Palestinian community. I have Palestinian identity, and I am considered the phronimos. I have identified that the research is part of my doctoral research, I have stated in the Facebook Page Wasatia.info in the about page, who am I and what I'm doing in the research at the start of conducting empirical work.

2. *Investigation phase* is the development of questions regarding the theme of the research. I used different data sources, social network site materials, people posts, and comments, which could be researched according to a netnography research design. The investigation phase involved attaining information that would assist in understanding the Reconciliation process from materials obtained from the online social network site and the Facebook pages. This phase was done when developing the empirical data, as published on the social network site and posted on the Facebook Page.
3. *An informational phase* involved gathering all information regarding the themes and the new emerging theme from exploring, collecting, and analyzing datasets. I considered the ethical consideration, which was raised as part of this phase and related to the conduct of ethical practice done in the research. All collected data were from the social network site and from the datasets developed from the Facebook pages, and these were in the public domain of the Internet.
4. *An initial interview stage* was where I searched and organized the social network site material according to its information. This phase was done in the developing process of the social network site and deploying the two Facebook pages. All information was gathered from the NGO Wasatia in Palestine regarding articles, book, journals, booklets, and news published online or in the median, considering the activities and workshop operated by the NGO. Also, I have used the Polling stages with participants and visitors. And I have also used the posts and comments as published in the Facebook Pages.
5. *Inspections phase* was where different sites were evaluated according to topics, person, groups' combinations, and interactions. This empirical work considered researching, exploring, and analyzing published material on the social network site, as conducted in Chap. 9.
6. *An interaction strategy* involved interaction, whether it was between me or the research and the participants or the interaction of the participants with each other. This was done by publishing online articles from the social network site (http://www.wasatia.info), posting these on the Facebook page of the NGO, and then boosting those posts to get more participants who would influence interactions on the Facebook page and to obtain more visits.
7. *Immersion in the data* refers to how the data are emerging from deploying the strategies conducted through the phases of netnography. In integrating the Facebook page with the social network site, with the approval of Facebook, which eventually created more visits and more interactions between visitors and members of the social network site and the Facebook pages. This led to developing new members for the NGO Wasatia because of the social network site. One example of emerging data are the trip to Auschwitz post; it had many

comments and hundreds of likes, it will be discussed further in the communication research.

8. *Indexing data collection* was where data were collected from a relevant variety of relevant resources that were in relation to the theme of the research. I used the deductive and inductive method when collecting and coding the data, as discussed in depth in Chap. 8. Using Nvivo for windows as an indexing software, and computer assistant software to collect and organize the data.
9. *The interpretive analysis* involved interpreting strategy for the netnography research, which included phenomenological, and hermeneutic methods, as described in Chap. 8 and practiced in Chap. 9.
10. A number of iterations were conducted by applying netnography through the multiphase mixed-method design, as described in Chap. 2.
11. *Penultimate phase* involved checking whether the phases were of the corresponding types of netnography, which included symbolic netnography, digital netnography, auto, and humanist netnography. ICT-MR Internet communication research was conducted within the phases of the three types of netnography: symbolic, digital, and humanist netnography, respectively.
12. Detecting and measuring the results of the netnography research occurred in Chap. 9, which examined the reliability and the validity of the results.

7.3 The Ethical Considerations

The ethical consideration was an important concept when conducting the research. And that is for developing research to be considered as academic research. According to Kozinets (2002, p. 65), there are concepts for adopting ethical conduct in netnography studies, and these are as follows:

> *(1) The researcher should fully disclose his or her presence, affiliations, and intentions to online community members during any research. (2) The researchers should ensure confidentiality and anonymity of informant. (3) The researchers should seek and incorporate feedback from members of the online community being researched. (4) The researcher should take a cautious position on the private-versus-public medium issue. This procedure requires the researcher to contact community members and to obtain their permission (inform consent) to use any specific postings that are to be directly quoted in the research.* (p. 60)

The ethical concerns involve the researcher, participants, and the degree of maximum and minimum harm, and the maximization of benefits of the research, within the concerns of the researcher and the participants. In addition to other aspects such as injury, vulnerability, and personally identifiable information.

The Association of Internet Research – AoIR offers some guidelines for conducting ethical consideration. The ethical decision making is best approached by applying practical judgment attentive to the specific context (Ethics Working Committee 2012, pp. 4–5). In a netnography, netnographers must balance the rights of subjects as authors, as research participants, as people. Moreover, they measure these with the social benefits of research and researcher's rights to conduct the research. The rights of human subjects may offset the benefits of research. The ethical decision-making in the research is a deliberative process, and researchers should consult as many people and resources as possible in this process (Ethics Working Committee 2012, pp. 4–5). Three concepts work when applying Internet research, and these are summarized as follows:

1. Human Subject Research: According to (Kozinets 2015, p. 135). Based on the Protection of Human Subject, the US Code of Federal Regulations Title 45 Part 46 (2009); is research in which intervention and interaction of persons to gather information for the research conducted. Therefore, the Netnography Internet communication research ICT-MR has fitted in this category. The ICT-MR Internet communication research is more oriented on topics discussed than it is on the personal participants or the researcher himself, and the discussion is in the open Internet space and is considered a social space according to (Basset and O'Riordan 2002), ICT-MR Internet communication research is an open social space of discussion on the topic of moderation, reconciliation, and democracy.
2. Public or private: The Internet is not a text or a place. It is an open space of interactions of culture capitals; it is not private or public the problem with that are the creations of digital doubles of identities and different narratives of ourselves (Kozinets 2015, p. 139). Anyone who is using the open space or the free domain is liable to the extent that his data or posts are not private and are open to the public. In the research, the researcher is accountable in accordance with the concept of contextual integrity according to (Nissenbaum 2010).
3. Data Text or Persons: Where there is no one, type of social interaction for one participant. The participant can change views and can post differently in a different context. In the ICT-MR Internet communication research, in accordance with the measuring of harm for participants and the research and the benefits of the study. The research possesses no actual harm to participants, as well as participants were acknowledged as being on a Facebook page. In addition, it is research conducted as part of the empirical work is done for a German university, and that is being posted on the Facebook page. In addition, their posts, photos, comments, and interactions are publicly shared on the Internet. According to Beckman (2005), the access to the ICT-Internet communication research is not restricted, and then it is not private but public (Langer and Beckman 2005, pp. 189–203).

The ICT-MR Internet communication research was in accordance with the ethical considerations, as stated above in the discussion.

7.4 Phases of Action in Applied Phronetic Netnography

The phases of action to develop a netnography research design for the ICT-MR Internet communication research, as these are stated in Table 7.1, with considering applying Netnography principles as we apply the phases of action.

Table 7.1 Netnography Research Design – phases of actions

	Phases of Netnography for the ICT-MR Internet communication research
Phase 1	Developing an online social network site, www.wasatia.info Gathering material on reconciliation and moderation for publishing on the social network site. Organizing workshops and activities about reconciliation and moderation for the community. Trying to recruit members for the NGO Wasatia through the online application Learning results: the social network site has visitors, but they are not registering, to become members, so we will create a Facebook page and integrate it with the social network site.
Phase 2	Developing Wasatia Facebook page (wasatia.info) Gathering material and posting them on the Facebook page Boosting posts on the Facebook page to create awareness. Integrating with comments on the page. Integrating the Facebook page with the social network site. Learning results: more visitors to the social network site and more members registering on the social network site. The need for distinguishing between Wasatia, as a concept of the way of life, and as a religious concept.
Phase 3	Developing Facebook page Wasatia Vision for the promotion of Wasatia values on Facebook, without the integration with the social media website. Learning results: it has followers more focuses on Wasatia as a religious concept. The need to distinguish between the social Facebook page for Wasatia.info and the religious concept for Wasatia Vision.
Phase 4	Promoting the site www.wasatia.info on social networks using Facebook boosting properties. Promoting awareness for moderation and reconciliation in the Palestinian society and recruiting new members for the NGO on the website Conducting reconciliation workshops and activities using online applications. Influencing interactions between members and visitors on the social network site and the Facebook page. Publishing more materials and journals about reconciliation and moderation. In addition, gathering news to be released about the activities of the NGO. The most important activities, which had a gradual influence on the society, included the trip to Auschwitz for Palestinian students in 2014[a]. Learning results: More members are registering to the social network site www.wasatia.info – and followers are becoming more on the Facebook page. In addition, more members of the social network site want to become more active within the Wasatia's activities on the ground. The influence of ICT for switching virtual members to the original member in the NGO and their enthusiasm to become active members in the work of the NGO.

(continued)

Table 7.1 (continued)

	Phases of Netnography for the ICT-MR Internet communication research
Phase 5	Developing reconciliation forums: Stage 1: A forum of open-ended questions for members of the social network site only to debate on crucial issues that the Palestinian community faces in the conflict. Most questions are related to reconciliation and moderation influencing the Palestinian society. Stage 2: An online forum for members and visitors of the site, to read and write comments, and integrating this forum with the Facebook page wasatia.info Conducting reconciliation videos and interviews about Wasatia NGO and the values the NGO promotes. Publishing the videos on the social network site and boosting them on the Facebook page. Boasting more materials about reconciliation and moderation on the Facebook page wasatia.info, and gather the interactions of members toward the posts on the page. Gathering online news about the influence of the trip to Auschwitz, which was conducted by the heart of flesh, not stone project, and publishing it on the social network site. Learning results: Some participants want to be part of the Facebook page and not to become active members, and there are participants who want to become members of WASATIA NOG, and to be active in the workshops and activities done by the NGO.
Phase 6	Applying qualitative content analysis on the internet communication research ICT-MR. utilizing a Netnography social structure is to explore and examine the new themes that emerged from the data. Then collect, analyze and interpret the data, developed from the social network site and from the Facebook pages Wasatia.info and Wasatia vision. The ambition is to examine the influence of ICT on the discourse for the reconciliation process in the Palestinian community that is in conflict.
Phase 7	Evaluation results of the website www.wasatia.info, and a Facebook page– wasatia. info – Wasatia vision. Conclusion remarks of the internet communication research ICT-MR Interpret and evaluate the effect of ICT on the discourse for the reconciliation process in the Palestinian community that is in the middle of the conflict.

[a]The trip was part of the Heart of Flesh not Stone Project. The trip was conducted by German professors, professors from Israel, and the Wasatia NGO led by Professor Dajani. The trip made an uproar in the Palestinian Society, which ended in torching his car and accepting his resignation at the Al-Quds University

7.5 Deploying the Phronetic Netnography Research Design

The initial attempt in developing a social network site involved gathering the domain name, Uniform Research Locator (URL); for this research, it was appropriate to choose http://www.wasatia.info as the domain name, which presented Wasatia to no governmental organization in Palestine (West Bank). The name had a significant meaning, which is 'moderation' as translated from the Arabic language, which also is the name of the Palestinian NGO (Wasatia). There should be three main elements to consider when developing the social network site: The nature of the domain URL, the material should always be current or dated on the website, and the information should corroborate truth and legitimacy. The URL contributes to the Wasatia Ngo, "*WASATIA*," as mentioned is the NGO residing in Palestine (West Bank) that aims to reclaim moderation and a centralist position in the conflict and to advance the Reconciliation process in the community.

The social network site needed to be current and regularly updated by the administrative staff and members, who were active in maintaining and developing the site, which made the site reliable. Participants could share their thoughts and ideas with each other. The social network site would help develop a culture that adhered to moderation and reconciliation in the community. It was a priority that all information was valid and verified because the purpose of the information was to uphold the real values of the NGO and to promote Reconciliation process in the site to the visitors and members. The gathering of information was based on the articles the NGO presented, which gave it legitimacy.

7.5.1 Introspection, Investigation, and Information Phase

Within the discourse of the empirical work to develop the netnography research from the Palestinian community, ICT-MR introduces the netnography social network design to reach an entire city, country, or continent around the globe due to its ease of access to information over the Internet. ICT platforms can bring people with seemingly unpopular political opinions together to generate new policies that were previously unthinkable and to promote peace and reconciliation or promote enmity and incitement against a common enemy (Richmond and Tellidis 2014).

The netnography social network site could play a significant role in the case of the Israeli/Palestinian conflict as an openness toward moderation, reconciliation, negotiation, and peace. Which is unpopular or stigmatized as seen as unfruitful. Thus, if through ICT platforms, one could bring citizens together who would believe in these seemingly unpopular ideas, these convictions could gain currency in the mainstream political sphere and help win over the undecided, to advance the Reconciliation process. Applying social network platforms to explore and examine Netnography research, could make some influence on the Israeli-Palestinian conflict, because if both sides can soften their attitudes toward one another, then a new political solution for the conflict might emerge. The ICT platform used to research a Netnography which is a useful medium for fostering democracy that is, by encouraging citizens to consider the merits of various arguments on equal terms to reach consensus.

This process of ICT platform to explore a netnography that might foster democratic political ideals – even unconsciously – by creating a space for citizens to express their freedom of speech and expression for their ideas, debate, and deliberate political issues as equals. In this sense, while often ICT platforms have functioned to separate opposing parties, it can be honed to bring people together and work toward shared, democratic political values. Because Israel and Palestine have remained such an excellent and heartbreaking political standstill for so long, the possibilities that the Internet and Communication Technology poses cannot be unheeded (Dahlberg 2007). Facilitating such an ICT platform for a netnography about the Palestinian community required oversight, guidance, and a commitment to ideals of moderation, reconciliation, democracy, and peace, that Wasatia could deliver. Harold (1951) stated the following:

> *We can perhaps assume that the use of a medium of communication, over an extended period, will to some extent determine the character of knowledge to be communicated and suggest, that its pervasive influence will eventually create a civilization in which life and flexibility will become exceedingly difficult to maintain, and that the advantages of a new medium will become such as to lead to the emergence of a new civilization.* (p. 34)

The procedures and the extent of the empirical work included information on activities, such as recruiting, training, and testing the actual work required. Wasatia needed the skeleton of an ICT platform, operating as a netnography design, to generate discussions and promote reconciliation within the Palestinian community. Furthermore, searching for improvement encouraged Wasatia as a way of thought and a way of life in the Palestinian community.

Launching an ICT platform of a netnography for the Wasatia NGO contributed to creating an online community that advanced moderation, reconciliation, and democracy in the Palestinian society. The social network site operated as a netnography design with standard "Home" and "About" pages that accompanied most NGO websites. There were links for activities, articles, news, videos, polls, applications, courses, and forums; online social network software for members to develop bases of interaction, and a Facebook integration with a Facebook page. Those links detailed Wasatia's recent efforts at facilitating moderation, reconciliation, and democracy between Palestinians and Israelis on the ground and online, such as through workshops, publications, conferences, and all online materials and activities. These also importantly contributed to the online participant's sense of Wasatia as a valid, credible, active, and responsible organization.

The social network site also had links such for articles that displayed links to articles, materials, and workshops that promoted the Reconciliation process. In addition, it provided more insight for participants who were online and wanted to further educate themselves on moderation and reconciliation in Palestine. The forums in the netnography design for the social network site hosted interactive discussions with its members to debate moderation and reconciliation efforts in Palestine.

The social network site had online courses, creating a breakthrough for the understanding of reconciliation and moderation online concepts endorsed by the Wasatia NGO; the courses taught reconciliation on an online basis. In this context, the informed public must become informed of what the word moderation, reconciliation, and democracy (Reconciliation process) meant in their communities. The social network site contributed to defining reconciliation in practice through articles, forums, and ICT programs; news feeds and web links; and other social media applications. The development of the ICT-MR Internet communication research expansion consequence is shown in Fig. 7.2.

The expanding progression diagram works in developing waves of influence to affect the community. The waves have more influence when moving from Phase 1

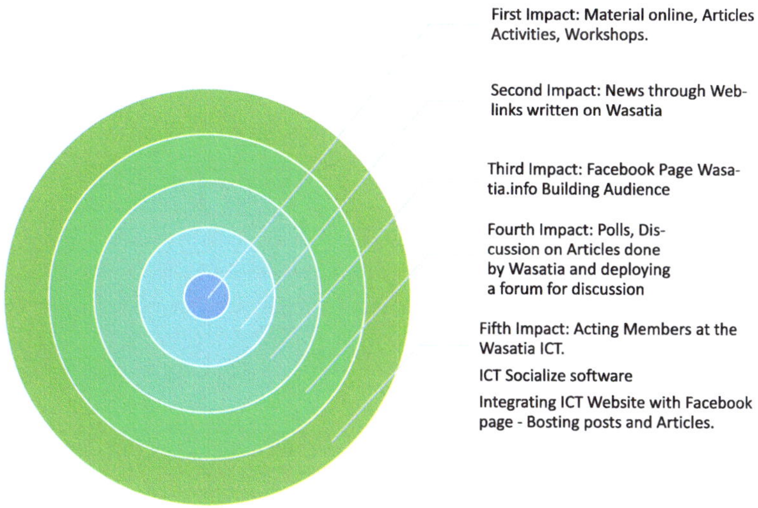

Fig. 7.2 The expanding progression diagram for the ICT netnography design

to Phase 5, as shown in Fig. 4.4. The expansion gets wider when creating waves of knowledge. These involve implementing the applications of ICTs on the social network site and influencing the community to spread the knowledge of the Reconciliation process in depth. The reason for this phenomenon was as Flyvbjerg (2001) illustrated, power is knowledge, and knowledge is a way to gain control (pp. 353–366). In this context, one can understand that knowledge is the virtues of the Reconciliation process, and the power to apply these to the online community was the ICT's goal in this ICT-MR research. As the phases were transgressed, more visitors were exposed and recruited through the wave of knowledge. Therefore, the social network site gained more members and more visitors. In that aspect, one could assure the Reconciliation process was being exposed to the open domain using ICT as the tool for that purpose.

7.5.2 *The Social Structure for the Netnography Research*

The domain name for the NGO was http://www.wasatia.info. Figure 7.3 shows the content of the social network sitemap for the adherence to the discourse of the Reconciliation process in the Palestinian Society.

The main map of the social network site took the visitor on a journey involved in understanding the concepts of the Reconciliation process and its methods. One of the essential features of the social media site was to give creditability and assure the reliability of information for the members and visitors, as well as to validate activities done by the NGO Wasatia supervision. The design also assures the validity and the reliability for the discourse of the communication research.

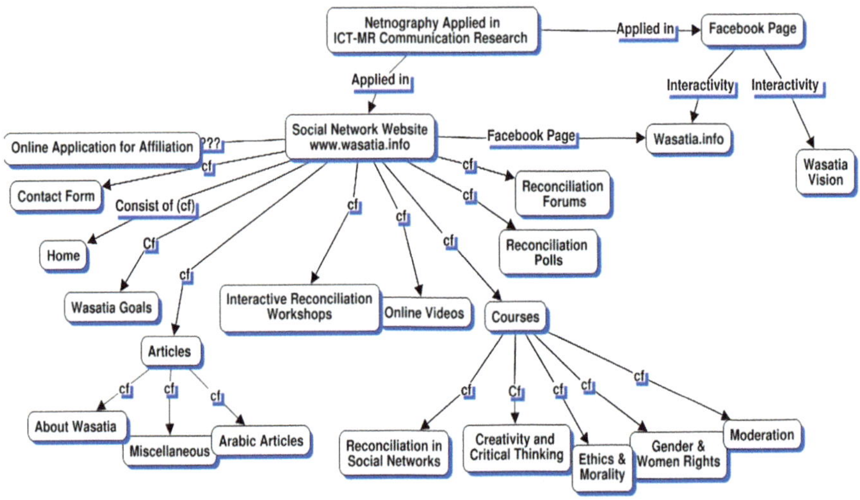

Fig. 7.3 Netnography in ICT-MR communication research. (Developed on 2 February. 2014)

7.5.3 *Interactive Reconciliation Workshops (Online and Offline), Immersion, and Data Collection*

The ICT-MR Internet communication research used features to advance toward a process that develops a moderate society within its online social network approach using different online application for that purpose. Table 7.2 illustrates the description for the online interactive reconciliation workshops.

The Content of the Front Page: Home The front page of the social network site has a logo and a slide show, which acts as a banner and is recurrent for activities done by Wasatia NGO. Links for the navigation of the site are illustrated in the main menu and accessible to all users and guests visiting the site. The slideshow illustrates all workshops, courses, and activates done by the Wasatia NGO (Wasatia NGO is referred to as NGO in the rest of the document).

The main homepage has a registration system for visitors to endorse them to become members of the NGO. In addition, it encourages them to be part of the activities and workshops done by the NGO. The home has an emailing list registration system to send newsletters to all members regarding all activities, courses, and workshops. The NGO website endorses concepts and principles of moderation and reconciliation in the community to contribute to publishing books for the NGO. A Facebook page is integrated with the social network site that appears on the right side of all pages of the website; moreover, social network software is installed on the social network that functions as a medium between all members. The Facebook is the leading social media network combined with the NGO's site to help spread the understanding of the Reconciliation process (see Fig. 7.4).

Table 7.2 Description for the interactive reconciliation workshops (online and offline)

Interactive reconciliation workshops online and offline.	Descriptions of the features used
Goals and missions of the NGO.	Promoting Wasatia Ngo goals and its mission.
Rules and regulations	Developing a moral standard of conduct for members and visitors on the site.
Articles in Arabic and in English written by Wasatia.	Publishing reconciliation articles online written by the Ngo Wasatia.
Intellectual articles	Publishing online reconciliation articles from a different professor in the field.
Article about "Wasatia" moderation	Articles about the Wasatia as a concept and as a religious concept and a movement.
Interactive Reconciliation Workshops Online & Offline	Publishing online interactive reconciliation workshops & activities (online &offline)
Reconciliation videos	Publishing online reconciliation teaching videos, its concepts, and theories – Debates about reconciliation.
Reconciliation forum	Developing an online forum for discussions done by members and publishing articles by members about reconciliation. And debating on the reconciliation articles
Web links	Gathering online information that is being issued by the NGO and its mission. In addition, gather all news articles about Wasatia and about reconciliation.
Social media online software	Social media software for developing social networking among online members, to develop interactivity between members and visitors.
Reconciliation polls	Reconciliation polls, to debate and develop surveys about reconciilation.
Reconciliation courses	Reconciliation courses for promoting the understanding of the reconciliation process. Including another course that teaches democracy and moderation.
Reconciliation news	Gathering all news published by different news agencies about the interactive reconciliation workshops and activities.
Online applications	Application form for registering new member using the online request form.
Contact application	Online contact form for ease of communication
Facebook connecting with social network website	This link helps users register with their user account on Facebook with the social network website.
Integration with wasatia.info Facebook page for interactive reconciliation debates	A Facebook page integrated into the used to boost reconciliation articles from the website into the Facebook page. This helps get more audience and target a wide spectrum of audience.
Facebook page Wasatia vision	Publishing reconciliation articles and also moderation and democracy articles.

Social network site

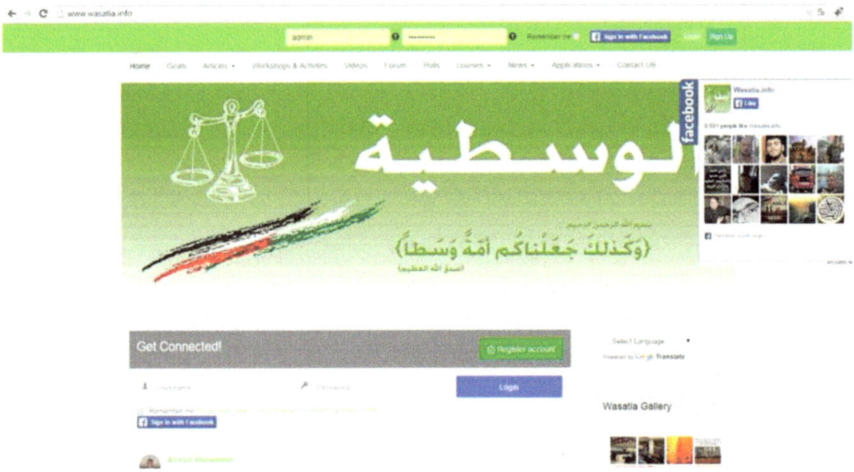

Fig. 7.4 Wasatia.info "Homepage". (Accessed on 13 Feb 2015)

Goals Anticipated Influencing the Community The NGO goals are elaborated on one web page illustrates the mission of the NGO and its objectives. The Wasatia movement was created in January 2007 as a non-governmental, non-profit, non-political initiative to change the trend to promote reconciliation and moderation in Palestine (Dajani 2011). The NGO declared goals are stated on the web page as follows (Dajani 2000):

1. To bring a broader and more rational understanding of Islam to Moslems as well as to non-Moslems.
2. To clarify the distortions to which Islam has been subjected at home and in the West.
3. To educate Palestinians on taboo topics such as the Holocaust taking a new humanistic approach.
4. To seek answers to the deeply religious, political, social, and economic crises inflicting the Palestinian society.
5. To strive and work for ending the Israeli military occupation through negotiations and peaceful means.
6. To spread and promote Islamic tolerant concepts, values, and principles within the Palestinian community
7. To encourage the practice of moderation among Palestinians to mitigate religious radicalism and bigotry and reduce political extremism.
8. To bring a message of peace, moderation, justice, coexistence, tolerance, and reconciliation to the Palestinian community through vocal civic leaders.
9. To teach creative and critical thinking and open-mindedness.
10. To empower the potential for leadership in their society. The goal of dialogue and education is to deconstruct mythologies, distortions, and misinterpretations and to promote knowledge and empathy for the other.

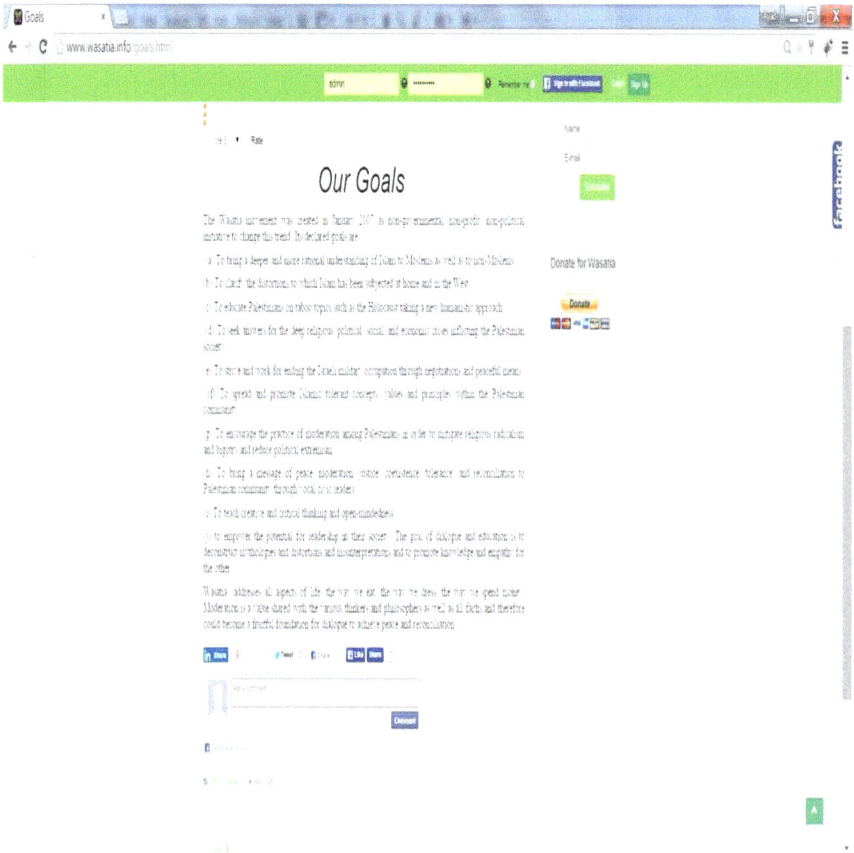

Fig. 7.5 Goals shared by the NGO, Wasatia.info "Goals". (Accessed on 13 March 2017)

Wasatia NGO addresses all aspects of life, such as the way one eats, dresses, and spend money. Wasatia NGO endorses moderation as a value shared by the various thinkers and philosophers, as well as all faiths; therefore, it could become a successful foundation for dialogue to achieve peace and reconciliation (see Figs. 7.4 and 7.5).

Rules and Regulations for the Site as for the Perspective of WASATIA The online social network site has standards and regulation for the control of the conduct for visitors and members of the social network site, as well as their interactions in the form of the site. The regulation and rules for the visitors and members conduct are as follows:

1. The dialogue in a moral manner with respect to Wasatia Community
2. Respect the narrative of the other.
3. Recognize the historical factors among members.

4. Part of the dialog should consider empathy to the other.
5. Keep it Clean. Please avoid obscene, vulgar, racist or sexually- oriented language.
6. Don't Threaten or Abuse. Threats of harming another person will not be tolerated.
7. Be Truthful. Don't knowingly lie about anyone or anything.
8. Be Nice. No racism, sexism or any –ism that is degrading to others.
9. Be Proactive. Let us know of abusive posts.
10. Shares with Us. We'd love to hear eyewitness accounts, the history behind the story.
11. Be Positive. Bring the best out of the person, rather than bring the worst. If you do not have something beautiful to say, do not say anything.
12. Be Rational. Avoid making emotional, unsubstantiated statements.
13. Be Creative and Original. Bring new ideas to the Dialogue forum.
14. Do not accuse or blame. Blame games are not productive.
15. Be respectful. Show respect for the values, ideas, and views of others.
16. Seek knowledge with respect and recognizing the different of narratives
17. Be Moderate. Seek the middle road in your views and in the way you express those views.

Online Interactive Reconciliation Articles Articles published online about Wasatia are essential, as these create more understanding of the concepts of moderation and reconciliation. The about page has articles about the Wasatia initiative and why it was created. It also explains what Wasatia NGO endorses and its mission for its contribution to the community (see Fig. 7.6).

Reconciliation Articles Published Online Visitors can navigate and read all articles published online by the NGO. The articles have a counting system that counts how many times articles are read by visitors and members. The count system indicates the social network site has been visited, and each article has its own counting methods (see Fig. 7.7).

Reconciliation, Moderation and Democratic Articles Published in Both Languages Arabic and English on the NGO's Website The need for articles in various languages backs the members in their social realities, as article are also written in their native languages, which enhance the chance of understanding the Reconciliation process. Participants are either native Arabic or English-speaking members. Therefore, the NGO has Arabic and English articles published on the website (see Fig. 7.8).

Intellectual articles are written by philosophers who support the concepts of moderation, reconciliation, and democracy, which are also endorsed by the NGO. Articles are published by other scholars who support the ideology of moderation and reconciliation between the Palestinians and the Israelis (see Fig. 7.9).

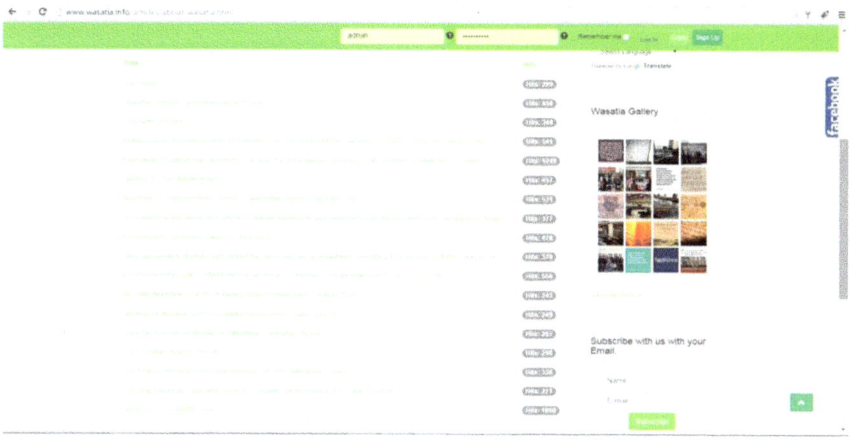

Fig. 7.6 Wasatia.info "About Wasatia". (Accessed on 13 March 2017)
Articles for the Wasatia NGO Published online including online hits, dated 20.5.2015

Fig. 7.7 Wasatia.info, "Articles". (Accessed on 20 April 2017)

Interactive Reconciliation Workshops (Online and Offline) Interactive reconciliation workshops can be online activities and offline activities that promote reconciliation, moderation and democracy, as illustrated on the social network site, to validate activities orchestrated by the NGO, which hold more credibility and truthfulness for the work done by the NGO (see Fig. 7.10).

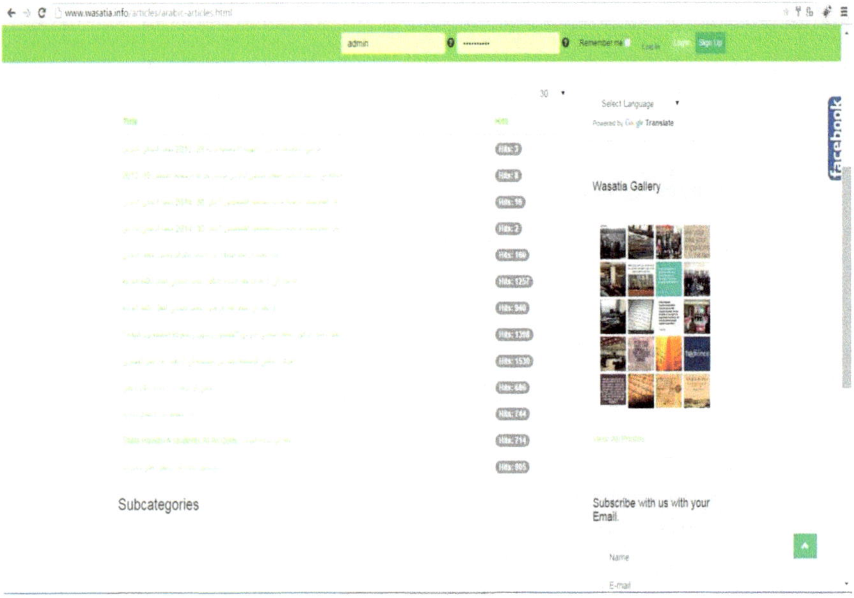

Fig. 7.8 Wasatia.info. **"Online Arabic Articles".** (Accessed on 20 April 2017)

Fig. 7.9 Wasatia.info "Articles by different scholar". (Accessed on 20 April 2014). [Offline] The website was destroyed by an online attacker "Hacker"

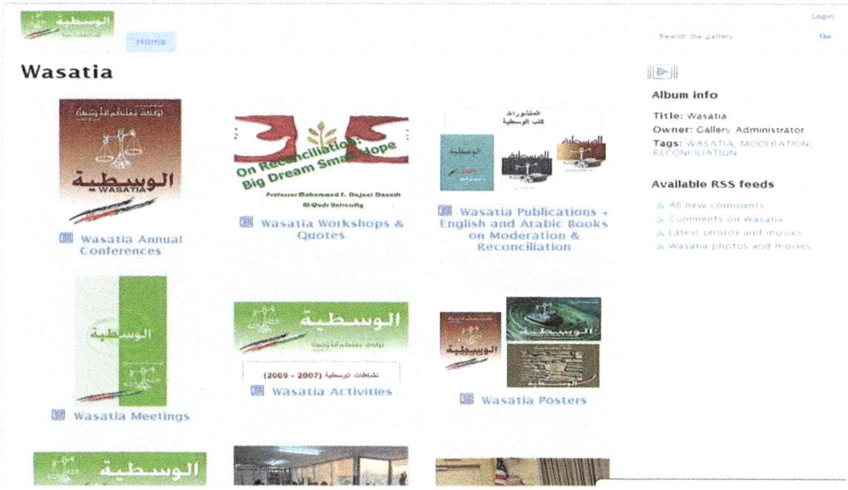

Fig. 7.10 "Interactive Reconciliation Workshops" [offline]. (Accessed on 20 May 2014)

The page was developed to illustrate all activities of Wasatia are done online and in the community but the webpage was destroyed by an online attacker, which was not identified, but then we used harder security methods.

Reconciliation Videos and Interviews Made by Scholars Who Teach the Reconciliation Process Videos spread the knowledge and the understanding of the principles of moderation, reconciliation, and democracy in Palestine. According to Rosensteel (2013), the human face is presented as a gathering point for information and credibility to influence more visitors and members to the social network site. It encourages members to participate and become an active participant of the NGO. It is easier to understand the message that the NGO has tried to convey to its visitors and members, such as *"a human voice speaking to us has an amazing way of converting information into meaningful content"* (Rosensteel 2013, p. 20). The body language in videos is more appealing to the visitors and members; it conveys the message that the NGO is trying to share in a more simplistic way for visitors and members to understand. The power of peripheral motions can grab the visitors and members with an emotional impression on themselves. It can express the concepts and principles that the NGO wants to spread (see Fig. 7.11).

Reconciliation Forum An Internet forum and social media software illustrate discussions and dialogues of members as influencing a type of online deliberation about topics in the social society. This forum is essential for the following reasons (Owais and Kiat 2013):

1. Networking with members of the NGO.
2. To gain knowledge about what the NGO promotes on its site.
3. A method to share information and experiences on the NGO's website.

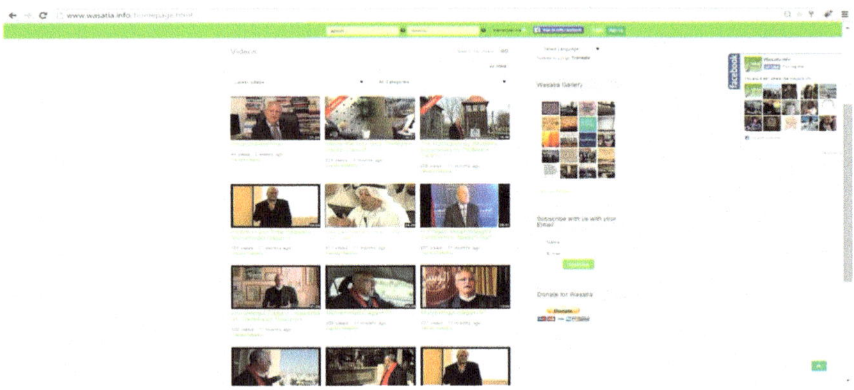

Fig. 7.11 "Reconciliation Videos". (Accessed on 20 April 2017])

4. To teach and give more details about subjects that the NGO searches on.
5. To build a relationship between NGO's members.
6. Share the experience with members.
7. A mine of data knowledge for all visitors that are in concern with the topics of moderation reconciliation and democracy.

There are two stages of implementing the Internet forums to advance a more reliable understanding of the concepts of the Reconciliation process. Stage 1 was to publish open-ended questions for discussions among members of the social network site. This act made more useful for the understanding of subjects considered as a taboo to discuss in the community. The interaction of members regarding taboo topics illustrated an openness for members to understand how the reconciliation process can be achieved. When arguing about taboo topics from different perspectives and the willingness to respect and understand other member's viewpoints and their narratives. The forum demonstrated a change within the members of the NGO, toward the Reconciliation process, through their interactions with each other. The problem with this stage was that it targeted only the members of the social network site, and new members could not interact in the forum because they were new to the discussion. In addition, the forum was not public; therefore, visitors could not interact, view, or become part of the discussion.

Some problems occurred in having rules for the participants; the rules made participants less open for discussions, which minimized the interactions between members and participants, but the use of standards was critical to maintain respect and gain the best behaviors by the participants and the visitors. However, asking for the best conduct online by members and visitors would minimize the influence that the primary aim of the forum should have gained (see Figs. 7.12 and 7.13).

The Analytics Shows Monthly and not Yearly, and it Was Eventually Destroyed by a Hacker The forum software was involved in Stage 1 and used http://vanilla-forums.com (see Fig. 7.13). The forum was destroyed multiple times until security

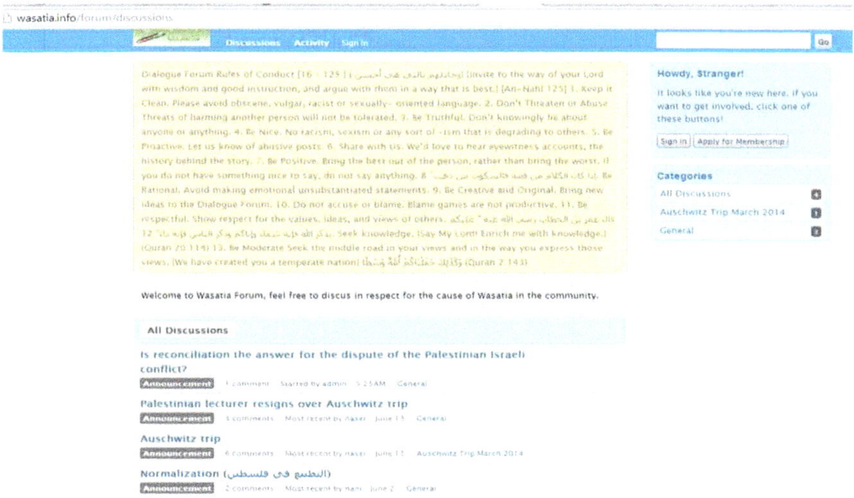

Fig. 7.12 Wasatia.info "Online Interactive Reconciliation Workshop". (Accessed on 30 July 2014)

Fig. 7.13 Wasatia.info "First Stage discussion Analytics". (Accessed on 30 July 2014)

measures were taken into consideration, and the developed a new forum application as described in Fig. 7.12, and later explained in Fig. 7.14.

Stage 2 involved creating an Internet forum that hosted all articles published by the NGO, exemplifying a more simplistic way for members to interact with each other. In addition, visitors could view and read the published articles and comment on them, creating a more exposed collaboration between visitors and members. Stage 2 had a broader influence on visitors and members alike than stage one in

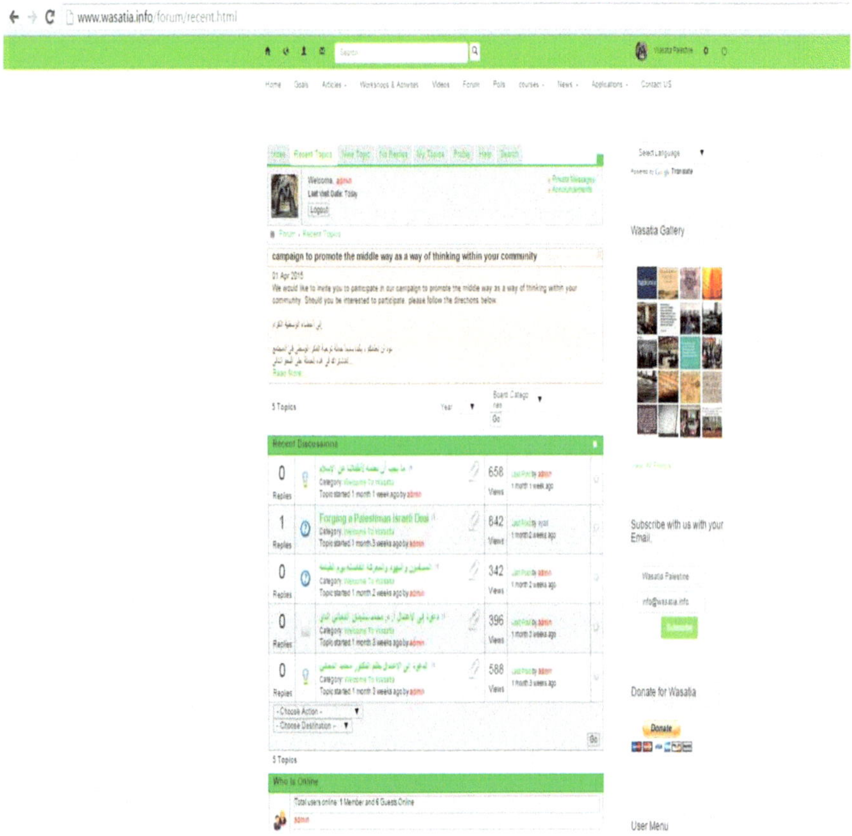

Fig. 7.14 www.wasatia.info – Online Reconciliation forum". (Accessed on 4 May 2017.)

promoting principles and concepts of reconciliation and moderation in the community (see Fig. 7.14).

The Web Links to the Site The feature of web links was preferred to bring traffic to the website, as web-links held all links that were published online about NGO's activities, and it showed some kind of legitimacy for activities of the NGO in the community. Part of the analysis was to gather all information about any activities published on the Internet about NGO's activities, including web articles, news, and any workshops published online. Web links were used to track the number of web articles, news, and activities exposed by visitors and members in the NGO website to create more recognition for the NGO website from visitors and members (see Fig. 7.15).

Online Social Network Approach The feature of using the social network software involved developing a community who comprehended the principles of Reconciliation process in the Palestinian community. The social media software

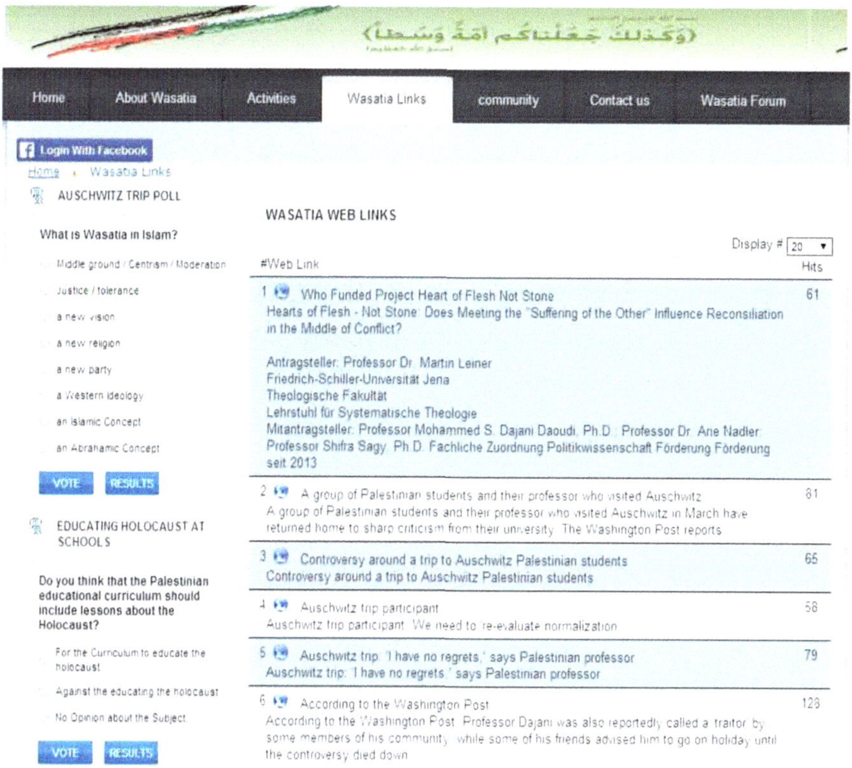

Fig. 7.15 www.wasatia.info "Online Reconciliation News". (Accessed on 30 June 2014)

held all profiles of members of the NGO who had confidence in moderation, reconciliation, and the NGO's initiatives. The software provided a backup for all discussions, videos, images, application, comments, and posts exposed on the NGO website, see Fig. 7.16.

Reconciliation Polls Introducing a polling system for the visitors and members exposed them to reconciliation and moderation concepts. Promoting polls one the website measured the acceptance of what the NGO was working and promoting in the society. Working with polls as part of the social network site occurred in two stages, as implemented in Fig. 7.17 to show Stage 1's direct question addressed to visitors and members of the website.

Figure 7.18 shows Stage 2's introducing polls for members of the NGO and giving them the opportunity to share their thoughts. This aspect promoted a more online deliberation process, which strengthened the understanding of the Reconciliation process.

Phase 2'a polling application had a vast, far-reaching influence on members and visitors. It provided members and visitors to demonstrate their thought on a public

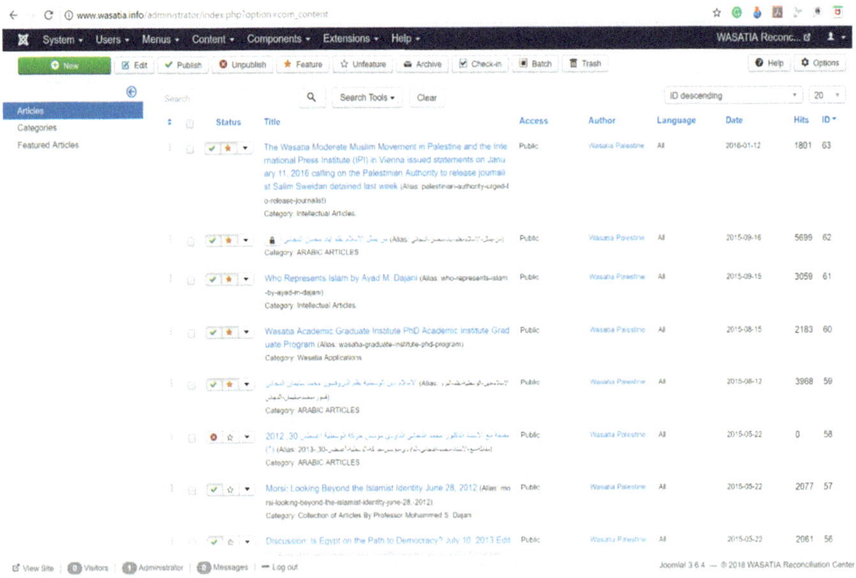

Fig. 7.16 Wasatia.info "Online Reconciliation Articles ". (Accessed on 30 April 2018)

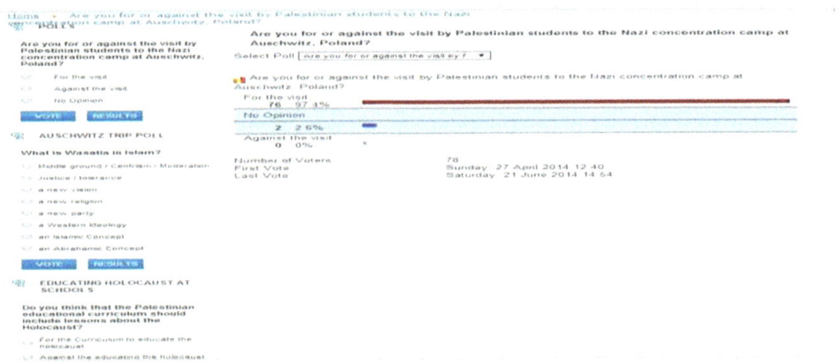

Fig. 7.17 Reconciliation polls, first phase start date 27 April until 21 June 2014. (Accessed on 24 May 2014)

campus. Which made it a challenge that members and visitors participated in validating topics and activities done by the NGO. Polls in the second stage had more influence, because visitors and participants were interacting with open-ended questions by only polling, which was an advantageous mechanism, For example, the polling about another trip to Auschwitz, which was also boosted on the Facebook page, gained on the Facebook page 501 likes according to the Wasatia Facebook page facebook.com/Wasatia.info dataset recorded and dated 10/25/2015 9:26:00 AM, but the polls majority for 40 participants were 80% for the Auschwitz

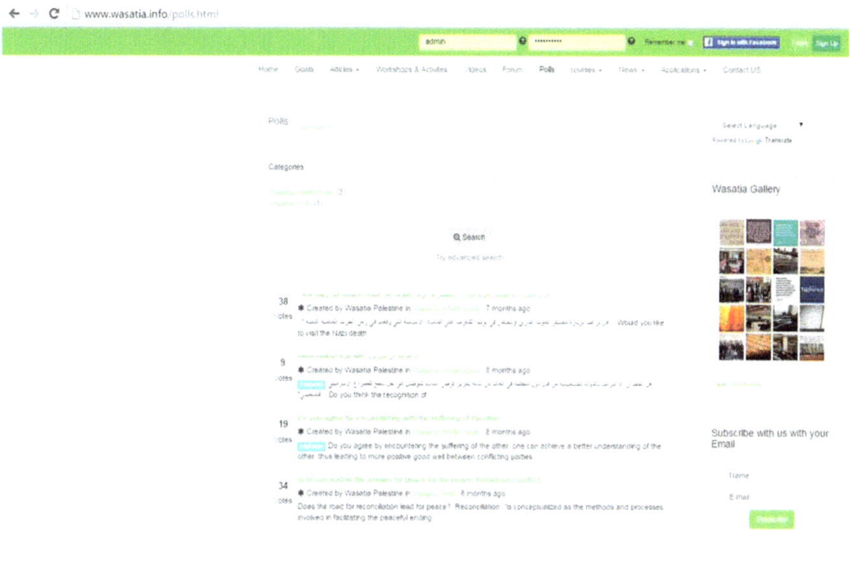

Fig. 7.18 Wasatia.info. "Reconciliation Polls Second Phase". (Accessed on 30 July 2016)

trip by Palestinian Students . Polling can cover a right spectrum of the community that is active online.

Courses Online courses for members and visitors on the website to educate members and visitors were about reconciliation and moderation concepts and ideologies (see Fig. 7.19).

Interactive Reconciliation Online News Articles News articles were published by other news agencies on the social network site to promote the concept of the Reconciliation process. In addition, these were published to show activities done by the NGO, thereby creating recognition for the NGO online and validating the NGO presence in the community (see Fig. 7.20).

Reconciliation Applications for Membership Applications included forms for new members to enroll in the NGO's activities, as well as for new visitors to participate and be part of the membership of the NGO (see Fig. 7.21).

An Online Contact Application The application was only sued for contacting Wasatia NGO; there was also an application form to make it easy to request information from the NGO (see Fig. 7.22).

Facebook Is Connecting with the Online Social Network Website A Facebook page was integrated with the NGO social network site to promote the NGO in social media and endorse the NGO Facebook page. It allowed visitors to become members

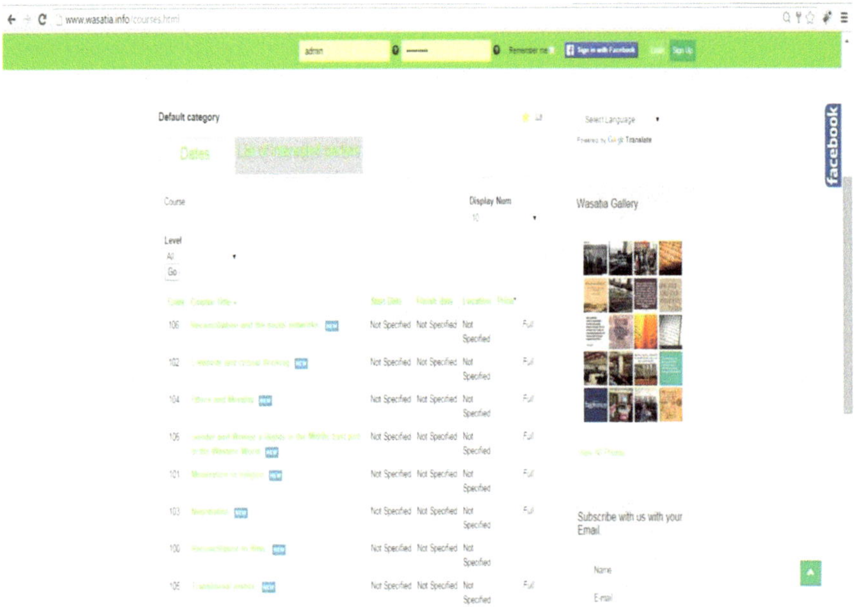

Fig. 7.19 wasaita.info "Online & Offline Reconciliation Courses". (Accessed on 30 July 2015)

by using their Facebook accounts, which made it easier for visitors to register as new members in Wasatia's society (see Fig. 7.23).

Using Web Software for the NGO's Website for Promoting Reconciliation and Moderation Online A statistics software was to track visitors to the website and to count the number of hits on the domain name, http://www.wasatia.info, had over the period of 3 years, including hits to its application and pages. The software created a chart of visits and hits on all files and applications for every year (see Figs. 7.24 and 7.25).

Web editor software (Joomla) was used for the websites to provide counts for all hits on articles published on the website that were read by visitors and members alike. The software illustrated that the visitors and members were reading the articles posted on the social network site (see Figs. 7.26 and 7.27).

In another phase, posting links for published articles on an online social network such as Facebook Pages "Wasatia.info" develops an integrating procedure between the Facebook Page and the social network website. The figure illustrates articles for topics discussed by participants in the forum; only participants can enter and engage in the articles discussed. This method illustrates spreading the knowledge about reconciliation, moderation, and democracy in the published articles (Fig. 7.28).

An online Social network software called "Joomla" was used to host the community of Wasatia in Palestine over the Internet. It hosted all members and

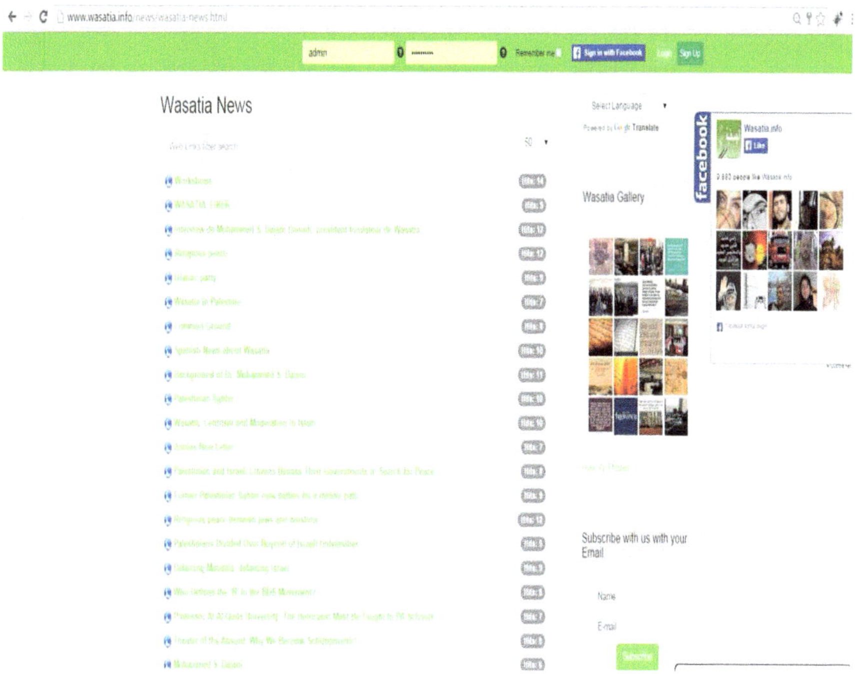

Fig. 7.20 Wasatia.info "news about reconciliation". (Accessed on 30 June 2016)

Fig. 7.21 Wasatia.info "Registration form, to register new members". (Accessed on 30 May 2017)

activities, photos, events, and was made by the WASATIA community online (see Figs. 7.29 and 7.30).

Figures 7.29 and 7.30 represent the software used as a social software for the site, and the registration area, the software is part of Joomla editor software, and it is free, used from the Internet. However, installing it, completing the setup, maintaining it and the extensions that were installed for online social debate to prove online deliberation was costly. These are called plug-ins and can be found on the Internet; these are called Joomla extensions.

The figure illustrates the member era for the online social network website.

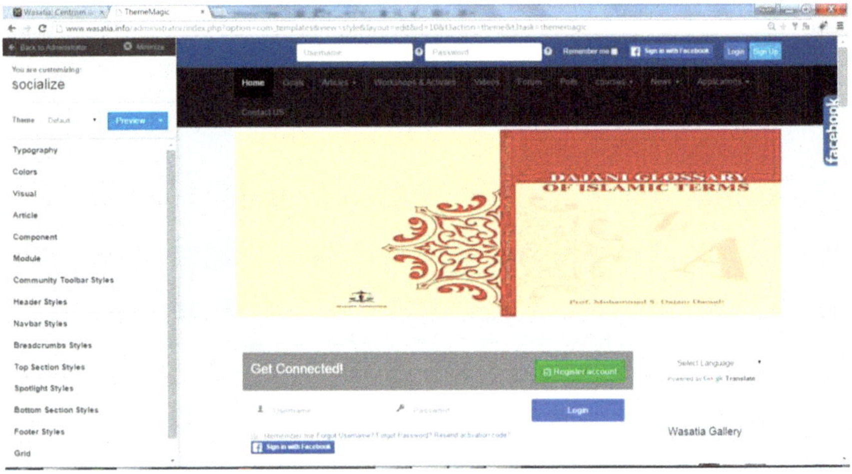

The Integration of a Facebook Page for the ICT-MR A Facebook page was relevant to integrate with the ICT-MR social network website. A high percentage of Palestinians have used Facebook; according to statistics published by *This Week* in Palestine published the article in 2014, over 1.2 million Palestinian have used Facebook. The usage of Facebook to expose concepts of reconciliation and moderation in the Palestinian community was essential, as Facebook reaches every house, business, and organization in Palestine. The figure below illustrates that articles were read either by the visitors or by the registered members.

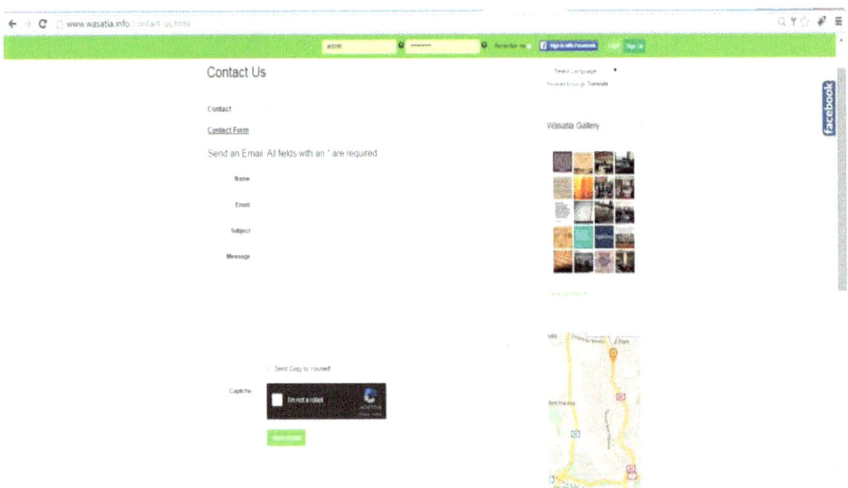

Fig. 7.22 Wasatia.info “Contact form”. (Accessed on 30 July 2017)

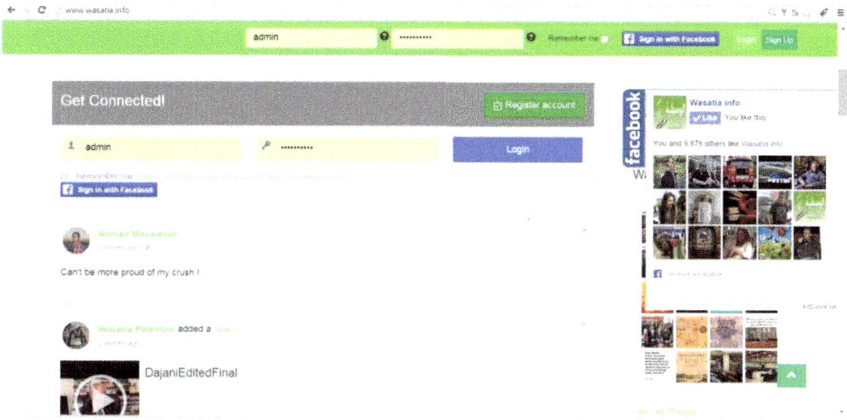

Fig. 7.23 Wasatia.info/home. Using Facebook accounts to become part of the memberships in the social network site (Accessed on 30 July 2017)

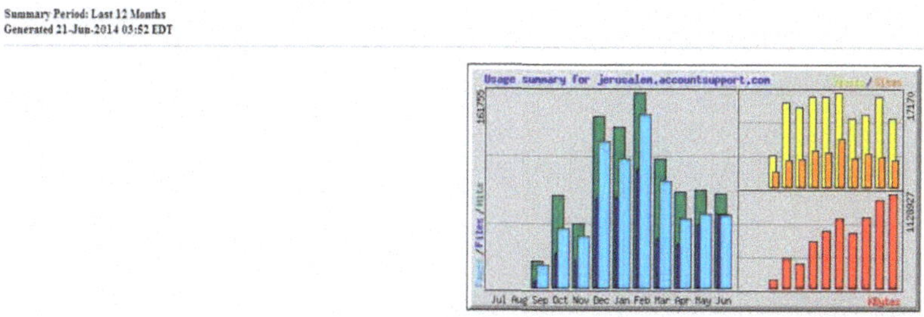

Summary by Month

Month	Daily Avg				Monthly Totals					
	Hits	Files	Pages	Visits	Sites	KBytes	Visits	Pages	Files	Hits
Jun 2014	3714	2893	2868	586	4763	1128927	12315	60247	60755	78012
May 2014	2620	1683	1960	530	5360	1049797	16431	60774	52202	81243
Apr 2014	2629	1207	1901	438	6123	845618	13140	57051	36232	78894
Mar 2014	3422	1299	2830	399	5139	664508	12382	87747	40288	106087
Feb 2014	5776	3497	5114	613	8705	830969	17170	143212	97922	161755
Jan 2014	4283	2397	3431	529	6290	684655	16401	106388	74311	132783
Dec 2013	4572	2365	3894	526	6582	561358	16326	120721	73321	141757
Nov 2013	1758	743	1426	482	5008	290847	14486	42803	22314	52740
Oct 2013	2466	903	1589	495	4751	363954	15367	49273	27995	76466
Sep 2013	1563	361	1316	412	2633	92180	5774	18434	5062	21893
Totals						6512813	139792	746650	490402	931630

Fig. 7.24 Wasatia.info "Statistics for visitors" [Online] between Sep. 2013 until June 2014–931,650 hits. (Accessed on 30 June 2014)

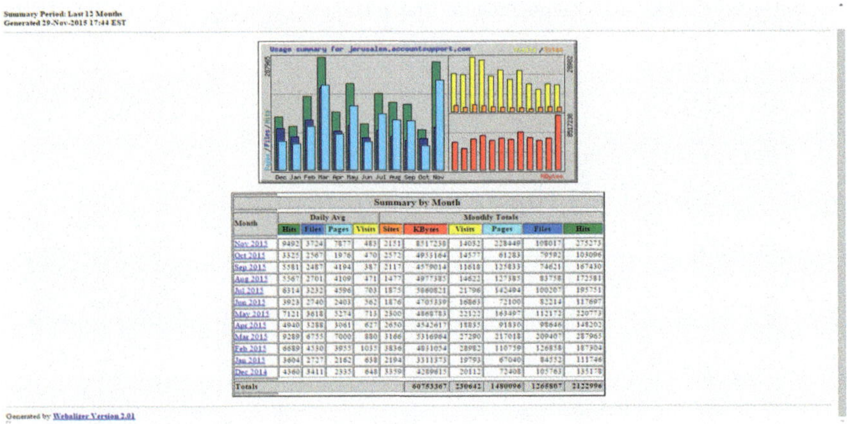

Month	Daily Avg				Monthly Totals					
	Hits	Files	Pages	Visits	Sites	KBytes	Visits	Pages	Files	Hits
Nov 2015	9492	3724	7877	483	2151	8517238	14032	228449	108017	275273
Oct 2015	3325	2567	1976	470	2572	4935164	14577	61283	79592	103096
Sep 2015	5581	2487	4194	387	2117	4579014	11618	125833	74621	167430
Aug 2015	5567	2701	4109	471	1477	4977385	14622	127385	83758	172581
Jul 2015	6314	3232	4596	703	1875	5860821	21796	142494	100207	195751
Jun 2015	3923	2740	2403	362	1876	4705359	16863	72100	82214	117697
May 2015	7121	3618	5274	713	2500	4868783	22122	163497	112172	220773
Apr 2015	4940	3288	3061	627	2650	4542617	18835	91830	98646	148202
Mar 2015	9289	6755	7000	880	3166	5316984	27290	217018	209407	287965
Feb 2015	6689	4530	3955	1035	3836	4831054	28982	110759	126858	187304
Jan 2015	3604	2727	2162	638	2194	3311373	19793	67040	84552	111746
Dec 2014	4360	3411	2335	648	3359	4289615	20112	72408	105763	135178
Totals						60753367	230642	1480096	1265807	2122996

Fig. 7.25 Wasatia.info "Online Statistics for visitors based monthly, 2-million visitors". (Accessed on 29 Nov 2015)

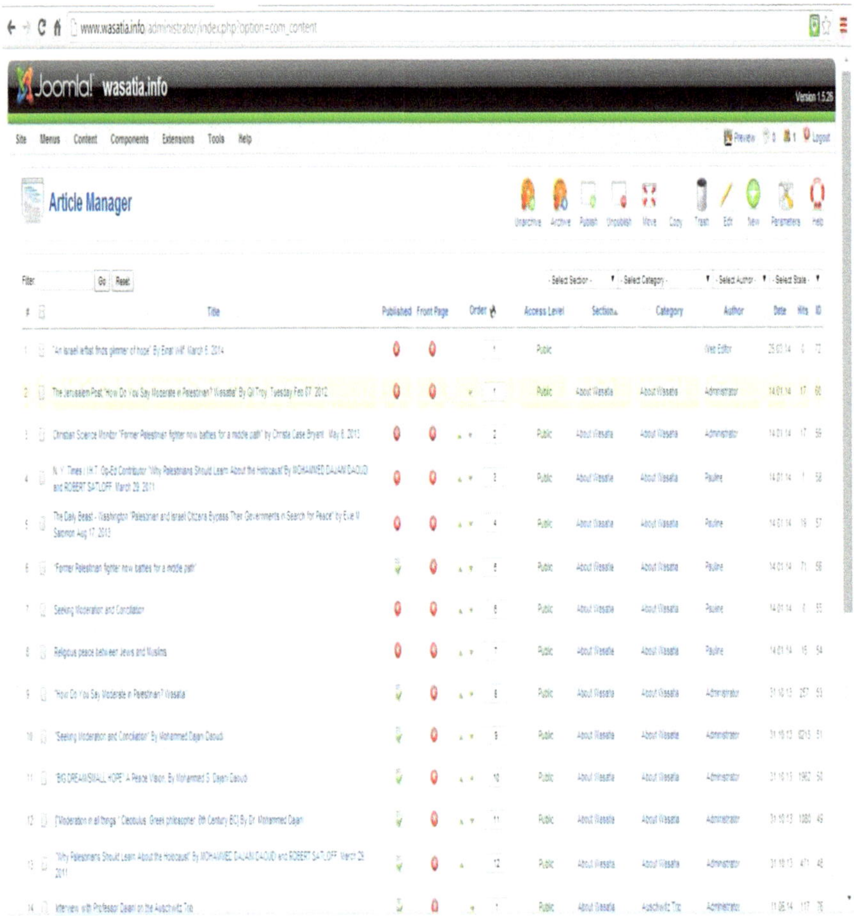

Fig. 7.26 Wasatia.info "Analytics for Published online news articles," "Articles about Wasatia" and "Articles about Auschwitz trip by Palestinian Students". (Accessed on 30 June 2017)

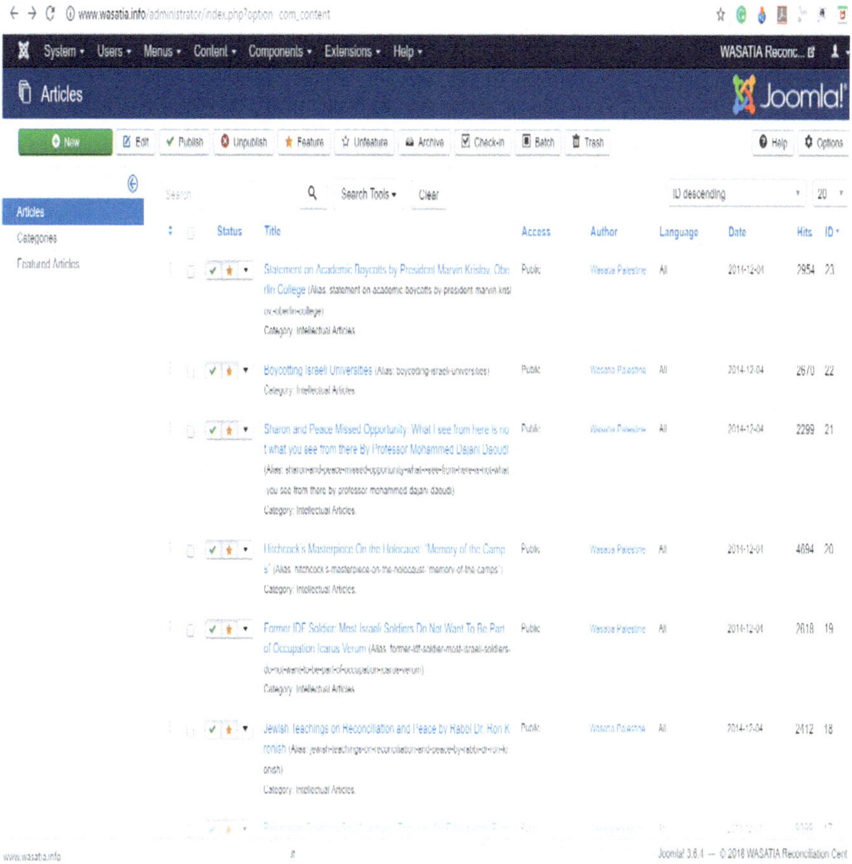

Fig. 7.27 Wasatia.info "Analytics for articles published by other scholars that support Wasatia". (Accessed on 30 June 2017)

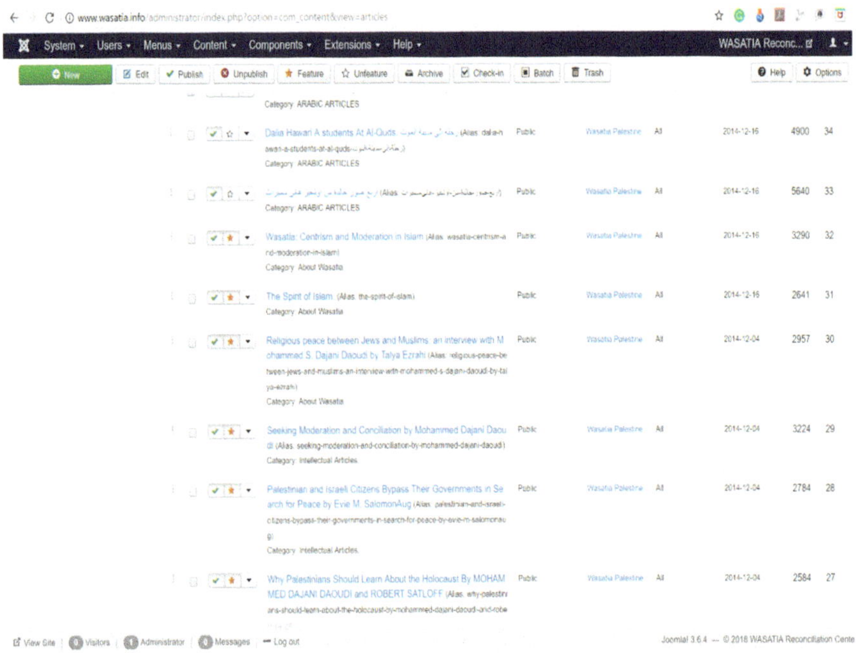

Fig. 7.28 Wasatia.info “Online Users Participating in the Forum of Stage 2” Accessed. (Accessed on 30 June 2017)

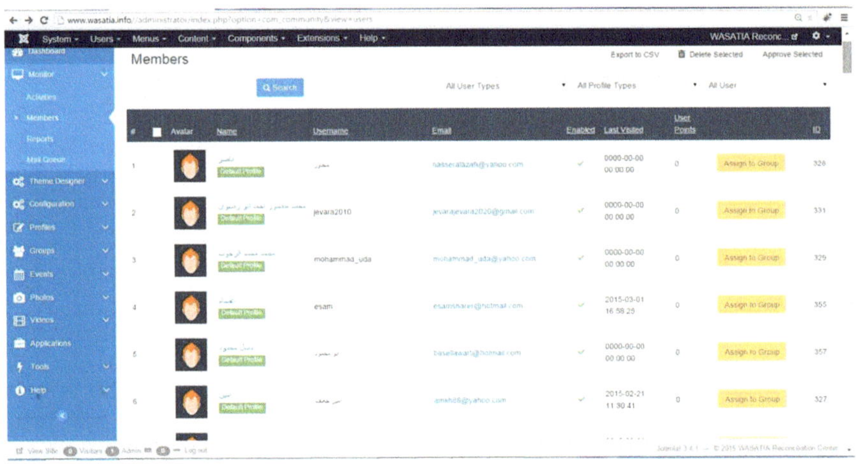

Fig. 7.29 Wasatia.info “Online Community 255 Members”. (Accessed on 30 June 2017)

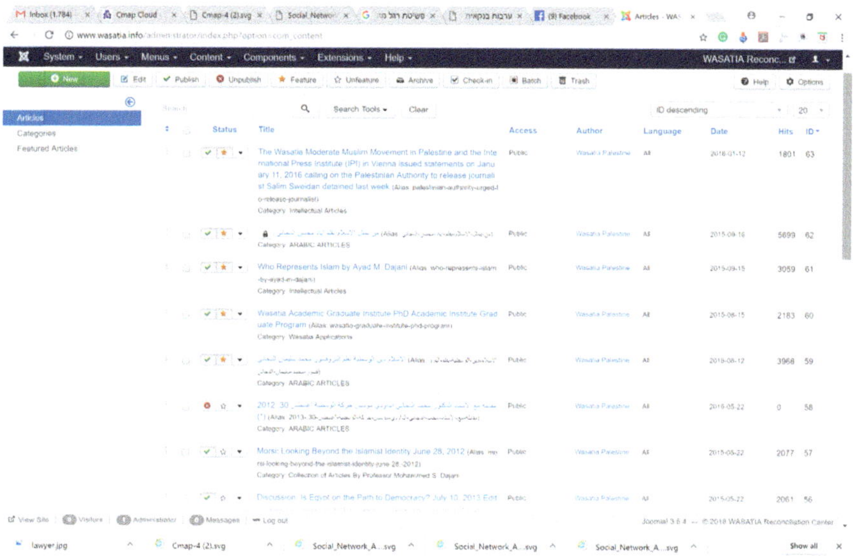

Fig. 7.30 Wasatia.info "Analytics for Wasatia. Articles including Hits from visitors and members". (Accessed on 30 June 2017)

Figure 7.31 illustrates the Facebook for Wasatial.info, which have 9000 members accredited on the data accessed at [31.Dec.2015].

The procedure for using the Facebook page (Wasatia.info) to illustrate a natural understanding of the concepts, moderation, reconciliation in Palestinian communities. The integration happens through posting articles and workshop activities of the NGO, and through photos and videos that adherence to the Reconciliation process. Posting and boosting the published articles and workshops included the videos and activities on the Facebook page, that are initially links to the main website "www.wasatia.info." The boosting and targeting audience for the Facebook page was a facility used by Facebook Inc., which was also used to target Palestinians using the Facebook page. Using the insight of the Facebook page provided statistics to determine the most related posts on the Facebook page, which showed what posts were most popular and the demographic influence of the ICT-MR in Palestine (Figs. 7.32 and 7.33).

The concept for integrating the Facebook page with the ICT-MR social network site was to attract more members to sign and register to become members of Wasatia. Members could use their Facebook accounts to enter the ICT-MR social network site www.wasatia.info.

The figures below illustrate the interaction of members in the site and in the social network Facebook pages, this resulted from the integration process between the articles published on the online social network site www.wasatia.info and posted as a link in the Facebook Page "Wasatia.info" (Fig. 7.34).

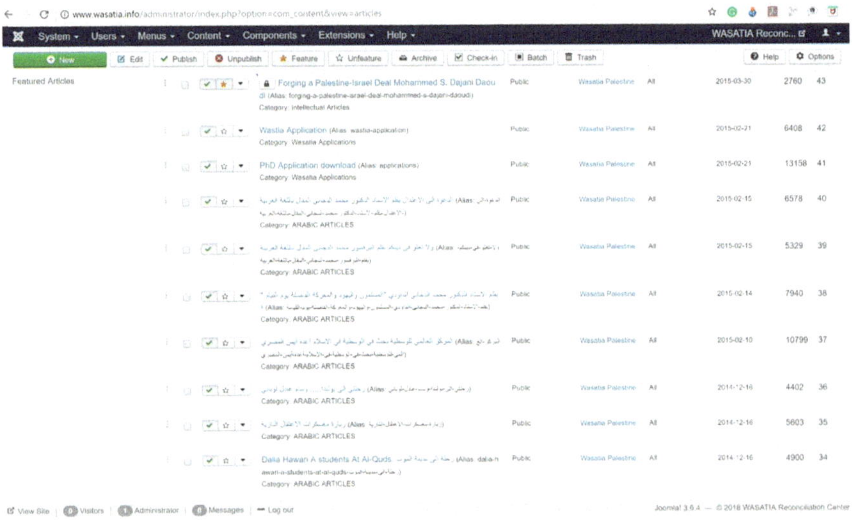

Fig. 7.31 Wasatia.info "Analytics for posted articles on Facebook Page and visited by members and visitors and Hits". (Accessed on 30 June 2017)

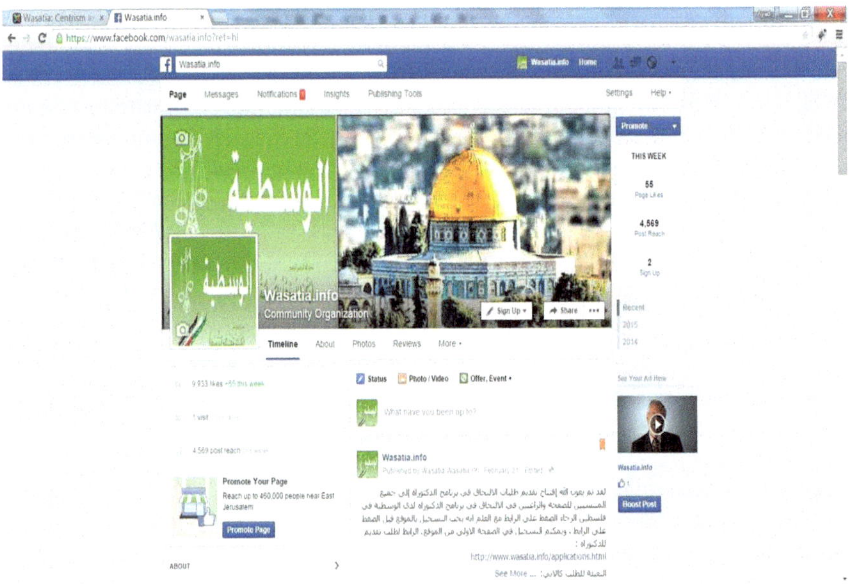

Fig. 7.32 Wasatia.info, "Facebook Page". (Accessed on 30 June 2017)

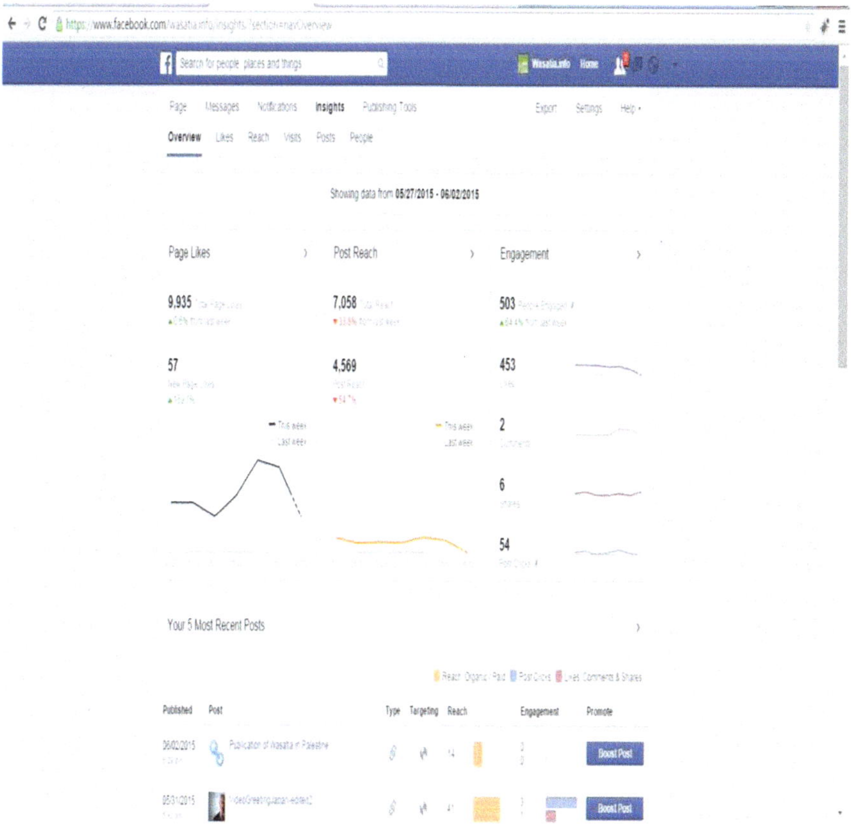

Fig. 7.33 Wasatia.info **"Facebook Page sample of analytics for all posts, comments, and likes".** (Accessed on 27 May 2015)

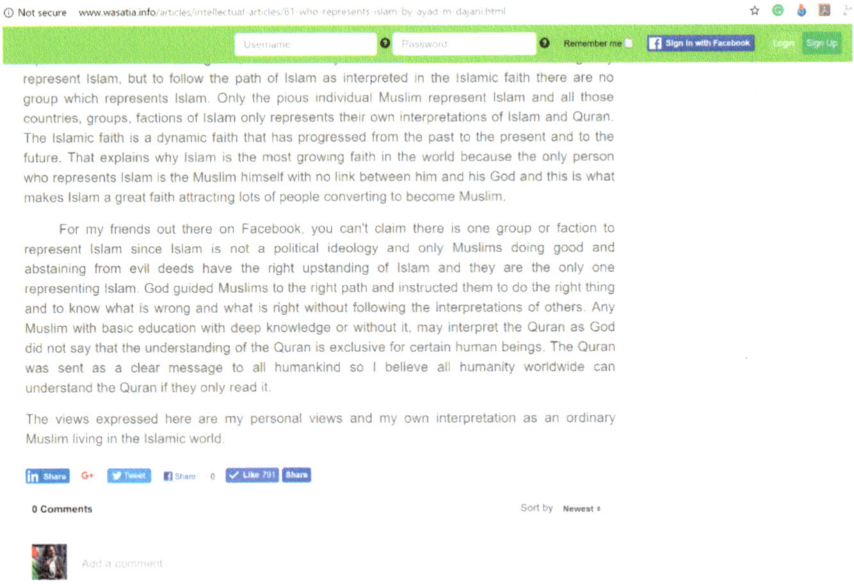

Fig. 7.34 Wasatia.info "Articles integrated into the Facebook Page". (Accessed on 30 April 2018)

The figure shows that interaction between the users and the published article, the article has "791 likes" the visitors are directed from the Facebook Page Wasatia.info towards www.wasatia.info, the visitors are being registered with different an addressee identifier, which shows no redundancy.

The results from using the integration detailed before created followers for the Facebook page (Wasatia.info); in a later stage, those followers might become active NGO members by registering on the social network website A lot of members entered the social network website and become a member also asked how they can become a social, active person to promote Wasatia in Palestine. This was considered as a successful initiative; the Facebook page was created in 2014 to encourage moderation and reconciliation, which the NGO in Palestine represented. Until the date of December 25, 2015, the Facebook page had precisely 10,000 followers and created 257 members for the Wasatia social network site (Figs. 7.35 and 7.36).

Some politician from the United States wanted to help the moderate vision of Professor Dajani and bring more support to moderation in Palestine; this vision was supported by different scholars worldwide, one of the most famous is President Barak Obama, who visited the American Studies Institute at the Al-Quds University in 2008 and supported Dajani's vision for moderation. President Obama was an American senator visiting a Palestinian University.

Fig. 7.35 Wasatia.info [Facebook Page] "Analytics". (Accessed on 30 May 2018)

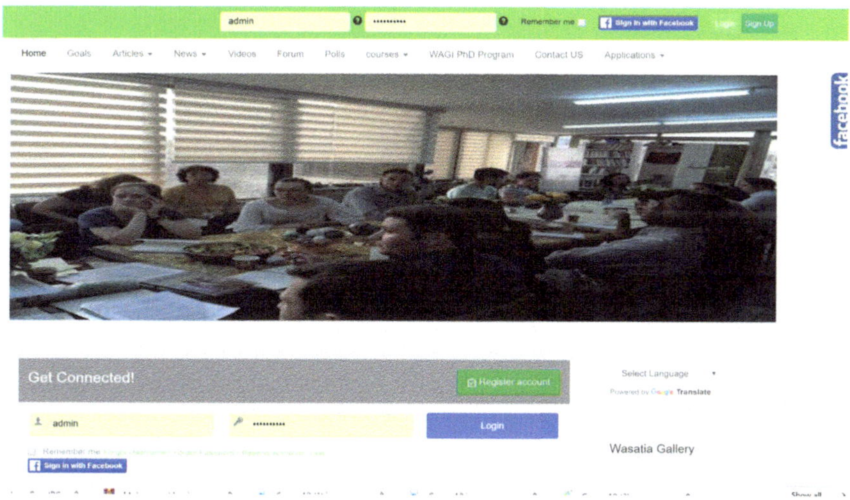

Fig. 7.36 Sharing activities in the social network sites. (Accessed on 30 June 2018)

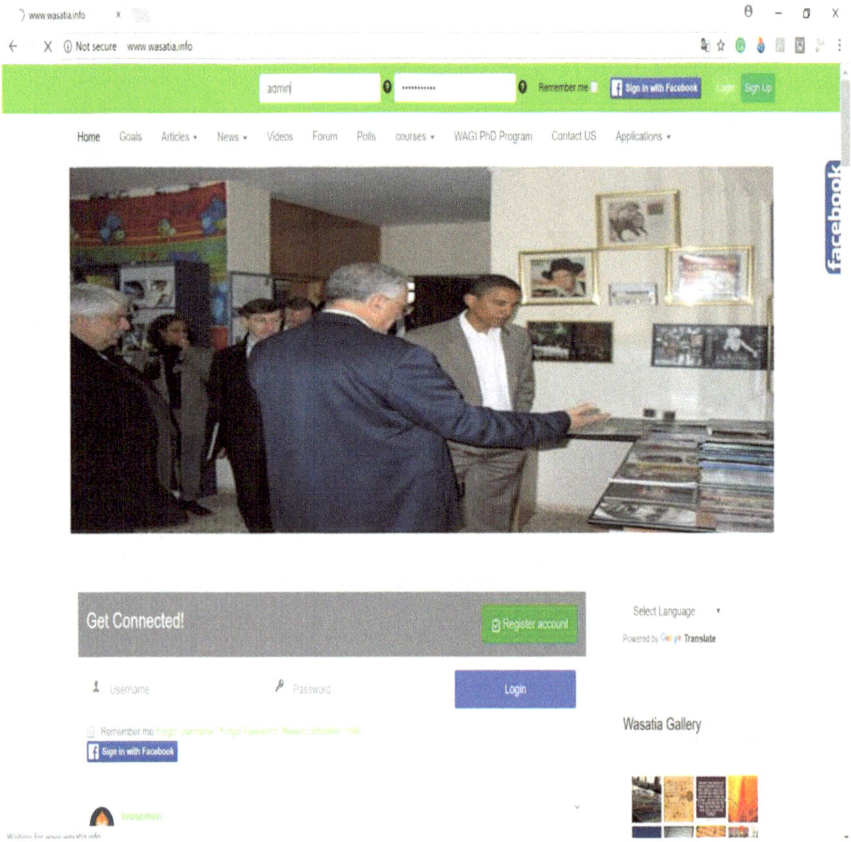

Other interaction between the online social network and the Facebook pages, is the interaction of participants on the posts on the Facebook page. They are described in the next section.

7.6 The Integration Between Facebook Pages and the Social Network Site

This ICT-MR integrated the Facebook page (www.facebook.com/wasatia.info) and the social network site (http://www.wasatia.info). The intent of this integration was to develop the Mixed-Method experimental design discourse. The datasets for this integration were used to **develop the online interactive reconciliation debates** about topics that help understand reconciliation. Another intent for the integration is to influence followers of the Facebook page to become members of the social network site and then integrate into the activities and workshops held by the NGO. The concept is that the followers on the Facebook page are directed toward http://www.

wasatia.info social network site, upon clicking on posts or titles in a post on the Facebook page. The link directed the user to the article published on the website, which gave information about the Reconciliation process. By understanding the concept and terminologies of Reconciliation process, the followers adopted those ideologies; eventually, these would influence their daily habits of interaction, perhaps directing their intentions toward moderation, reconciliation, and democracy in their daily communications.

Another purpose of the integration was when the user decided to become part of the Wasatia membership site; the user created an account or could integrate his or her user account on Facebook within the Wasatia membership account on the website (http://www.wasatia.info). The user registered as a Wasatia member, as well as a Facebook member. This strategy helped to recruit as many members from the Facebook page as well on the website as members of Wasatia. The number of members registering in the social network to become active members in the Wasatia NGO in Palestine proved that ICT exposed the Reconciliation process. The complete dataset consisted of 473 dataset rows acquired from the NVivo 10.

According to Fisher (2001), this develops a **social-psychological process** in an **Interactive Conflict Analysis Reconciliation workshop**, and that is the intent of having two Facebook pages. This kind of approach also enhances the process of conflict analysis from using Facebook pages and deepen the process of dialog in implementing a constructive confrontation in both virtual communities that are active in the Facebook pages.

7.7 Interactive Online Reconciliation Workshop Process

One of the most interactions in the research towards reconciliation is the interaction of the online Palestinian community after Auschwitz Trip that took place in March 2014 by the Heart of Flesh Not Stone Project (HoF). A new post is exploring whether Palestinians want to have another experience as such an Auschwitz Trip done by the project (HoF). The new post about Auschwitz trip reached (52,368) thousand viewers and had over (497 Likes), it shows that Palestinian who are the most users of the Facebook page, want to understand their own occupiers, and understand the history of the other, this is a process for reconciliation, where it is known as "***rediscovering the self and the other***" accentuated by (Rasmussen 2001), which means that ICT, in this case, has reached a breakthrough in advancing towards the reconciliation process. These the figures below illustrate a discussion about the Auschwitz trip, after the Auschwitz trip that took place in "March 24 to 30th 2014" came back.

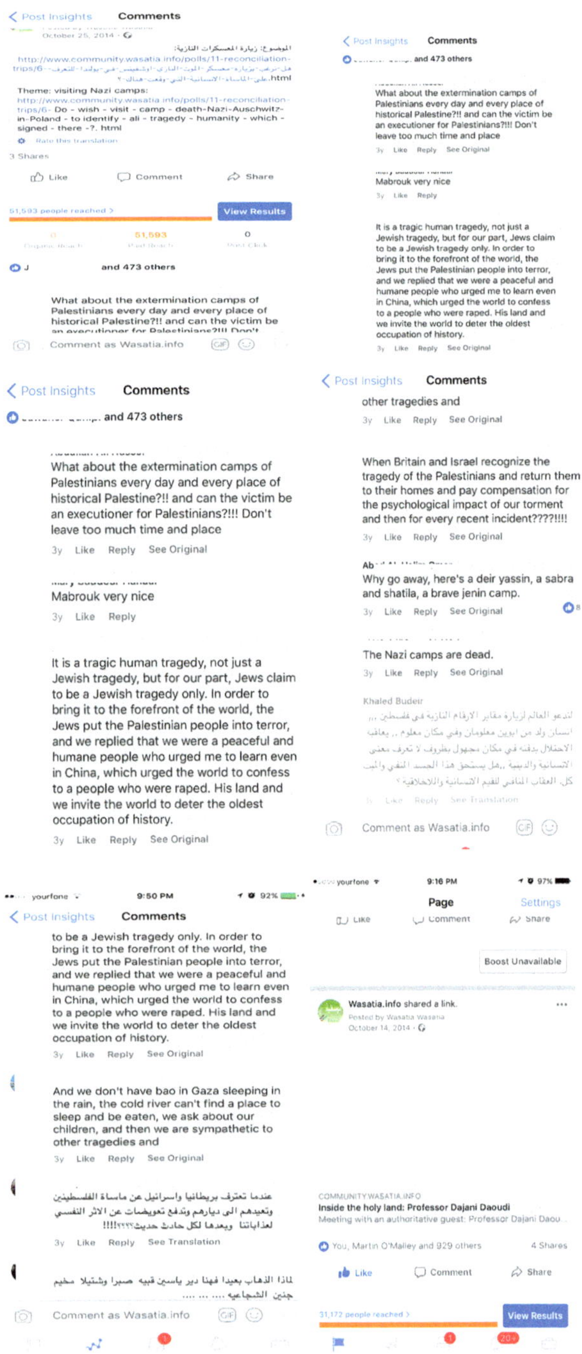
Post Insights
Comments
October 25, 2014
Theme: visiting Nazi camps:
http://www.community.wasatia.info/polls/11-reconciliation-trips/6- Do - wish - visit - camp - death-Nazi-Auschwitz-in-Poland - to identify - ali - tragedy - humanity - which - signed - there -?. html
3 Shares
Like
Comment
Share
51,593 people reached
View Results
51,593
0
and 473 others
What about the extermination camps of Palestinians every day and every place of historical Palestine?!! and can the victim be an executioner for Palestinians?!!! Don't leave too much time and place
Comment as Wasatia.info
3y Like Reply See Original
Mabrouk very nice
3y Like Reply
It is a tragic human tragedy, not just a Jewish tragedy, but for our part, Jews claim to be a Jewish tragedy only. In order to bring it to the forefront of the world, the Jews put the Palestinian people into terror, and we replied that we were a peaceful and humane people who urged me to learn even in China, which urged the world to confess to a people who were raped. His land and we invite the world to deter the oldest occupation of history.
other tragedies and
When Britain and Israel recognize the tragedy of the Palestinians and return them to their homes and pay compensation for the psychological impact of our torment and then for every recent incident????!!!!
Why go away, here's a deir yassin, a sabra and shatila, a brave jenin camp.
The Nazi camps are dead.
Khaled Budeir
yourfone
9:50 PM
92%
And we don't have bao in Gaza sleeping in the rain, the cold river can't find a place to sleep and be eaten, we ask about our children, and then we are sympathetic to other tragedies and
3y Like Reply See Translation
9:16 PM
97%
Page
Settings
Boost Unavailable
Wasatia.info shared a link.
October 14, 2014
COMMUNITY.WASATIA.INFO
Inside the holy land: Professor Dajani Daoudi
Meeting with an authoritative guest: Professor Dajani Daou...
You, Martin O'Malley and 929 others
4 Shares
31,172 people reached

According to Fisher (2001), this is what he defines as Interactive reconciliation workshops (IRW) workshops, but as an evolvement towards the reconciliation process, I have called them the Online Interactive Reconciliation Articles (OIRA), ICT can develop such IRW using the online social network website and Facebook Page and other online application software such as reconciliation forums, reconciliation chats, and reconciliation polls. ICTs can also develop OIRA that exposes virtues of the reconciliation process. Also, OIRA can be implemented as a sequenced element towards the reconciliation process. In this research, the OIRA is developed by internet communication research.

Another Post is about Teaching Empathy and Reconciliation, the figure below shows that it had (325likes), reached (15,984), this post also proves that the Palestinian community has a fertile ground to reconcile with the enemy, according to Leiner (2016), **reconciliation starts with the minority and the minority becomes majority**, here the post has reached a significant amount of followers and proved that ICT could advance the community that is in conflict towards the reconciliation process.

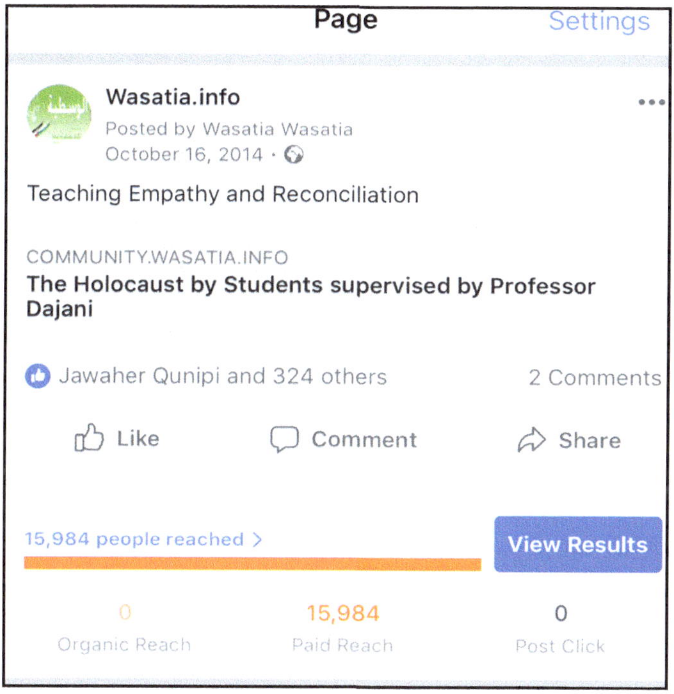

I also observed, from the dataset that is provided from the Auschwitz post, that ***Auschwitz trip by Palestinian Students** post* had more impressions and interactions than taking part of the course, and the ***Teaching Empathy and Reconciliation**,* post; part of these posts involved taking students to Auschwitz, which indicated

Palestinians wanted to be educated about the Holocaust but not in the context of a political agenda or even an educational syllabus.

Another interaction online towards the reconciliation process, is a post about whether Reconciliation is the way to solve the Israel-Palestinian conflict, as you the figure below illustrates that (733) likes the idea, six comments, and three shares, according to (Hollenbeck 2001), the community that is online is in a state of reconciliation, as she states **that reconciliation is a state were neither parts harbor animosity and hatred to each other and agree to behave in a manner that will not threaten their enemy or their rights**.

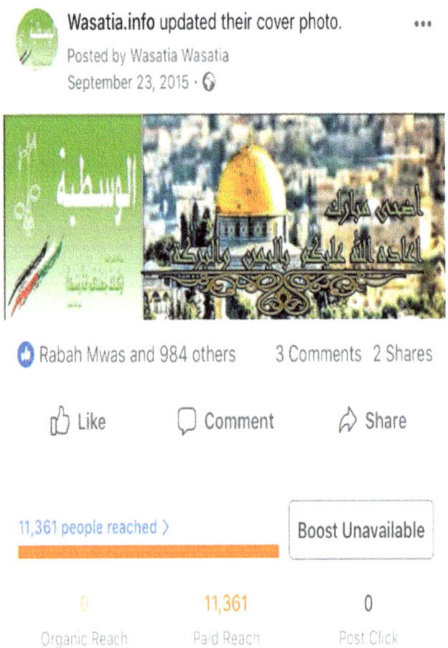

Another Post that is of significant, it was posted on September 23, 2015, about Wasatia in Palestine, the post (984) Likes, three comments and two shares, this proves that the Palestinians are open for Moderation in Islam and as a way of life. **Having accepting to become a moderate as an Islamic idealist, this is part of the Wasatia Cycle**, which proves that the Wasatia Cycle discussed in Chap. 5, using ICT to develop social transformation, has been accepted by the online participants. As discussed in Chap. 5 moderation leads to reconciliation and reconciliation leads to the aspects of democracy. This post shows that ICTs can also work towards the reconciliation process.

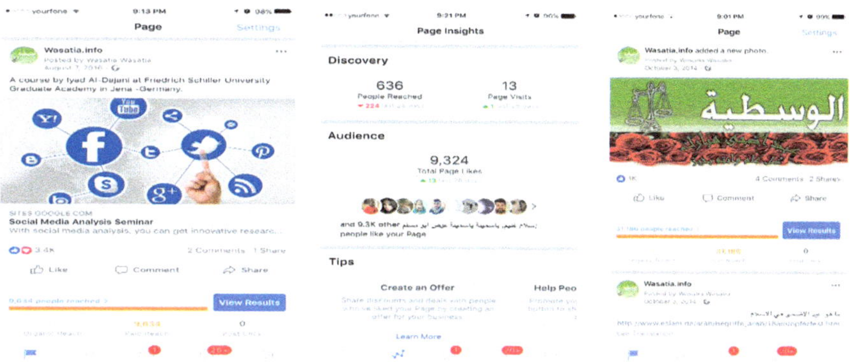

The Wasatia Facebook page had 10,000 followers as of December 12, 2017. The social network website had 257 members that want to be part of Wasatia activities in the community. This indicates that the virtual community is interactive and that the ICT-MR has developed Interactive Reconciliation Workshop according to and the virtual community transformed into a real community that wants to help the community, with having members registered in the social network website.

The use of the Facebook pages influences Interactive Reconciliation workshops (IRW) and Online Interactive Reconciliation Articles (OIRA) that are exposed in Reconciliation Forms, Reconciliation Chats, Reconciliation Bogs, Reconciliation Facebook Pages. Such as “IRW – Auschwitz trip (2014)[1] – which illustrates how the trip affected online users and visitors of the social network’s website and the Facebook Page Wasatia.info. The figure below proves that the IRW exposed online members and visitors to what happened in the Auschwitz Trip that happend in 2014, participants were israeli, German, Palestinian professors and Ph.D candidates and forty Palestinian students.

[1] Described in Chap. 5 under section Auschwitz Trip (2014).

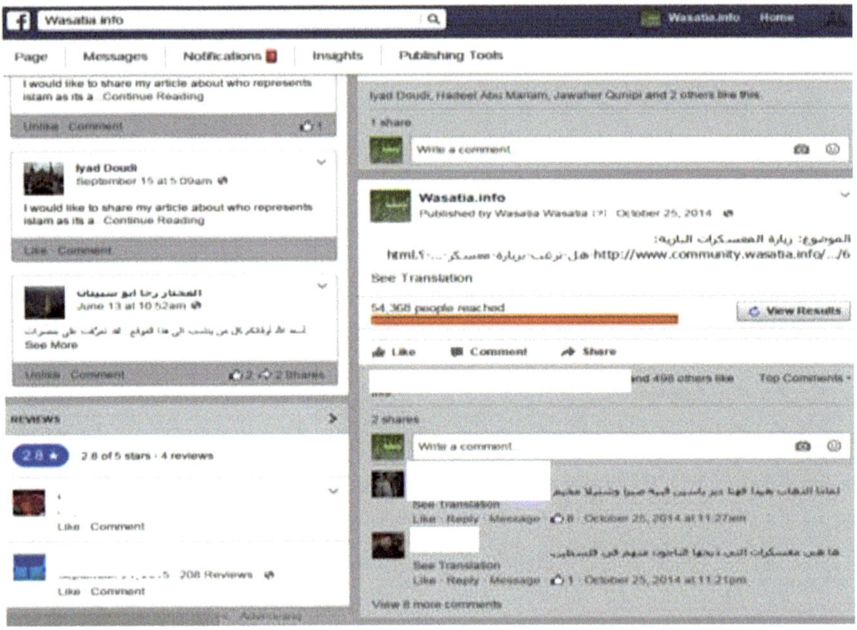

IRW are posted online via the social network website and in the Facebook Page, such as Posts concerning the visit to Auschwitz dated 25. October.2014, the post is accessed [20.5.2015]. The figure illustrates that 498 participants like the idea of visiting Auschwitz, and the post reached 54,368 people; comments against the trip are demonstrated below; for ethical considerations, the user's names are removed.

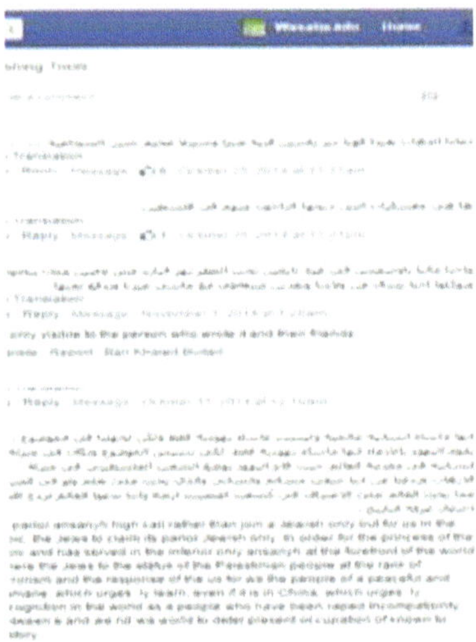

The discussion was in Arabic and in English by members and visitors, the interaction is dated form October 2014. For ethical consideration, no names are illustrated in the discussions.

The profiles are real profiles, according to the Internet research association, as people have published their comments online in an open domain; the users know that their comments are not considered private, and they are part of a study of research this was discussed in the ethical consideration in Chap. 6. Anyone who is using the open space or the open domain is liable to the extent that his data or posts are not private and are open to the public. In the research, the researcher is accountable in accordance with the concept of contextual integrity according to (Nissenbaum 2010). Anyone that visits the Facebook page Wasatia.info will find their posts and comments with their names. Users and followers of the Wasatia Facebook page do not own their comments, as they posted their comments on a page conducted by Wasatia Nog. Therefore, they know about the exposure of their comments in that context, Wasatia has the right to disclose the information or not. Concerning the courtesy of conducting the ethical consideration, the usernames were cut off from the figures, and the discussions are bleary.

This proves that IRW and OIRA expose visitors and members to the virtues of the reconciliation process, and give the members the opportunity to interact together and debate on reconciliation issues and practices in regards of their own enemy in the context of the Israeli Palestinian conflict. According to Fisher (2001), this is part of the reconciliation process. Therefore IRW & IORA are fostering the reconciliation process; the next other chapter goes more in-depth in proving the hypothesis of the communication research.

7.8 The New Tactic: Phronetic Mixed-Method Netnography Approach (PMMN)

According to Kozinets (2002, pp. 61–71), netnography can be applied to the qualitative content analysis (QLCA). In the ICT-MR communication research, I applied the research to QLCA, according to professors in the field (Krippendorff 2013; Saldana 2013; Schreier 2012); their techniques and methods are discussed in the next chapter. This ICT-MR applied the QLCA to the Mixed-Method design, as discussed in Chap. 1, which involved collecting data developed from the netnography approach, referred to as the Phronetic Mixed-Method Netnography Approach (PMMN). Therefore, I explored a new approach, which involved a discourse to conduct a qualitative and quantitative content analysis using NVivo for Windows.

Big data emerged from applying the PMMN, which were abundant with qualitative and quantitative content from applying netnography and the open-ended questions from the netnography techniques. These aspects made PMMN adaptable to a variety of researchers, which led to its endorsement as a method to conduct the research. The netnography procedure is commonly ethnographic procedures to

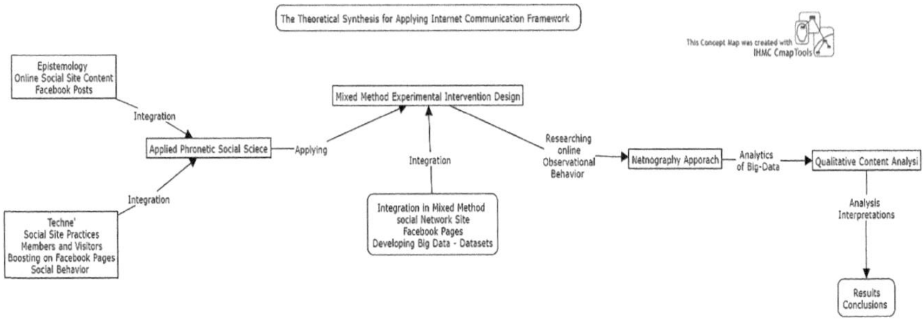

Fig. 7.37 Phronetic Mixed -Method Netnography research design

shape researchers' online participant observations, and it includes making cultural entry, gathering and analyzing data, ensuring accurate interpretation, conducting ethical research, and providing opportunities for culture member feedback (Kozinets 2002, p. 63). The PMMN procedures were implemented in this ICT-MR Internet communication research using principles of netnography, which were as follows (Kozinets 2002, pp. 63–65):

1. Making a cultural entry from the ICT-MR platform involved the case study phenomena of ICTs, which could influence communities during the conflict by endorsing values of the Reconciliation process through interaction between the members, the social network site, and the two Facebook pages.
2. Gathering and analyzing data was part of the ICT-MR platform, which should be identified regarding the phenomena in the research. The gathered data were from the social network site www.wasatia.info and the Facebook pages of Wasatia.info and Wasatia Vision.

The methods of collecting data from the ICT-MR platform were as follows: The data derived from observations and interactions of the online community, and its members' interactions and the meanings, including posts, comments, shares, and photos. The data were also collected from published articles, videos, and images on the social network website www.wasatia.info. Those collected data developed from datasets from both Facebook pages and the social network website to be analyzed to examine the hypothesis of the communication research. This ICT-MR used the Nvivo 10 for Windows. This software can create qualitative analysis packages, and help expedite coding, content analysis, data linking, data display, and theory building functions (Richards 1994). The data analysis was conducted using patterns matching approach and hermeneutic interpretations. During PMMN, data collection and analysis was used to examine whether the data were legitimate or trustworthy by conducting reliability and validity concepts using qualitative content analysis methods, on the ICT-MR Internet communication research.

Figure 7.37 describes the discourse of the Internet communication research as well as the discourse for implementing the PMMN.

The next chapter explores the method, and practical procedures of the PMMN applied to the research, in theory, referencing to professors in the field of content analysis. Moreover, the chapter discusses the historical approach for qualitative content analysis as well as definitions for it and exploring methods and procedures of qualitative content analysis to apply for analyzing PMMN approach.

Chapter 8
Internet Research Analytics (Theories and Methods)

This chapter introduces the theories and methods for exploring and analyzing the analytics evolved from conducting the Netnography approach, the figure below serves to build up the theoretical foundations for the research design in researching theories and methods applied in the qualitative content analysis for the online social network to investigate the empirical work in the research. The chapter explores the analytics from datasets collected from online social network site www.wasatia.info and the Facebook pages Wasatia.info and Wasatia Vision. (Fig. 8.1).

8.1 Introduction

This ICT-MR was interdisciplinary research that involved conducting and exploring mixed data using the Phronetic Mixed-Method Netnography approach (PMMN). This chapter introduces how to explore, collect, analyze, interpret, and examine the dataset developed from the approach. Applying PMMN developed a method for Internet communication analysis, which was applied to social networks. The content analysis procedures included a combination of concepts and theories written by different professors in the field of qualitative content analysis. The utilization of PMMN techniques analyzed big data for social Internet networks, such as Facebook, Twitter, and YouTube. This chapter explores the discourse for the methods used to examine the significant data engineered from the netnography approach in Chap. 7.

This chapter contains the theory behind analytical procedure for qualitative and quantitative content analysis for the ICT-MR Internet communication research. The chapter combines the aspects of applying content analysis to datasets triggered by applying the PMMN research design. The chapter contains research about the essential characteristics, advantages, and limitations of qualitative data analysis in concept. In addition, the chapter includes a Phronetic mixed-method coding

I. M. AlDajani, *Internet Communication Technology (ICT) for Reconciliation*, Lecture Notes in Social Networks, https://doi.org/10.1007/978-3-030-41203-6_8

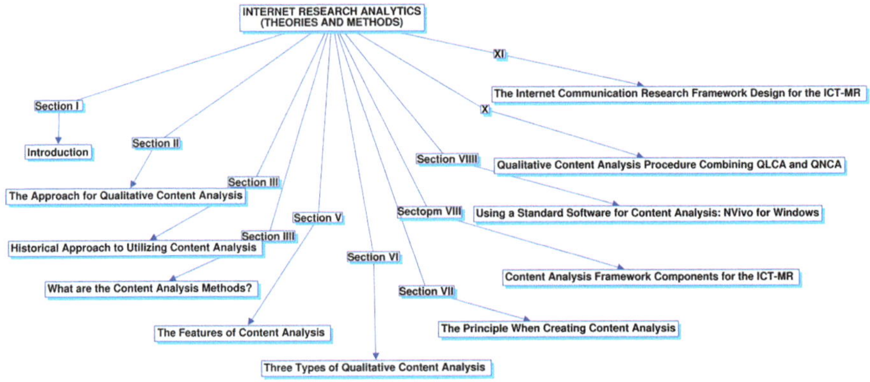

Fig. 8.1 Internet research analytics

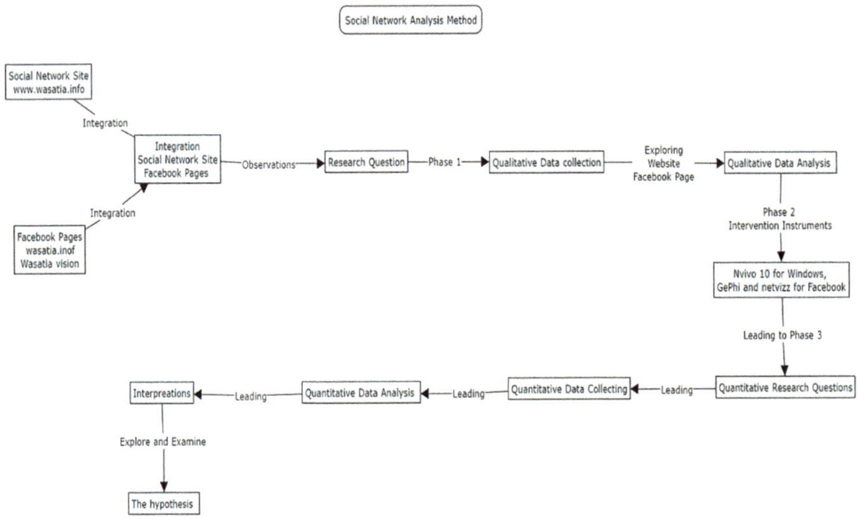

Fig. 8.2 The qualitative content analysis discourse in Mixed-Method experimental intervention research design

approach for the analysis of ICT-MR Internet communication research, as described in Chap. 2. The content analysis procedure is used to combine quantitative and qualitative elements; the data analysis is conducted to create patterns to match approach and hermeneutic interpretations. The chapter contains a focus more on the theories and methods to conduct qualitative and quantitative content analysis for social networks, thereby utilizing those methods for conducting functional analysis for the ICT-MR Internet communication research content in the next chapter.

Figure 8.2 is also illustrated in Chap. 2 of this research, as a discourse for the content analysis procedure; however, in Fig. 8.2, the integration is between the online social network site and the two Facebook pages. This discourse develops the complex Mixed-Method experimental (intervention) design, as discussed in Chap. 2 (Creswell and Plano Clark 2018).

8.2 The Approach for Qualitative Content Analysis

The term *content analysis* was introduced in 1961. Krippendorff (2013) defined it as "*analysis of the manifest and latent content of a body of communicated materials through classifications, tabulations, and evaluation of its keys symbols and themes to ascertain its meaning and probable effect*" (p. 1). Content analysis has three characteristics, as described by Krippendorff (2013). Content analysis is "*empirically grounded method, exploratory in the process, and predictive or inferential in intent*" (pp. 1–5). The content analysis uses words from Greek concepts, such as a *sign, significance, symbols,* and *logic,* to conduct normative conceptions of persuasive argumentation in any type of empirical work. I conducted content analysis to examine data, printed materials, images, or sounds from today's videos to understand the implications these had and the meaning these portrayed. Content analysis "*transcends traditional notions of symbols, content, interpretations, and intent*" (Krippendorff 2013, p. 1). I used content analysis to plan, execute, communicate, reproduce, and quickly help in evaluating the results that included collecting, analyzing, and interpreting data and testing the hypothesis of the research.

8.3 Historical Approach to Utilizing Content Analysis

Berelson (1912–1979) provided the first integrated image of content analysis; moreover, the evolution of content analysis happened because of the massiveness of available communication, which continued to attract many scholars from different disciplines. Berelson[1] (1912–1979) stated, "*Content analysis is a research technique for the objective, systematic, and quantitative description of the manifest content of the communication*" (as cited in Schreier 2012, p. 20). Gerbner (1969) developed *culture indicators* used to analyze a 1-week fictional television programming per year; this project took two-decades of research to complete. Gerbner's goal was to establish violence profiling for different networks and to track various groups that were portrayed on U.S. television. From this development, a new trend was created for content analysis of the media.

Content analysis was introduced for different disciplines, beginning with psychologists who used it in four primary areas, as in the inference of motivational, mental, or personality characteristics through the analysis of verbal records. They

[1] Bernard Reuben Berelson (1912–1979) was an American behavioral scientist, known for his work on communication and mass media.

used content analysis to gather verbal data in the form of open-ended interview questions, focus group conversations, and verbal responses to various tests. They used it as a process of communication, in which content analysis was a vital integral part. They applied content analysis took in the form of generalizations for the measures of meanings over a broad range of situations and cultures.

Anthropologists used content analysis techniques in their studies of myths, folktales, and riddles. In addition, they introduced ethnography. From this introduction, emerging ethnographers have started to rely on content analysis methods to apply to their research. Moreover, historians have started to analyze historical documents using systematic ways of analysis, where data are numerous and statistical accounts may help. Social scientists have also recognized the use of content analysis as a suitable technique to gather educational materials for any research that focuses on the rich source of data and gathering data from the process of reading. Scholars have begun to apply the techniques of content analysis to societies' larger political, attitudinal, and value trends.

In 1955, the Social Science Council Committee on Linguistics and Psychology sponsored a conference on content analysis. Their contribution shifted content from analyzing content in communication to drawing inferences about any discipline in a research study. In addition, the council shifted the discourse of analysis from measuring the volume of a subject matter to counting single frequencies of symbols and then creating co-occurrences in content analysis.

8.4 What Are the Content Analysis Methods?

The content analysis included gathering content and examining the logic of meaning for this content. As stated by Krippendorff (2013), *"To enable researchers to plan and examine the logic, composition critically, and protocols of research methods; to evaluate the performance of diagnostic techniques, and to estimate the likelihood of particular research designs and to contribute to knowledge"* (p. 5). Content analysis represented a research technique required in this Internet communication research for analyzing content. Researchers have described it as one of the most critical methods for analysts to view data, not as physical events, but as texts, images, and expressions that were created to be shown, read, interpreted, and acted on to derive meanings (Krippendorff 2013, p. xii). Content analysis became an important science to understand because it assisted social researchers in adopting natural scientific methods of inquiry, which addressed what mattered most in everyday social life. Content analysis was useful in Internet communication research because it created a sense of mediating between social members of a community and a tool for comparison of meanings through developing an understanding of textual matter, symbols, messages, information, mass media content, and technology (Krippendorff 2013, p. 55). Content analysis involves

Table 8.1 Types of content analysis

Analysis	Goal
Content analysis	Identify keywords, paragraphs of themes
Discourse analysis	Identify the central themes and test the way they were expressed; also, we examine the actual words being used in the text. And new themes will emerge in the discourse of analysis.
Relational analysis	Identify concepts, keywords, and paragraphs of themes and explore the relationships between the concepts

Table 8.2 Krippendorff's (2013) qualitative approach

Analysis	Goal
Discourse analysis	Focusing on how themes are being represented for analysis; to understand how realities are constituted in human interactions whether in language or writings or studies how emotions are conceptualized. New themes emerged, and the analysis procedure is continued.
Rhetorical analysis	Focuses on how messages are received, and what are their intent or their actual effects. This kind of analysis concentrates on the style of argumentation, speech acts, and the like.
Conservative analysis	Recording variable interactions in natural settings and its aims to analyze the transcripts of record toward a collaborative construction of conversations.

> *Referring to any technique for classification of sign-vehicles which relies solely upon judgments which theoretically may range from perceptual discrimination to sheer guess of an analyst or a group of analysts as to which sign-vehicles fall into categories, provided that the analyst's judgments be regarded as the report of the scientific observer.* (Krippendorff 2013, p. 127)

This study's content analysis had three types of analysis that were applied to the discourse of qualitative content analysis in the ICT-MR Internet communication research, as seen in Table 8.1 (Schulz 2012).

Krippendorff (2013) described content analysis in a more qualitative approach to three types of analysis, as illustrated in Table 8.2 (p. 22).

The primary procedure for conducting social networks analysis for the ICT-MR communication research involved the qualitative content analysis approach, as defined by Krippendorff (p. 22). Furthermore, I utilized computer software that assisted in conducting these three kinds of analysis techniques. The applied discourse for discourse analysis, rhetorical and conservative analysis to develop relations between concepts and theories of analysis were applied to the data collected from the PMMN procedure to examine or support the phenomena of the hypothesis in this research. The software was utilized to assist in the qualitative content analysis is NVivo 10 for Windows (i.e., Computer Assistant Qualitative Content Analysis Software [CAQCAS]); Krippendorff (2013) presented NVivo for Windows as one of the best qualitative content analysis software. The software also supported developing a qualitative data analysis (QDA) (Krippendorff 2013, pp. 208–214).

8.5 The Features of Content Analysis

Content analysis has three key features. Krippendorff (2013) described these as theories, which analysts should comprehend when applying the content analysis in any research. Table 8.3 indicates those characteristics used in this ICT-MR communication research.

The characteristics were used to examine the logic, composition, and research methods critically, which contributed to the legitimacy of the study and developed a well-designed analysis to help and assist in understanding the phenomena for the hypothesis of the research. The quantitative content analysis (QNA) and qualitative content analysis (QLCA) are discussed in this chapter.

QNA can be defined as *"a research technique for the objective, systematic and quantitative description of the manifest content of communication"* (Bernhard 1952, p. 18). This kind of analysis transfers the focus on the objectivity of the content that is being analyzed and is maintained by a general quantitative content (Kaefer et al. 2015). The delinquency of applying quantitative alone was that the research would be focusing on the quantity, where data might be ambiguous when conducting the analysis of significant data. For that reason, a follow up with the qualitative analysis technique was more desired. This kind of approach was signified as the explanatory Mixed-Method approach (Creswell 2014), which was later developed into the complex-mixed-method experimental intervention design (Creswell and Plano Clark 2018).

QLCA treats context as a *"central to the interpretation and analysis of the material"* (Kohlbacher 2012, p. 25), which transfers the content from the focus on objectivity, as maintained by traditional quantitative content. This aspect allows for *"the subjective interpretation of the content of text data while maintaining the systematic classification process of coding and identifying themes or patterns"* (Hsieh and Shannon 2005, p. 1278).

Table 8.3 Content analysis characteristics

Characteristics of content analysis	Description of the characteristics
Empirically grounded method	Exploratory in the process and predictive or inferential in intent. Example: Words, signs, significance, symbols, and logic, examining data, printed data matter, images, or sounds and text to make an understanding of what those data would mean for people and how they enable or prevent a situation and what information conveyed by them do.
Contemporary content analysis transcends traditional notions of symbols content and intents	How does the development of the media technologies shape the attention of its users to communicate and its role in influencing the culture in assigning significance to what is being analyzed?
Contemporary content analysis has been forced to develop a methodology of its self.	It enables researchers to plan, execute, communicate, reproduce and critically evaluate results through analysis

Using QLCA has much more advantages compared to the QNCA. Its openness for analysis and more theory guided the process in this paper; in addition, in QLCA, analysts could combine QNCA within the QLCA in its process (Kaefer et al. 2015). Krippendorff (2013) also stated the following:

> *All reading of the text is qualitative, even when specific characteristics of a text are a letter converted into numbers. The fact that computer process enormous volumes of text in a brief time an represent those volumes in ways someone can understand does not remove the qualitative nature of the text being analyzed and the algorithm used to process them.* (p. 22)

QLCA can include all elements of QNCA characteristics. For example, QLCA can apply to word frequency analysis, which allows focusing on quantitative components of the research data to emerge themes from a qualitative analysis perspective. That would be most relevant to the research in question (Kaefer et al. 2015). The characteristics of QLCA allowed me to combine both QLCA and QNCA elements, as illustrated in the ICT-MR content analysis procedure in the next chapter.

One could use QLCA to apply double coding as a means to ensure analytical reliability (Kaefer et al. 2015). QLCA can *"preserve the advantages of quantitative content analysis as developed within communication sciences and to transfer and further develop them into qualitative interpretive steps of analysis"* (Mayring 2000, p. 1). The qualitative content analysis is systematic, reliable, and valid. In applying quantitative analysis, analysts do not acknowledge their contribution to the research (Krippendorff 2013, p. 17). The epistemological approach by Krippendorff (2013) in qualitative analysis contributes to legitimizing values, truth, and reliability of the analysis (p. 27). The epistemological approach involves conducting content analysis in such a way others can use it, and the analyzed material is useful for other researchers, as well. It is *"the primary reason for conducting a content analysis that acknowledges that all text is produced, read by others are expected to be significant to them, not just to the analyst himself this is what makes a content analysis epistemological"* (Krippendorff 2013, p. 27). In the analysis, Wasatia, moderation, reconciliation, democracy, and peace were considered. Krippendorff (2013) introduced an understanding of the characteristics of QLCA, which were used for conducting the qualitative content analysis in this ICT-MR communication research (see Table 8.4).

8.6 Three Types of Qualitative Content Analysis

The types of qualitative analysis, suggested by Mayring (2000, pp. 1–20) were considered when conducting qualitative content analysis, as shown in Table 8.5.

Table 8.4 Characteristics of qualitative content analysis for internet communication research (ICT-MR)

Openness and ability to deal with complexity	All data can be represented – It can be easily validated and reliable "*synthesize openness*" complexity: analysis can be dealt with in a systematic process.
Theory-guided analysis	The center of the analysis is to regularly compare the theory with data and end with a combined theory that would represent the data and the theories. It also gives the ability for a secondary data analysis, which is critical. "*Theory-guided analysis also offers the chance to compare and complement the primary data collected within the research project with secondary data.*" (Kohlbacher 2012). This helps to examine the Hypothesis
Integration of context	The context is also essential to the interpretation and analysis of the material, meaning. It can take a quantitative data and explorer it into qualitative measures, like asking a question from "what" and "how much" changing it to "why" and "how" to get the emphasis on understanding processes as they occur in their context in each research. This can apply to PMMN in the ICT-MR research.
Integration of different material and evidence	Materials can be any data, from text to videos to survey, recorded communications, transcripts of interviews, protocols of observation, videotapes, written documents, in general. All those data can be integrated and analyzed in one qualitative analysis. It can integrate diverse types of data material into one kind of analysis, and it would be a theoretical guided way including all specific evidence collected for the data. As the Social Network Site "*www.wasatia.info*" in the ICT-MR research.
Integration of quantitative steps for analysis	Integration of classical quantitative content analysis, and thus includes quantitative steps for analysis, meaning both quantitative data can be used to reach for qualitative analysis. In addition, creates a more profound insight gain and a more general view of the scientific research deeper insight and a more general view of the object of research

Table 8.5 Types of qualitative content analysis

Type	Description
Summary: Referring to a reduction of the material to the essential content to make it more manageable	Meaning gathering all material and taking the essentials that we need it for our research
Explication focuses on explaining, clarifying and annotate the material	Creating meaning for the material such as explaining it and gathering materials that have the same interpretations with each other.
Structuring: Seeks to extract a structure from the material in questions	In other words, classifications of the material that are gathered from the research.

8.7 The Principle When Creating Content Analysis

Schulz (2012) showed the most important principles to apply when conducting content analysis, as illustrated in Table 8.6.

8.7.1 Inferences, Deductive, and Inductive Approaches

The meaning of Inferences was the human process of coding that it was conducted in this analytical procedure. It appeared in my interpretations of the findings, as suggested by researchers (Krippendorff 2013, p. 41). Inferences were distinguished by deductive procedure and inductive procedure. The three types of inferences that analysts could apply to their research, as stated by Krippendorff (2013), are described in Table 8.7 (pp. 41–42).

8.7.2 The Procedure for Inductive Approach

The conduct of the inductive approach, as designated by Schulz (2012), was used to define elements of the stages for content analysis. These were applied when conducting an inductive approach. Table 8.8 contains a description of the components of the procedure for conducting the inductive approach, while Fig. 8.3 contains a depiction of inductive content analysis.

8.7.3 The Procedure for Deductive Approach

Schulz (2012) illustrated the deductive approach that analysts needed to conduct when applying QLCA. Table 8.9 contains a description of each element of the procedure for the deductive approach.

Table 8.6 Content analysis principles (Schulz 2012)

The principle of creating themes	Always allow themes and new ideas to emerge while building content analysis.
Systematic	Following the procedures of content analysis in the case study or research
Transparency	The content of analysis should be described details
Coherent	Ideas and patterns should reflect the themes of the research study.

Table 8.7 Inferences in content analysis discourse used in the communication research

Inferences	
Deductive inferences	Logically conclusive, the process is from generalization to (Krippendorff 2013) Less Structured Data (Schulz 2012).
Inductive inferences	Generalizations about similar kinds, meaning we have a generalization of an element to a type, from to generalizations (Krippendorff 2013). More structure data (Schulz 2012).
Deductive & inductive inferences	Proceed across logically distinct domains meaning from of one kind of another. Contributing conditions between to different. By applying qualitative and quantitative approaches as they are described by (Creswell 2014), in Chap. 1, and in Chap. 6, introducing the PMMN.

Table 8.8 Inductive procedure

Stages	Description
Unitizing: Units of analysis	Breaking the content into separate chunks of data: Words, phrases, sentences, and paragraphs.
Sampling: Sample of data.	One or two words of summary,
Recording	Open coding: Describing the meaning of the data. Closed coding: Categories codes into groups, grouping coding. Listing all codes. Constant comprising
Reducing	Looking for similar codes and redundant codes and separating them, reducing the list of codes to a smaller list, which is easier to manage.
Inferring	Themes should reflect the purpose of the research and answer the research questions. That data should be comprehensive, and sensitive to what is in the data itself. Themes will merge into many types such as common themes, and unexpected themes, hard to classify themes, major and sub-themes. Examine ideas that made themes and subthemes look for evidence for relations between overarching themes. Constant comparing between themes.

In a deductive approach, the theory comes first. Analysts can create a hypothesis and test the hypothesis to validate the theory, known as the *Top-down Approach* (Schulz 2012; see Fig. 8.4).

8.8 Content Analysis Framework Components for the ICT-MR

To conduct a qualitative content analysis for this ICT-MR, framework components for the research are explored in Table 8.10. Those framework components are considered by (Krippendorff 2013, p. 35) as used to conduct a qualitative content analysis for any research.

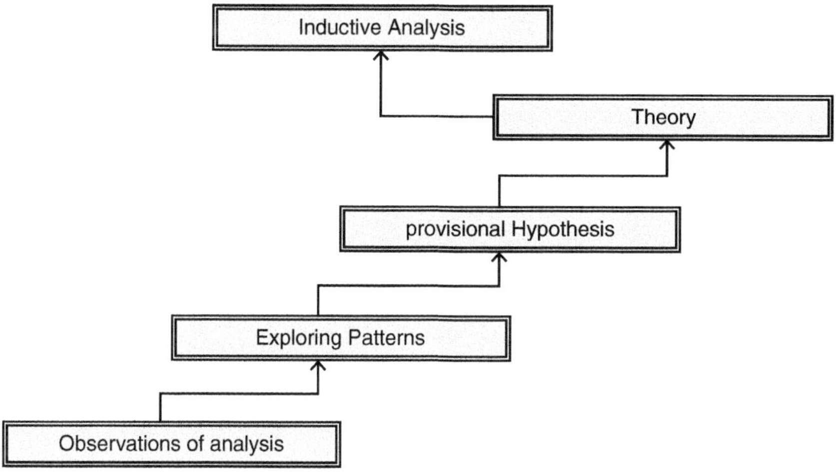

Fig. 8.3 Inductive content analysis

Table 8.9 Deductive content analysis

Stages	Descriptions
Stage 1: Unitizing	Creating a set of themes, gathering literature about the theory or a theoretical concept
Sampling	Breaking up data into separate chunks of evidence. Writing sentences and paragraph that are sensitive to the theory itself
Recoding	Allocating and labeling the chunk of data
Reducing	Gathering all individual quotes and themes separately then gathering all topics and quotes that have similar meaning together
Inferring	Examining all themes, and searching how themes and sub-themes relate to each other
Final stage	Construct a narrative to support the theory through the themes and quotes that were gathered, and discover relations between the themes to support the theory in question.

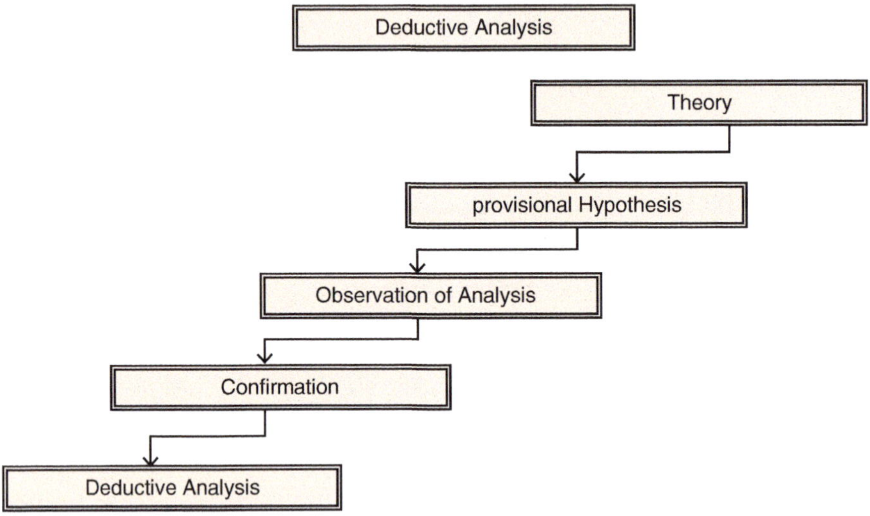

Fig. 8.4 Deductive approach

Table 8.10 Content analysis framework

The component of the ICT-MR	Description of using the components.
The primary element of the research (http://www.wasatia.info), and two Facebook pages. Wasatia.info and Wasatia vision.	A body of the research that contains text, images, videos, and other research components are gathered in one element. A haven for all elements of the research.
Can internet communication technology work for the advancement of the reconciliation process during conflict (Israeli- Palestinian) conflict.	A research question that we seek to answer by various examination using the primary component of the research and the other elements affiliated with the main component.
Software and components utilized in the research like articles, forums, polls, post interviews, news, like, comments, workshop activities, applications, images, and videos.	A context of the analyst's choice in which to make sense of the body of text.
Using QLCA software CAQCAS NVivo for Windows to explore inferences.	Inferences that are intended to answer the research question, which will constitute the core conclusions of the content analysis of the research
Validating the evidence and reliability of the research	Implementing validation elements to and reliability elements to the scientific research, to gain a content analysis regarding those perspectives.
Research results report.	Gaining answers to the research questions; in the form of data files, summaries, statistical accounts, propositions of specific nature recommendation for actions, all regarding their potential validity, compelling results that the anticipated readers of the research result can understand.

8.9 Using a Standard Software for Content Analysis: NVivo for Windows

There was standard software available to assist in qualitative content analysis. The assistant software was referred to as a Computer Assistant Qualitative Content Analysis Software (CAQCAS). This included NVivo 10 for Windows, which was used to support the analysis and not conduct the analysis. The software was used to create an analytical process that was more flexible, transparent, and legitimate, thereby making it more trustworthy.

CAQCAS can accommodate complex data for analysis, whether it was qualitative or quantitative data, and could combine these to create a qualitative content analysis of the research; in that case, researchers have called QLCA, homogenous content analysis (Kohlbacher 2012). One of the principal reasons for considering NVivo for Windows was that *"NVivo is more than a mere qualitative data analysis program; it is a vital resource for planning, team coordination, literature, research design, data analysis, and reporting",* stated by Chris Atchison, a Professor at the Department of Sociology and Anthropology, Simon Fraser University (Atchison 2014). CAQCAS gathers qualitative data analysis software package for QDS. CAQCA S assists in a fast and more efficient method to manage data being used by numerous scholars who work on the same research. The advantages of creating QDS using packaging software, such as NVivo 10 for Windows is shown in Table 8.11 (Kaefer et al. 2015).

The use of CAQCAS assisted in managing the data and organizing these by querying data, graphically modeling ideas and concepts, and reporting forms of data. In addition, the searching and modeling tools of CAQCAS, such as NVivo 10 for Windows, developed data visually in ways not possible with the standard manual methods, allowing the research to have new insights and reflections on the research (Kaefer et al. 2015). NVivo 10 for Windows was used to combine qualitative content analysis and quantitative content analysis to create a more efficient reliable, validated analysis for the research (Penna 2013; see Fig. 8.5).

NVivo was used to develop qualitative content analysis; it concentrated on the coding of text, audio, video, images, data sets, and web pages to develop social network datasets. In addition, it was used to work on open-ended surveys and interview responses. In addition, it was used as a tool to explore and help in audio and video interpretations. NVivo was also used to develop quantitative analysis, as it provided a focus on descriptive statistics to create inferential statistics and develop meta-analysis of the data explored. NVivo was used to import and export text data that were developed within and between group analysis coding and text analysis.

> NVivo developed Kappa coefficient analysis, exploring items such as physical findings, radiographic interpretations, or other diagnostic tests often rely on some degree of subjective interpretation by observes. In addition studies and measures the agreement between two or more observers that include a statistic, that takes into account the fact that observers agree or disagree, merely by chance. The kappa statistic (or kappa coefficient) *"is the most commonly used statistic for this purpose. A kappa of one indicates perfect agreement, whereas a kappa of 0 indicates agreement equivalent to chance. A limitation of kappa is*

Table 8.11 Using packaging software

Benefits of using software NVivo 10
The ability to handle large volumes of data and complex data gathered from the research
Capacity to reduce the complexity of the data collected in the research
Capacity to improve the methodological of the research.
Capacity to have consistency in the research
Capacity to have analytical transparency

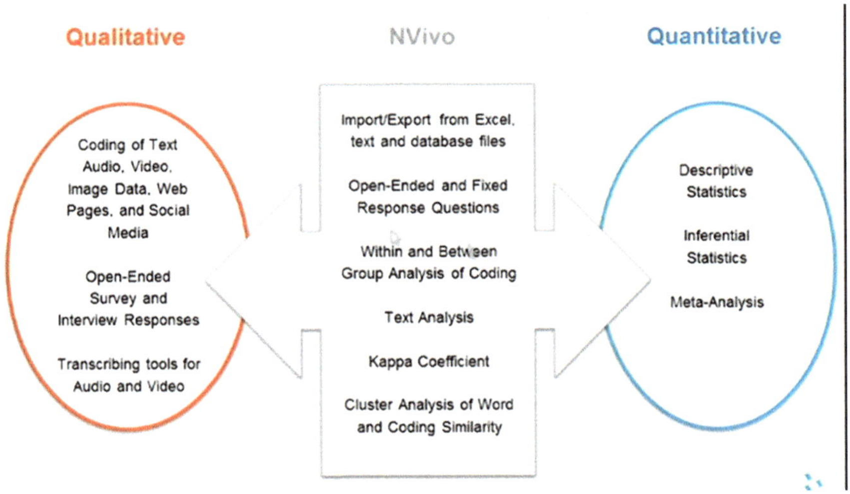

Fig. 8.5 NVivo for Windows (Penna 2013)

> *that it is affected by the prevalence of the finding under observation. Methods to overcome this limitation have been described"*. (Viera and Garrett 2005, p. 360)

NVivo was used to develop a cluster analysis of word and coding similarity. Figure 8.6 illustrates the stages of using the NVivo 10 for Windows software that creates the QDS for any Internet communication research (Penna 2013).

Figure 8.6 indicates ways in which NVivo 10 for Windows explores the content data for analysis, organizes data, and makes it ready for qualitative content analysis. NVivo 10 sorts data into three categories. All the sources of data for analysis should be organized as internal data, such as data sources and literature; external data, such as data on the web; and memos used to record observations and findings. The source classification included bibliographic information and node classification, such as demographic information. Node codes involved coding that was structured and

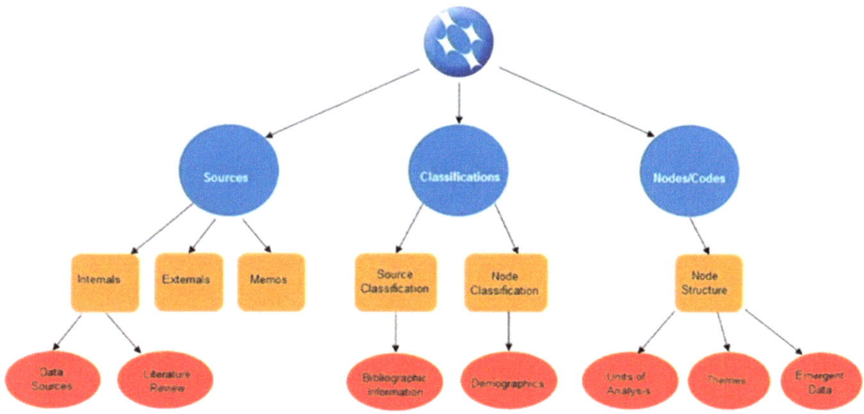

Fig. 8.6 Using NVivo to develop QDS (Penna 2013)

creating themes by combining units of analysis, themes, and emerging data. NVivo was used to develop the qualitative data structure. Furthermore, it was used to gather and work with web pages and online PDFs using NCapture. I also used it to import, Facebook datasets, YouTube videos and online comments, and discussions. In addition, I used it to work with pictures, datasets, tables, spreadsheets, audio files, videos, and automatically code social media data to visualize results easily.

8.10 Qualitative Content Analysis Procedure Combining QLCA and QNCA

The QLCA procedure for this ICT-MR was homogeneous only when combining QLCA and QNCA in the content analysis procedure, according to Krippendorff (2013, p. 86). The QLCA procedure for this ICT-MR combined quantitative content analysis and qualitative content analysis to reach a more desired understanding of the research in question; this described a more homogenous content analysis (Kohlbacher 2012). This ICT-MR developed a theory to conduct qualitative content analysis for digital social networks; this theory derived from combining and integrating theories for analysis from different professors in the field of analysis. The intent of integration was to merge datasets from new themes, as part of the Mixed-Method procedure described in Chap. 2. The procedure to apply this procedure in content analysis, it was done through stages. The first stage involved conducting Krippendorff's (2013) stages for QLCA: unitizing, sampling, recording, reducing, and inferring, as illustrated in Table 8.12.

The second stage involved applying Schulz (2012) to infer data, a procedure conducted respectively, explication, structuring, summary. This stage involved

Table 8.12 Stage 1 for content analysis

The question in the Internet communication research	Utilizing ICT for the advancement of moderation reconciliation and democracy in an online community.
Framework design	Exploring elements of the framework design (Krippendorff 2013, p. 35):
Unitizing	Creating units for analysis, according to data emerging from applying the Netnography
Sampling	Gathering samples to be obtained as part of the units of analysis. Samples are from the datasets and from the social network site.
Recording	Recording results from querying the data in question. Using NVivo for windows to develop query questions to examine, and test emerging themes that examines the hypothesis.
Reducing	Registering the most valuable data out of the recorded results. And collecting data in accordance with developed themes.

Table 8.13 Stage 2 in content analysis

Procedure	
Explication	Clarifying data and its relation to the question of the research. Developing emerging themes from data.
Structuring	Organizing the data – To make it more meaning full to answer the research question. Coding of data and observations. And collecting merging themes that lead to the same interpretation together.
Summary	Examining conclusion remarks of the content analysis of the research in question. Organizing and developing memos from observations and coding.

exploring the data that were registered from the recorded material by using the inductive and deductive methods, as shown in Table 8.13.

The third stage involved gathering reflection and observations out of the analyzed material. In addition, this stage was used to explore emerging themes that supported the hypothesis of the communication research. Moreover, I related the analysis results to the research in question, which meant interpreting and examining whether the research gave a precise answer to the hypothesis or if the conduct of analysis was not sufficient, needing more deep analysis. The results of QLCA indicated a sufficient understanding and interpretation of the research question under examination. Figure 8.7 illustrates the combination for all stages of social network analysis for the communication research for online social network analysis.

8.11 The Internet Communication Research Framework Design for the ICT-MR

A few conceptual components should be considered when creating a framework designed for Internet communication research, as illustrated as follows (Krippendorff 2013, p. 35):

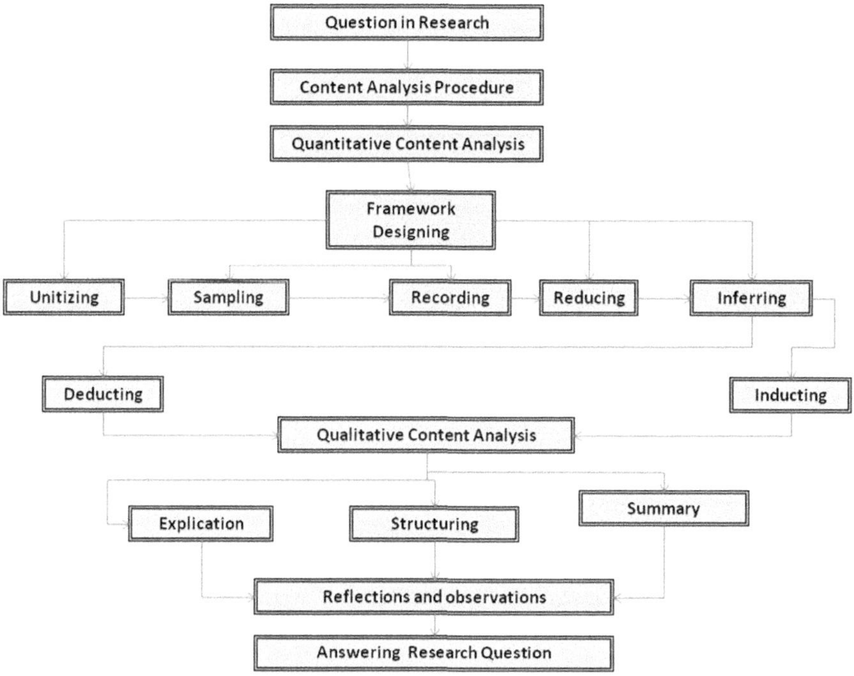

Fig. 8.7 Social network analysis procedure

Table 8.14 ICT-MR framework design

Data type(s)	Units of analysis	Variables	Time framework	Thematic framework
Website: Articles, article news, polls, members, website hits, articles hits video interview, recorded interviews, forum, courses, application, workshops. Social media: Facebook pages, comments, posts, photos, likes, members	Articles News articles Datasets	Gender, age, city, education Visited counts Article hits likes – Boosting (reaching) Comments shares Posts	2014–2018	Reflections, explanations summary, themes of reflections and observations

1. The body of the text, which the content analyst has the material to begin his analytical efforts. In this ICT-MR the materials are content from the social website www.wasatia.info, and two Facebook pages. Wasatia.info and Wasatia vision.
2. Research questions or themes, the analyst, seeks an answer or explore in the body of the text. The first phenomena are ICT can foster moderation, reconciliation, and democracy for directing intentionality's in the community from enmity and hatred to less intentionality.

3. Sampling, which the context of the analyst choice, that would make sense of the body of the text. The samples would be from the content of the social website, and the datasets developed from both Facebook pages.
4. An analytical construct that would operationalize what the analyst knows about the framework of the content to be analyzed.
5. Inferences intended to explore the themes of the research question. Exploring themes that would support the primary hypothesis of the communication research.
6. Validating the analysis, which is the justification of the analysis.

Table 8.14 indicates a framework design used in NVivo 10 for Windows to organize and manage data for conducting analysis on the ICT-MR.

Chapter 9
Applied Internet Research Analytics from Theory to Practice

The chapter introduces the practical methods and theories for analyzing the online social networks; the figure below serves to build up the theoretical foundations for the research design in applying procedures for the theories introduced in the previous chapter, the chapter introduces the practice for conducting qualitative content analysis for the internet research design.

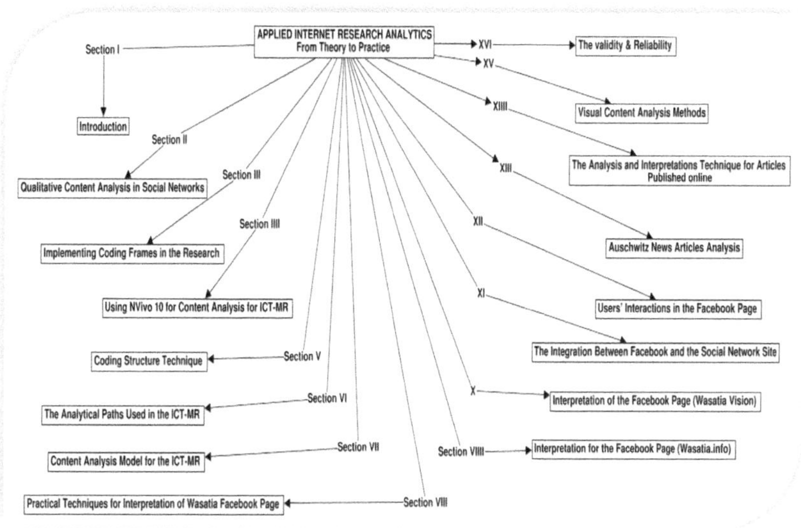

I. M. AlDajani, *Internet Communication Technology (ICT) for Reconciliation*, Lecture Notes in Social Networks, https://doi.org/10.1007/978-3-030-41203-6_9

9.1 Introduction

This chapter explores and examines in practice, the PMMN approach for social network structure analysis, as discussed in Chap. 7. The methods for developing a qualitative data structure (QDS) for this ICT-MR platform is used to examine the hypothesis: *ICT can be a means for conflict transformation for communities in the midst of conflict. This case study is about the Israel/Palestinian conflict.* Moreover, the chapter applies social structure network analysis for the ICT-MR platform by combining Krippendorff methods of content analysis in practice and implementing the deductive and inductive methods for Schreier 2012.

The method of implementing qualitative content analysis is advantageous when attempting to analyze the social network site and social networks platforms, such as Facebook pages, Twitter, YouTube videos, and Instagram. According to Krippendorff (2013), *"On-going changes in information technology allow the nature of content analysis an increasingly attractive method for understanding social phenomena"* (p. 24). Furthermore, the chapter describes ways in which to incorporate theories of content analysis into practice, for example, examining and applying qualitative content analysis on the ICT-MR platform to develop inferences and to emerging themes, which helps indicate the evidence about the ambiguity of the hypothesis of the Internet communication research ICT-MR. George (1959) stated, *"We employ the term 'none- frequency' to describe which uses the presence or absence of specific content characteristics or syndrome act as content indicators in an inferential hypothesis"* (p. 8).

9.2 Qualitative Content Analysis in Social Networks

Siegfried (1952/1953) stated that QLCA was more reliable to use than QNCA. The dependency on quantitative content analysis leads to neglecting of qualitative exploration of the content, which reduces the accuracy of the analysis. It excludes qualitative considerations that might arise in the Internet communication research, such as new themes emerging from the quantitative data. In addition, the explorations of the communications research can be developed only if; the analysis is shifted from qualitative (Web Materials) to quantitative (from Collected Web Materials) and then to qualitative analysis, lastly interpretation and examine the hypothesis. There are some compelling reasons to use QLCA method (Schreier 2012) when dealing with big data that require in-depth interpretations. One of those reasons is the emerging of verbal-visual data and variable data. That can develop the emerging of new themes that would need interpretation, to examine the hypothesis.

QLCA is useful for collecting data that has been sampled from other sources such as the Internet, surveys, questionnaires, documents, journals, images and videos, etc. In addition, when collating rich, complex data from the research itself. One of the primary advantages of using QLCA is due to its precise and flexible nature; it can appropriately reduce data to develop a more compelling broad interpretation that is feasible for research or the primary hypothesis (Schreier 2012).

The QLCA approach was executed via systematic steps to use when applying the implementation of QLCA in research, according to Schreier (2012). The following steps were utilized as guidelines for generating an actual content analysis procedure (Schreier 2012):

- Deciding a research question: Can ICT be a means for conflict transformation toward moderation, reconciliation, and democracy (Reconciliation Process) in the middle of a conflict? Can ICT develop an ego-centered social network in the community in conflict using its core values from the Reconciliation process?
- Selecting material, which came from http://www.wasatia.info, Facebook pages, http://www.Wasatia.info, and Wasatia Vision.
- Building material, which derived from social network site's articles, images, videos, posts, comments, likes, and counts from Facebook pages, social observations.
- Implementing the coding frame for the ICT-MR Platform, as described more in detail.
- Dividing material into units of coding occurred through the social media website, articles on moderation and reconciliation, news articles, forum articles, datasets from Wasatia.info Facebook page, and datasets from Wasatia vision Facebook page.
- Coding procedure always depends on the hypothesis explored and examined, which is referred to it in this communication research as the Reconciliation process.
- Coding schemes are
  - Open coding "Open *coding is the interpretive process by which data are broken down analytically*" (Corbin and Strauss 1990, p. 12).
  - Axial coding asserts that the hypothesis is under investigation during the research and keep gathering information about the hypothesis in question (Corbin and Strauss 1990).
  - Selective coding, "*coding is the process by which all categories are unified around a "core" category, and categories that need further explication are filled-in with descriptive detail. This type of coding is likely to occur in the later phases of a study*" (Corbin and Strauss 1990, p. 14). The procedure of coding is conducted sequentially.
- Evaluating and modifying the coding frame, which is discussed in further sections in this chapter.
- Analyzing and interpreting the findings involved illustrating the results of the ICT-MR platform.

When analyzing data, the need to achieve a reliable and valid analysis remains, which also means that it is partly driven, and any researcher can continue the research work after conducting the analysis and receiving results that are more accurate. By limiting data to just the relevant elements, the coding was at the higher level of abstraction compared to the more concrete information in the material. In the process of coding, I classified all accurate information as instances of a higher order; meaning, that it reduced data in an ordered, various category to develop new information about the data, thus creating a new degree of interpretation inclusive of new information, taking in consideration of applying three sequences of coding procedures.

9.3 Implementing Coding Frames in the Research

Researchers stated, "*Coding frame is a way of structuring your material; it consists of main categories for each main category specifying relevant meanings concerning these aspects*" (Schreier 2012, p. 61). To develop a coding frame for the research, firstly the main aspects of focus should be highlighted in the analysis. The research focuses on developing, and ego-centered social network within the online community in the Israeli Palestinian conflict. With this centered social network develops that values for moderation, reconciliation, and democracy in the middle of the conflict. From applying Internet communication technologies, such as the social network site (http://www.wasatia.info) and social networks platforms, such as Facebook pages. The focus begins with concentrating on analyzing the NGO Wasatia's online social network site www.wasatia.info and then analyzing the Facebook pages, http://www.facebook.com/wasatia.info and http://www.facebook.com/Wasatiavision. Coding frames were part of creating frequencies that were compared with different frequencies from various coding frames from the datasets to develop an analysis that examined the hypothesis (Fig. 9.1).

9.3.1 The Features to Analyze Social Networks

The focus of the analysis was to tackle various aspects of the hypothesis of the research. Such aspects were of prominence throughout the research and could be referred to as dimensions of the research (Schreier 2012). The main category for ICT-MR was about ICT frequencies and inferences, and its influence on conflict transformation toward more intentionality's for moderation, reconciliation, and democracy in Palestinian communities that were in the midst of conflict, as these adopted social change toward the Reconciliation process. The dimensions for the analysis of the research ICT-MR included the following:

1. Whether Wasatia NGO promotes moderation and reconciliation via ICT such as the social network site and the social network's platforms such as Facebook pages?
2. How were the discrepancies accounted for using online social media? How did they affect the hypotheses?
3. What are the political intentions for consciousness-raising from conducting internet communication research?

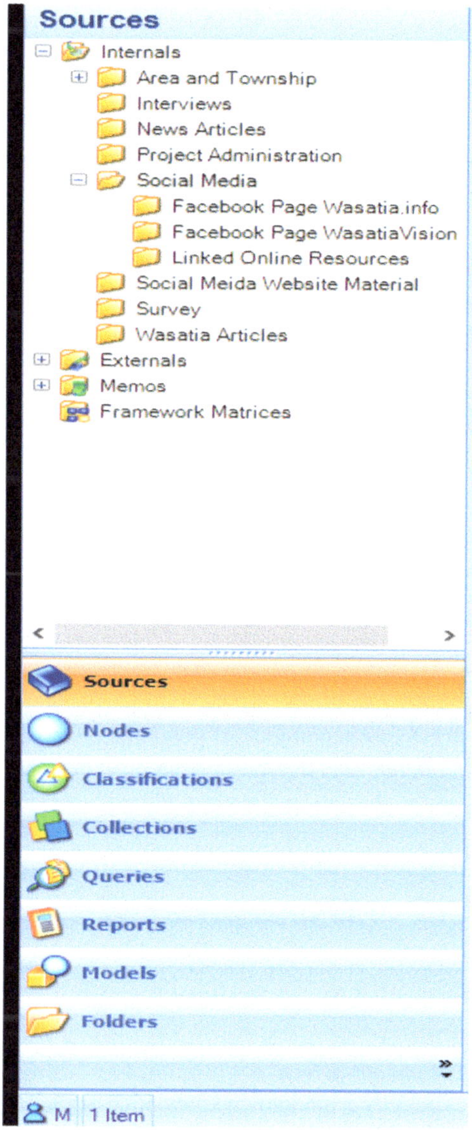

Fig. 9.1 Collecting materials and organizing them dataset collected and analyzed [2 June 2016]

4. Did the use of ICTs change intentionalities From Hatred and Animosity to adapt towards the virtues of the Reconciliation process?
5. Did the online social network deliberation correlate with the virtues of the reconciliation process?
6. Explain how does the Wasatia Cycle concept in phronetic social science transforms the conflict into the objective side of the conflict.
7. Did the online social network interaction correlate towards a reconciliation process in the Palestinian community?

Those questions are answered in Chap. 11, which is the evaluation chapter for the research.

9.3.2 What Are the Dimensions for Creating the ICT-MR Coding Frame?

When creating the coding frame for ICT-MR platform, we should use Unidimensionality mutual exclusiveness and exhaustiveness in mind (Schreier 2012, p. 71). There are three definitions for creating the coding frame in ICT-MR, according to Schreier (2012):

1. *Unidimensionality* "*means that each dimension in your coding frame should capture each only one aspect of your material*" (Schreier 2012, p. 20). When creating subcategories in ICT-MR, we must emphasize each subcategory and identify one unit of coding; helps in the analysis and that the analysis is not redundant. In other words, no double coding.
2. *Exhaustiveness* means including all aspects of the subcategories in a unit of coding; means that all unit of coding should imply to all subcategories without any redundancy, considering all elements or aspects in the subcategories.
3. *Mutual exclusiveness* "*means that each unit of coding only can be assigned to one subcategory within a given dimension*" (Schreier 2012, p. 20). When creating subcategories, it is good practice to assign a theme, which is considered as a coding unit that describes one subcategory within a given dimension. In addition, the subcategory would represent subjects to consider.

9.3.3 The Strategies to Create a Coding Frame

There are three aspects of strategies for creating a coding frame as categorized by (Schreier 2012):

1. ***Concept-driven*** coding involves using the deductive method that emerges directly from known knowledge or theories prior to analysis; those theories create subcategories that emerge as coding units to analyze data in the research.
2. ***Data-driven*** coding involves basing data categories from the materials analyzed for creating units of coding as it emerges from the material itself.
3. ***Combining concept driven and data-driven coding*** refers to the deductive and inductive methods as best used in QLCA. These are implemented in the startup of analysis as a concept-driven and within the procedure emerges the data-driven procedure in creating the categories that create the units of coding. Mason (2002) stated "*The development of an original theory is not always a necessary outcome of the qualitative inquiry, but acknowledge that pre-existing theories drive the*

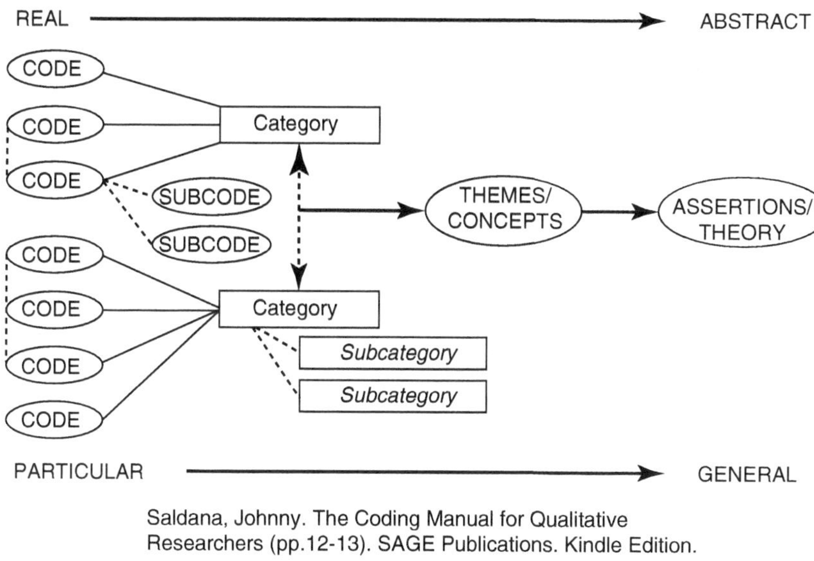

Fig. 9.2 Saldana coding procedure to develop coding frames (pp. 12–13)

entire research enterprise, whether you are aware of them or not" (Saldana 2013a, b, p. 12).

Figure 9.2 demonstrates the combining concept driven and data coding, according to Saldana (2013a, b). This method of coding was implemented in the ICT-MR coding process.

9.3.4 Creating Subcategories Coding Scheme for ICT-MR

Coding scheme specifies what interprets the themes and emerging themes in the research for the main category (Schreier 2012). In that sense, I used the deductive and the inductive methods to create subcategories for the ICT-MR from the main categories. Figure 9.3 illustrates how deductive and inductive methods are used in the ICT-MR, which is the process of building a frame code in ICT-MR.

This ICT-MR explored the primary hypothesis for the research by examining subcategories and coding to develop inferences and themes by considering the phenomena in question. The subcategories were the social network site (http://www.wasatia.info), Facebook page for Wasatia.info, and the Facebook page for Wasatia Vision. The combination of concept and data-driven for creating subcategories meant the main category for coding was concept driven from previous knowledge, and subcategories were data driven from material within ICT-MR the various online platforms.

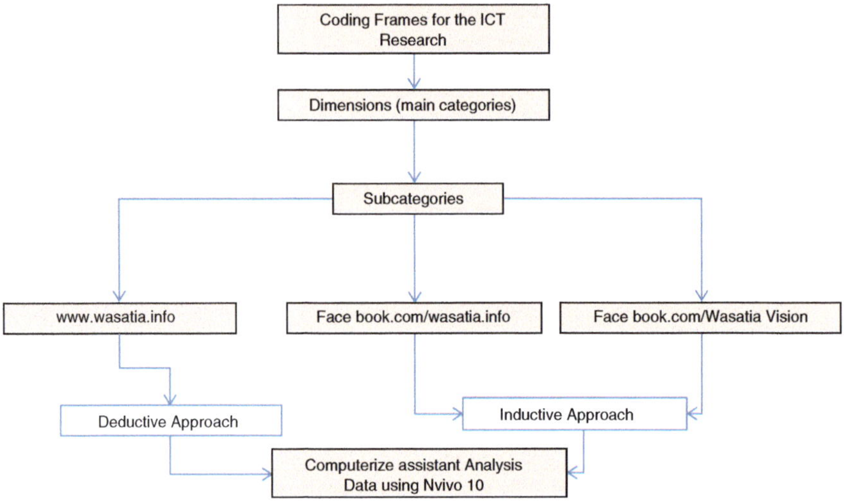

Fig. 9.3 The coding frame for ICT-MR

9.3.5 *Defining Subcategories for Creating a Coding Frame for ICT-MR*

According to Schreier (2012), subcategories for coding include the main category and subcategories. The main category included ICT inferences and its influence on conflict transformation toward more intentionalities for the Reconciliation process on a netnography developed in the Palestinian communities in the midst of conflict. Subcategories for the coding frame from the ICT-MR included the social network site (http://www.wasatia.info), including all indicators and attributes. Indicators were online web materials, such as articles, images, videos, online applications that were part of the analysis and the attributes, which were the tools used by the indicators to analyze the influence of the social network's site in the community to help interpret the inferences. In addition, two Facebook pages were used; one was integrated with the social network site, http://www.wasatia.info. The other Facebook page was Wasatia Vision, which was part of exposing Wasatia education and concept in the Palestinian community and was not integrated with a social network website. Both Facebook pages showed ways in which social networks, such as Facebook pages, could affect to expose the Reconciliation process in a virtual community in Palestine.

Table 9.1 Qualitative description (QD)

Design issues	Design specifics
Philosophy	ICT frequencies and inferences, to dictate its influence on conflict transformation toward the Reconciliation process.
Sample	Taken from the social network site: www.wasatia.info, Facebook page: Wasatia.info – Wasatia Vision.
Data collection	From Wasatia Articles, Wasatia News, Wasatia Auschwitz News. Facebook Datasets from Wasatia.info and Wasatia Vision Pages.
Analysis	See where the term occurs in content. Identify frequently occurring terms in content. Search for content depending on how it is being coded. Cross-tabulation for coded terms. A tool used are Word Counts, Word Cloud Analysis, Tree Maps, Correlations units of analysis and clustering analysis. Matrix Analysis.
Outcomes	To have a straight description coding method of the data organized in a way that "fits" the data findings of the analysis. Developing an emphasis on the phenomena that is in question.

9.3.6 Qualitative Description (QD) Design Strategy for the ICT-MR Communication Project

Table 9.1 shows QD design issues, as proposed by Sandelowski (pp. 334–340). Table 9.1 shows the ICT-MR Internet communication research.

Figure 9.4 describes the coding frame of ICT-MR when considering the aspects and elements described by Schreier (2012) and Saldana (2013a, b); the coding frame was the essential part of the analysis. Figure 9.4 demonstrates the methods for conducting the analysis method of ICT-MR.

9.4 Using NVivo 10 for Content Analysis for ICT-MR

Representing source classifications for the research in NVivo for Windows, which will present the influence relations between the units of coding and the units of analysis to conclude the themes of analysis (Bazeley and Jackson 2013). Table 9.2 shows the source of classifications for the research in NVivo 10. Table 9.3 shows the nodes of classifications.

Nodes of classifications involved the attributes of units of coding that were active in this ICT-MR, which examined the influence of those attributes on the ICT-MR to reach the observations of the units of analysis (Bazeley and Jackson 2013).

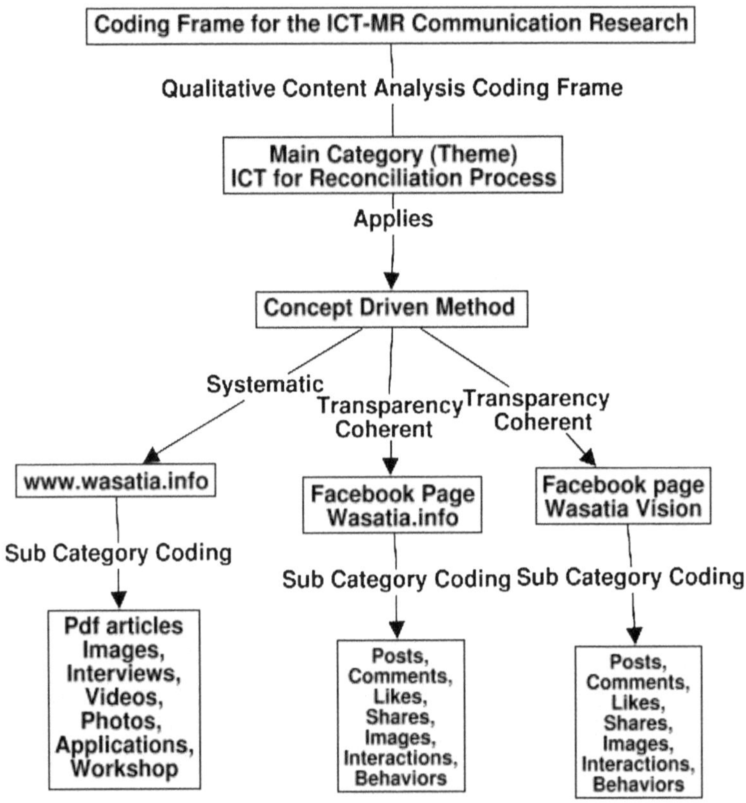

Fig. 9.4 Coding frames for online Social Network

9.5 Coding Structure Technique

I used the analytical representation techniques stated by Krippendorff (2013, pp. 189–208). These included counts as word counts that have a significant meaning for the statement in the unit of analysis that would be an indicator for the interpretation. For example, in ICT-MR, it would be the word "WASATIA," "Moderation," "Reconciliation," "Democracy," "Peace." the frequencies of using those words, and their influence on the community. The words would only be discussed if and only if the knowledge of the terminology exists. In addition, charts were used for the representation of frequencies. I used charts for frequencies, as generated by CAQCAS coding nodes developed from using NVivo 10 for Windows, to compare both the Facebook pages and social network site coding nodes of analysis. Moreover, I used cross-tabulations, associations, and correlations to compare results of coding nodes and using the cross-tabulation techniques that could be generated by NVivo 10 for Windows. I used factor analysis and multidimensional scaling to scale results by generating factored analysis for interpretations, such as comparing members'

Table 9.2 Sources of classifications in NVivo

Classification	Attribute	Attribute values
Interview	Interviewer	Wasatia interviewers
	Date of interview	Date
	Type of interview	Phone, Skype, in-person video
	Location	City or state
News article	Newspaper	The news article was done about Wasatia
	Newspaper location	City or where it was published
	Date of article	Any date
Electronic article	Type	Journal article, communication research, book
	Date of publication	Any date
	Author	Name of author
	Visits	Number of visits
Social Media – Facebook page	Facebook page names	Names of the Facebook pages
	Likes	Number of likes
Social media – Facebook page – posts	Shares	Number of shares
	Comments	Impact of comments
	Likes of posts	Number of likes
Applications	Polls	Kind of polls
	Visitors	Number of visitors
	Article visits	Number of article visits
	Website visits	Number of visits

activities and creating initiatives by comparing and scaling using Nvivo for Windows (CAQCAS). I used images, portrayals, semantic coding nodes, and profiles to create profiles for analysis from coding, with the assistance of NVivo 10 for Windows, which created datasets of packages that generated the units of analysis for interpretation. Furthermore, I used contingencies and contingency analysis, which was a technique that enables me to infer networks of associations from patterns of co-occurrences in the text (Krippendorff 2013). In this ICT-MR, the analytical representational techniques were conducted and presented by using the Nvivo for Windows program; moreover, all coding was generated using that software.

9.5.1 Creating Coding Frame for Facebook Page http://www.wasatia.info for Analysis

Defining a coding frame for the Facebook page http://www.wasatia.info can be illustrated in Fig. 9.5; moreover, the subcategories were used to analyze the content of the Facebook page. This coding method was the first cycle coding method, as

Table 9.3 Nodes of classifications

Classification	Attribute	Attribute values
Person	Gender	Female, male
	Age	Any integer, or ranges (16–29, 30–49)
	Occupation	Cook, teacher, IT professional
	Location	Palestine, East Jerusalem, Ramallah, Gaza, Nablus, Jenin, Bethlehem,
Organization	Location	East Jerusalem, Ramallah
	Members in Wasatia	Any integer, or ranges (1–14, 15–49, 50+)
	Sector	Academic, government, non-profit
Community	Population	Any integer or ranges (<10,000, 10–30,000…)
	Location type	Palestine, East Jerusalem, Ramallah, Gaza, Nablus, Jenin, Bethlehem, Tukaram
Facebook pages	Likes	Number of integers
	Members	Number of members on the Facebook page
	Dates	Any date
Facebook page posts	Likes	Number of internets
	Comments	How did the comment influence?
	Shares	Number of integers
References	Type	Journal, book, article
	Article visits	Integer number
News articles	Name	Newspaper name
	Seen	Number of visits
Polls	Type	Type of the polls
	Name	Title of the polls
	Agree	Some integer
	Disagree	Some integer
Website	Visits	Number
	Date	Any date
	Type pages	The name of the page
	Members	Number of members in the Wasatia website
	Impact	The page being connected with a Facebook post or not
Workshops	Attendance	Number of attending
	Type	Name of workshop
	Impact	Whether workshops influence the community

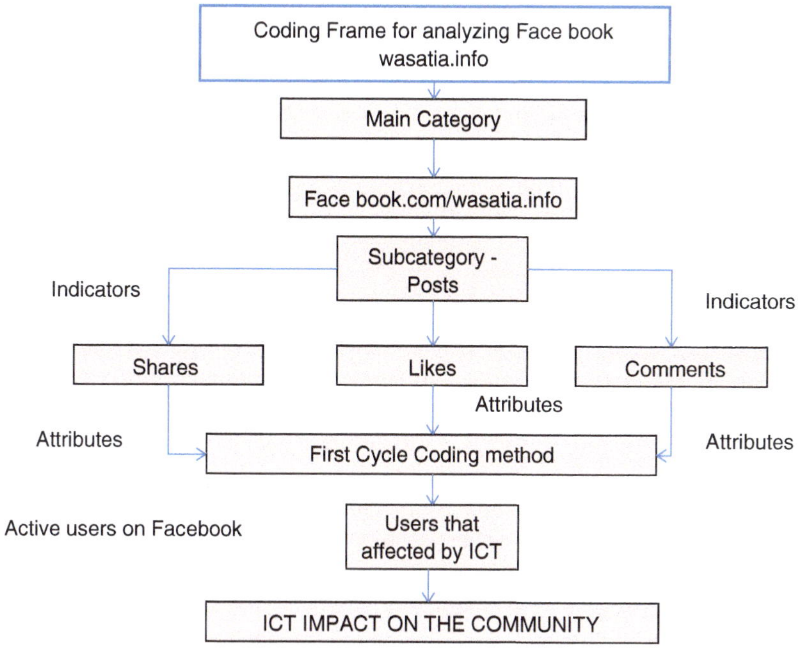

Fig. 9.5 Coding frame for Facebook pages

illustrated by Saldana (2013a, b), and the analysis of coding was data-driven coding (Schreier 2012), as materials emerged from the Facebook page.

9.5.2 *Creating a Coding Method Profile for Analysis of the Facebook Page*

The profile for the coding method of the Facebook pages for analysis was illustrated by Saldana (2013a, b), and these contained the following elements:

1. Source of the material from ICT-MR.
2. Description for each coding of the Facebook page.
3. Application of the strategies, grounded theories, or historical inquiries in creating the coding for the themes of analysis.

According to Kohlbacher (2012), to implement data analysis scientifically for the Facebook page, the coding always must be categorized to develop a profile for each type of coding for analyzing the ICT-MR datasets (p. 19). According to Schreier (2012), the coding method was parallel to the coding frame. Moreover, when generating the representational visualized data results from the ICT-MR

(Schreier 2012), analytical development for analysis was implemented, as presented by Krippendorff (2013, p. 41).

9.6 The Analytical Techniques Used in the ICT-MR

The logical path and techniques according to (Krippendorff 2013, p. 188) for the ICT-MR are as follows:

1. ***Counts*** refer to the following: "*Counting is justifiable only when the resulting frequencies mean something, or have something to do with the context of texts, including selecting quantitative answers to research questions*" (Krippendorff 2013, p. 189).
2. ***Word Cloud:*** The bigger the size of the word in the cloud, the most current term is used in a comprehensive model.
3. ***Cross-tabulations, associations, and correlations*** included finding relations between terms that are utilized in the content, which interpreted the relationships within the content used in the study.
4. ***Images, portrayal semantic nodes, and profile*** included creating a profile for the associated content in the analysis that was in the perspective of the coding nodes that were set up through the portrayals and images in the content.
5. ***Contingencies and contingency analysis refer*** to "*techniques that enable researchers to infer networks of associations from patterns of co-occurrence in the text*" (Krippendorff 2013, p. 189).
6. ***Clustering, which*** "*operationalizes of something humans do most naturally; forming perception holes from things that are connected, belong together or have common meanings while separating them from things whose relationships seem accidental or meaningless*" (Krippendorff 2013, p. 189).

The figure below illustrates the techniques used for the internet communication research analysis procedure.

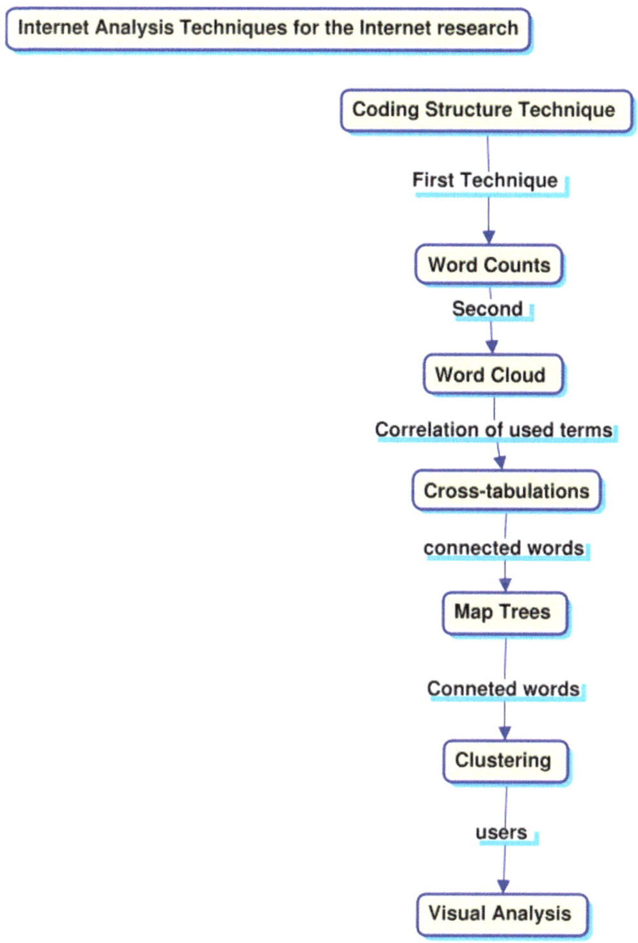

Qualitative Content Analysis Model for the ICT-MR

9.7 Qualitative Content Analysis Model for the ICT-MR

Figure 9.6 illustrates the influence of ICT procedure on the developed netnography that leads to conflict transformation by bounding the use of concepts adopted by the Reconciliation process that present the ideologies respectively to affect the community. According to Krippendorff (2013, p. 335), this kind of model creates evidence of structural validity, as it establishes relations between structural correspondence from available data and theories with modeled relationships within ICT-MR.

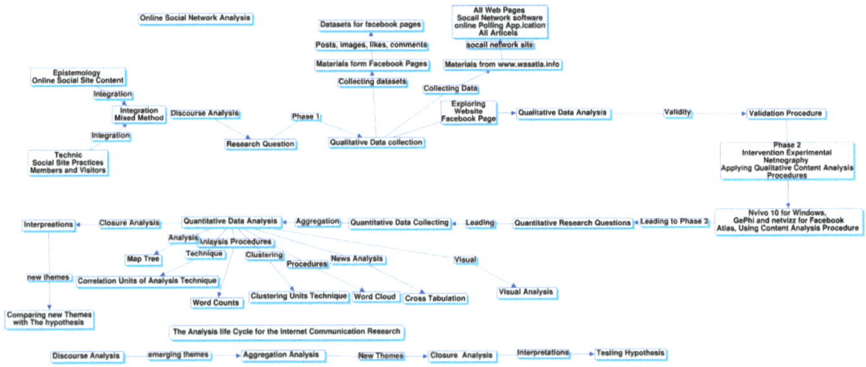

Fig. 9.6 Qualitative Content analysis procedure model for the ICT-MR Internet Communication Research Design

The Fifth part of this communication research explores the research evaluation, including, the evaluation from conducting and implementing the Phronetic Social Science, theoretical framework, the online interactive reconciliation workshops analysis, which explores and describes the analysis, the interpretation, and the results from conducting the ICT-MR communication research.

Part V
Research Analysis, Interpretations & Evaluation "Exploiting The Phronetic Approach"

Chapter 10
Research Analysis, Interpretations & Evaluation

10.1 Applied Internet Communication Technology (ICT) in Phronetic Social Science

The significance of this ICT research was to explore and investigate bridging a path between nations in the middle of a conflict by the influence of Internet Communication Technologies such as online social networks websites, online applications, and Facebook Pages. To adhere to communications on topics that are related to the reconciliation process as an ethical virtue to apply the process of moral practices towards reconciliation in the middle of the conflict.

Martin Luther King described a similar instance a few decades ago:

> *Man hates each other because they fear each other, they fear each other because they do not know each other, and are not aware of each other because they can't communicate with each other. They can't communicate with each other because they are separated from each other.* (Vrasidas et al. 2009, p. 205)

The figure below illustrates a research design discourse for the evaluation process of the ICT-MR communication research.

I. M. AlDajani, *Internet Communication Technology (ICT) for Reconciliation*, Lecture Notes in Social Networks, https://doi.org/10.1007/978-3-030-41203-6_10

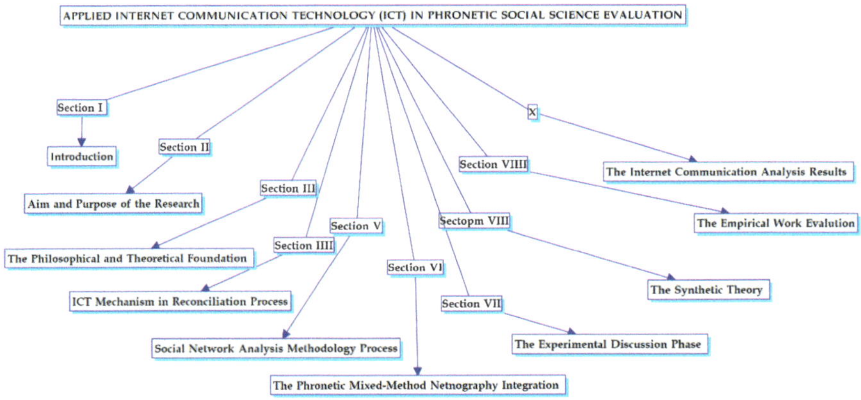

10.1.1 The Evaluation Process for ICT-MR

ICT can help build the bridge of communication between both parties in conflict to destroy the ambivalence of hate that was imminent between two nations in conflicts.

This Internet communication research, referred to as ICT-MR, was a broad interdisciplinary study. For the first time, this study brought together disciplines from different fields. These included philosophy, social science, religion, political science, communication science, computer, and information sciences. The study presented this in a coherent fashion, highlighting the intense sequences of theories and practices in different areas of the research, representing a synthetic approach for The Phronetic Mixed-Method Netnography approaches.

The ICT-MR started with the hypothesis that ICT could be an instrument used to foster concepts of moderation, reconciliation, and democracy in the middle of the Israeli/Palestinian conflict, an adaptation for this hypothesis is that ICT can foster social change towards reconciliation process in the middle of the conflict. To explore, collect, analyze, interpret, and investigate the hypothesis, the NGO Wasatia was chosen for its work to encourage and improve an understanding of moderate Islam and reconciliation in Palestine. This Internet communication research embraced the aim of promoting and fostering the Reconciliation process.[1] According to a study conducted on the transformation of Dajani Daoudi from *The Heart of Stone to the Heart of flesh: The Evolution Journey from Extremism to Moderation,*[2] the Wasatia circle elaborates that moderation leads to reconciliation. The Wasatia approach paves the way for democracy, thus aiming for creating a political system characterized by political stability, economic development, and social prosperity.

[1] Reconciliation Process, is the process that promotes moderation, reconciliation, and democracy for social change in the middle of conflict.

[2] Barakat, Z. (2017). From Heart of Stone to Heart of Flesh: Evolutionary Journey from Extremism to Moderation. München: Herbert Utz Verlag GmbH.

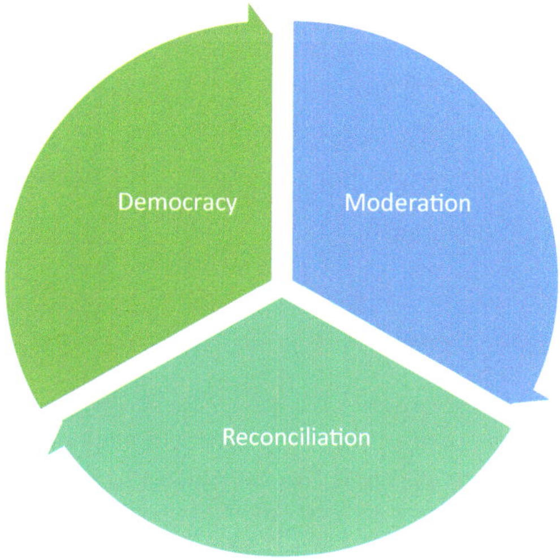

Fig. 10.1 Reconciliation process to be adepts as a discourse for the Wasatia Circle: Defined in more detail in Chap. 5

This aspect would allow respect for human rights and the rule of law and would put an end to authoritarianism, corruption, and prejudice within the community, leading to equality in the pursuit of happiness, resulting in a democratic society.

Moderation (Wasatia), as Dajani (2014) reviewed it, created an environment of trust, empathy, tolerance, acceptance, and temperance with the desire to conduct open-minded dialogues, while respecting the narrative of the other. Reconciliation employs the willingness to seek innovative solutions that are enhanced by moderation, which enhances dialogue, trust, and the understanding of the other. It also bridges relations with different narratives and paves the way for negotiations in good faith (Fig. 10.1).[3]

Internet communication technologies are instruments used for influencing conflict transformation within communities in conflict. It investigates the influence of ICTs on the Palestinian community. The ICT-MR Internet communication research objectives were to clarify, the adaptation from theory to practice, how to reach the Palestinian community and influence them positively toward the Reconciliation process (Fig. 10.2).[4]

One of the primary results that benefit the Wasatia NGO is developing a social network site and a Facebook page. Both social networks became a haven for promoting the Reconciliation process in the Palestinian community. It resulted in evolving an ego-centered social network,[5] which has at its core the values of the Wasatia Circle that transformed into the cycle by the impact of the use of ICTs.

---

[3] See Chap. 5 for details.

[4] See Chap. 5 for the adaptation of the Wasatia Cycle.

[5] Ego centered network is "personal networks or ego-centered networks, an ego-centered network is the network surrounding one particular individual, meaning, usually, the individual surveyed and his or her immediate contacts. The individual surveyed is referred to as the ego and the contacts as alters" (Newman 2010, p. 44).

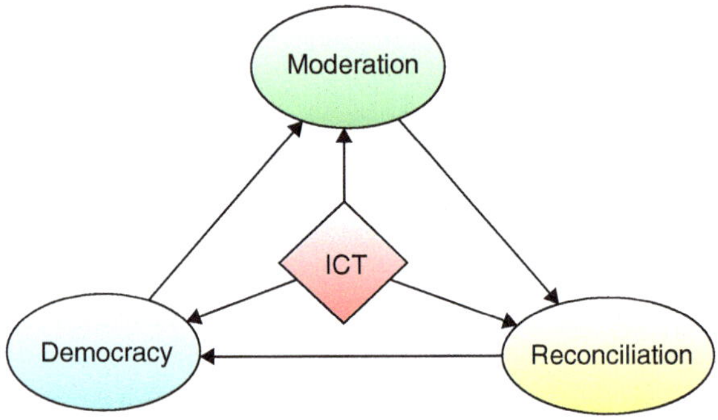

Fig. 10.2 The Wasatia Cycle: The adaptation from theory to practice by applying ICT

The funding for the Internet communication research was funded from the project Heart of Flesh Not Stone Project funded by the German Science Foundation (DFG). This was a significant benefit for the Wasatia NGO as it gave financial support to develop the research and to endorse the Reconciliation process in the Palestinian community by applying ICT in Phronetic Social Science for social change (see Chap. 2). The Internet communication research explores the exposure for the Reconciliation process within the community on the concept and the practical level by applying Phronetic Mixed-Method Netnography Approach (PMMN).[6] Moreover, Internet communication research demonstrates that a more extensive influence can be achieved with much higher financial and institutional support. The Internet communication research is a method that utilized Mixed-Method Experimental-Intervention Research design[7] on Netnography[8] to develop a virtual community, developed from http://www.wasatia.info and two Facebook pages: www.Facebook.com/Wasatia info and Facebook page Wasatia Vision in Phronetic social science.[9] Then implement qualitative content analysis methods and procedures to investigate and explores emerged themes that are related to the hypothesis in question.[10] With sufficient resources, the influence is scalable to reach a broad spectrum of the community. The Palestinian community has a need for understanding and learning of the Reconciliation process before trying to achieve peace. Such research is scarce and sometimes dangerous because it endorses the Reconciliation process in a hostile environment. For example, the posts on the Wasatia Facebook page reached a significant amount of people in the Palestinian community (see Fig. 10.3).

Figure 10.3 illustrates that on November 5, 2014, the Wasatia Facebook page reached (186,560) people, posts received (950) likes, (833) Likes on the post, and (117) shares. Interactions between visitors and members were visible through

[6] See Chap. 7.

[7] See Chap. 3.

[8] See Chap. 7.

[9] Described in detail in Chap. 3.

[10] The procedure in described in Chap. 8.

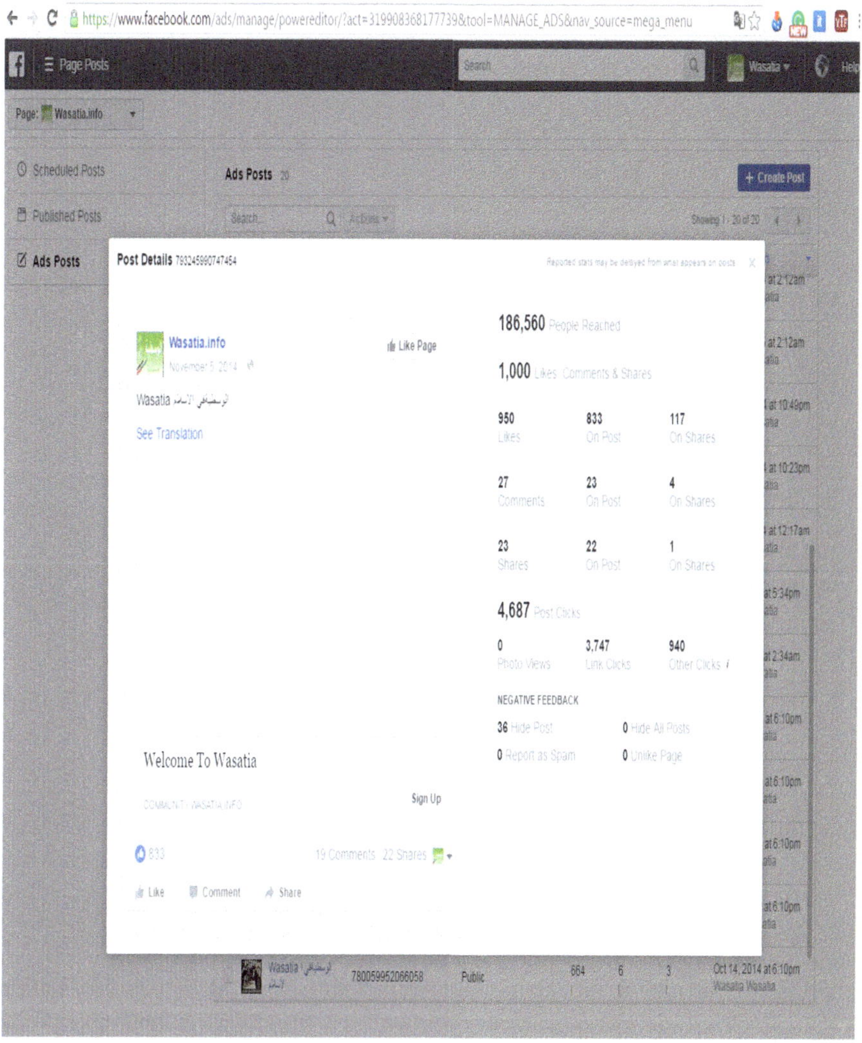

Fig. 10.3 Wasatia.info Facebook page reached a broad audience. (Source accessed [5, Nov. 2014])

comments and shares. (4687) post clicks and (3747) link clicks on the post and 940 other clicks, which directed the audience to the social network site www.wasatia.info.

Figure 10.4 illustrates the discourse for the communication research "*The Internet Communication Technology for Reconciliation: Applied Phronetic Netnography in Internet Research Methodology*." Also, see Qualitative Content analysis model in Chap. 9.

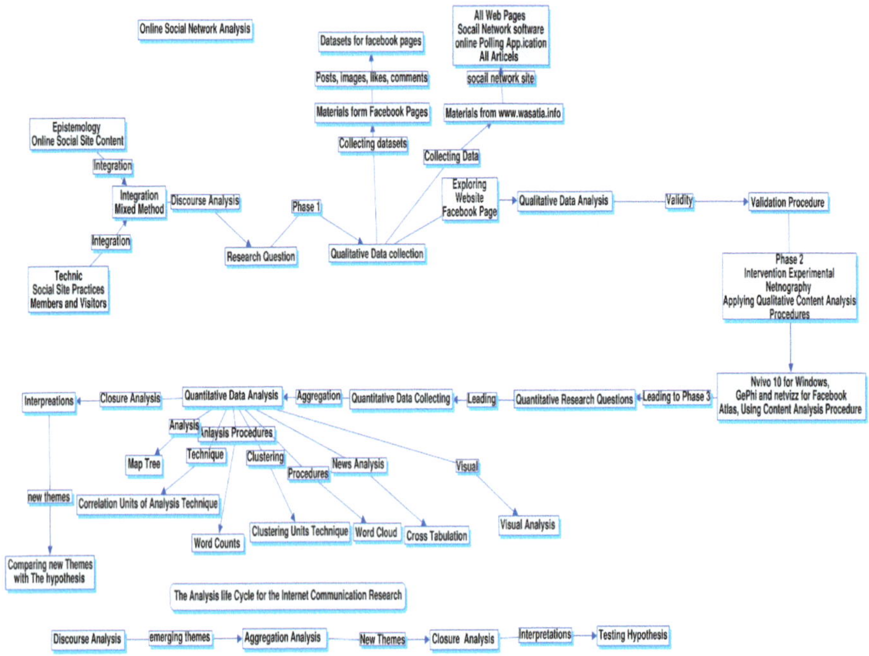

Fig. 10.4 ICT for Reconciliation Process in the applied content analysis

10.2 Aim and Purpose of the Research

The aim and purpose of this research were to generate an understanding to foster reconciliation, moderation, and democracy using ICTs as a tool for that purpose. The hypothesis of this research was that ICTs could foster moderation, reconciliation, and democracy in the middle of the conflict. The purpose was to understand and discover the social change in intentionality's from hatred and animosity to empathy, reconciliation, moderation, and democracy from the influence of online social network to develop an online community – Virtual Communities – and its impact on communities realities in conflict. The intent of the study is to explore, collect, analyze, and interpret datasets developed from online exposure to Reconciliation Process from a developed virtual Palestinian community.[11]

[11] More details about the aim and the purposes are discussed in Chap. 3.

10.3 The Philosophical and Theoretical Foundation

The philosophical foundation in the research evolved from Flyvbjerg (2001) in his phronesis approach to develop applied Phronetic social science using ICT as the element for social change. The adaptation of Phronetic approach to the ICT-MR communication research, developed the concept, The Phenomenology of ICT in Phronetic social science, the idea behind it, is to develop applied Phronetic approach by using ICT for social change towards the practices of moral issues to apply ethical values, in the case of this research the ethical values are the Reconciliation Process. The Applied Phronetic approach in ICT develops a social site that keeps on running forever to develop social change towards the reconciliation process in the middle of the conflict. To achieve these results, the researcher had to find the element of power that resided in the social science research to be active. In this research, ICT was the core and the element of power, according to principles stated by Flyvbjerg (2001). The element of power must always be ultra-dynamic which means switching from one participant to another, which makes the core always dynamic and changing and not dependent on one situation or one person.

The accreditation for the theoretical foundation is to Flyvbjerg (2001) for understanding and philosophical definition for "*Phronesis*" and the applied framework in applied Phronetic social science to (Flyvbjerg et al. 2012) (Fig. 10.5).

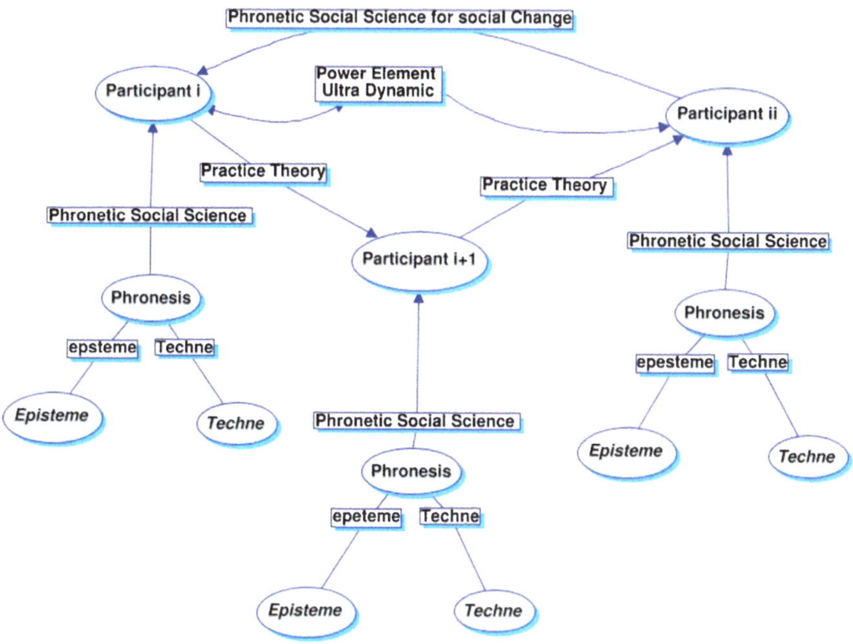

Fig. 10.5 Phronetic social science for social change

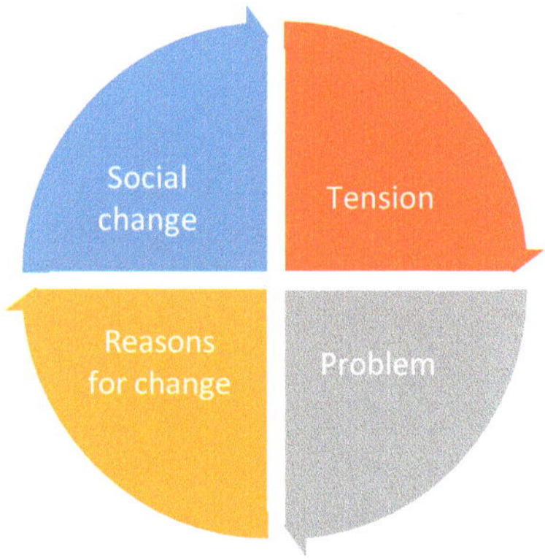

Fig. 10.6 Social change from using ICT in Applied Phronetic Social Science

ICT is the element of power, strengthening this element develops the recursive influence on society that leads to social change. The element of power should be ultra-dynamic, which means it changes over time, and from one entity to another, where no participant has the ultimate power, and it switches from environment to another. Where ICT can be in recursive influence between participants, the positive and the negative, for examples the stakeholders and participants in the site. This develops friction of power from the practice of participants, which is social science develops tension, and that tension leads to a problem, and the problem leads to reason and leads to social change that must happen to relieve the tension. This recursive ultra dynamic approach develops a social change in the society, and this is the core of the phenomenology of ICT in Phronetic social science (Flyvbjerg et al. 2012).

Figure 10.6 illustrates how to measure social change, within the community. The communication research influenced social change because there were tensions developed from the research, one of those tensions are illustrated in the debates in the Facebook pages, between Hamas members, who were questioning the intent of the social network website and the Facebook pages and the intent of Wasatia as a Ngo. The reason it is distinguished as Hamas members because they posted the flag of Hamas (a green flag – with the Islam versus of declaration), and they debated critically on topics related to the Islamic religion. Another tension evolved between participants about the Auschwitz trip by Palestinian Students that happened in March 2014 (more in detail in Chap. 5). The other tension also happened when the site was hacked and was destroyed by an anonymous user, and I had to start using hard measure and security software to protect the work being done online. The researcher used anti-hacking tools to remove any malware attacks against the site. Such as sitelock.com. The other tensions that happened were the debate about the resignation of the Professor who is the Palestinian leader of the trilateral project (HoF), who was one of the leaders who lead the Palestinian students' trip to Auschwitz Trip. In conclusion, many visitors became members of the Wasatia.info and Wasatia Vision. Also, they started wanting to be members of the social network website which developed more tension between members and participants. The problem evolved from the exposure of reconciliation in the middle of the Israeli

Palestinian conflict. This problem developed a reason for the importance of the exposure of reconciliation within the use of ICT in the Palestinian community. Finally, Wasatia as Ngo had been perceived wrongly, which the work of the Ngo, was thought to be normalization in Arabic (Tatbiea) and not reconciliation (Musallaha), which made Palestinian community against it.

10.4 ICT Mechanism in the Reconciliation Process

Before researching the approach of ICT for Reconciliation Process, the term reconciliation was vague in the Palestinian community. With the use of ICTs, the adherence developed an influence to understand and adapt why reconciliation process is needed in the path of resolving conflicts. ICT can be the tool that can apply the reconciliation mechanism; we have stated some of those approaches in Chap. 3 of this research. ICT-MR investigates and explores whether ICT did really adhere to reconciliation in the middle of the Israel-Palestinian conflict, by applying the mechanism stated in Chap. 3. In this research the focus was more on the Palestinian side, so the mechanism for reconciliation is being explored by ICT in the Palestinian context.

There is a working definition when adopted social change towards reconciliation process, the definition is referenced by (Moaz 2004) in "*Social Cognitive Mechanism in Reconciliation*", he defines reconciliation to have a working definition as a "*cluster of cognitive and emotional process through which individuals, groups, societies, and states come to accept relationships of cooperation, concession, and peace situations of former conflict.* (p. 225). I can also confirm that ICT brings a bursting political and cultural information that can operate on the level of individuals and groups in conflict, that help for the transition from conflict to cooperation and understanding of the narrative of the other.

ICT-MR applied the mechanism for reconciliation using ICT to investigate the mechanism of the reconciliation process, and they are described as follows in different phases:

1- Phase of Justice and Truth:

ICT-MR explored this phase when posting reconciliation polls about the Holocaust Trip that took place in (March-2014), which was the first trip organized by the Heart of Flesh Not Stone Project (Hof). Some Palestinian in the online debate about the trip comprehended that there is the truth about the Holocaust and the atrocities that happened in the camp. But the Palestinian participants debated online that it was exaggerated. But what was astonishing is that more Palestinian over (500) clicked "like" on the Facebook page post of the Auschwitz trip by Palestinian Students, on the subject of visit the Auschwitz concentration camp. The reason was that Palestinian participants want to understand and study the tragedy that happened to the Jewish People, making a distinction between the Jewish people and Zionists that they are the enemy. The truth aspect of considering Israel, the land of the Jewish people, the notion was not accepted. Palestinians participants debated that Israel doesn't present the Jewish People, they believe they are against a Zionist regime that is hiding under the Jewish faith. Some Palestinian in the group wrote about "*Nakba*," catastrophe that happened to the Palestinian people after the Arabs lost the six-day war. Some Palestinian argued that it is justice to teach the "*Nakba*" as we are also posting about the Holocaust trip, that the "*Nakba*" also should be addressed. The phase of justice and truth was debatable by the activities and posts done on the social network site www.wasatia.info and on the Facebook page wasatia.info. Articles were published online on the social network site, that had too many readers one of those articles where "*Why Palestinians need to learn about the Holocaust by Dajani.*"

2- Apology and Forgiveness

The Wasatia community didn't reach forgiveness or to give apologies for what other Palestinians did, their narrative that they have the right to fight the occupation, this was obvious, some online debates where against the work of Wasatia as a moderation from the West and not moderation of Islam. But they were a very fund of moderation "Wasaita" as a concept they would accept to live with and adopt in their lives. The online debate was between the researcher and several religious clerks that accused the social network site of spreading the moderation of the west and not moderation of Islam. After debating online what Wasatia as a concept and as movement and its mission, some apologies were applauded to Wasatia and that this concept should be taught more than the Jewish suffering. The apology happened after understanding that Wasatia is not promoting any western ideology but promoting an Islamic concept of moderation. This illustrates to adopt the apology concept we should have empathy, and understand the narrative of the other. As apology being part of the reconciliation mechanism, Palestinians are willing to apology but after they understand, but they feel they didn't get justice and that their acts are justifiable in reason being against the Israeli occupation. In this phase, ICT had a role in deploying the mechanism of reconciliation.

3- Phase of Acknowledgement and Recognition

Palestinian acknowledged the atrocities done in Auschwitz but didn't recognize that Jews had the right to come to the Palestinian lands and build the state of Israel on their lands, according to online debates in wasatia.info. Palestinians were ready to acknowledge Israel as a state, but they also wanted recognition of them as Palestinian people and to recognize the existence of Palestinian state beside Israel.

6- Culture Performance for reconciliation

In the reconciliation process, all parties must be integrated into the process, and it must be promoted in public sphere, that it must inquire and investigate into the exact needs of all parties concerned in the conflict, and that the preparations for reconciliation must be considered before entering the process (Leiner 2018, pp. 180–181). ICT-MR made the process in the Palestinian community a public matter for debate and discussing the reconciliation as a debate between Palestinians.

The debate continued whether reconciliation is normalization and accepting Israeli without forcing the recognition of the Palestinian people.

7- The cognitive biases and Subjective construal between Groups as barriers to the reconciliation process (Moaz 2004, pp. 227–231)

The ICT-MR explored two phases that develop barriers for the reconciliation process, that is important to mention, the first one was ingroup favorability phase, Palestinians in the wasatia social network website, were always favoring the Palestinian narrative and taking into consideration that occupation is part of the problem in their answers in the online debate. The second was the negative perceptions, and the brutal image of the other, that the conflict is persisting because of Israelis are cruel and not willing to compensate with what they have, but they want more, being greedy and building a settlement and processing the occupation.

With more effort and work, we can understand that ICT can be a heaven for debates, that can affect the community and can also provide political and cultural information that can alter a social change in the community. The study illustrates that ICT can develop social change, and in this study, we were able to direct the social change towards a reconciliation process.

The figure above illustrates the before and after the position of the community, the Wasatia, didn't have that exposure without the social network site and the facebook page, but after the Wasatia as a moderate movement and as a concept was being debated and accepted among some of the Palestinian communities. The figure illustrates that after having a social network site and a Facebook Page, it has (9400) likes on its social network Facebook Page and 3.8 percentage of acceptance of 5, The figure Accessed Online [06 of May 2018]. The notion that ICT can be a tool for social change has been explored.

10.5 Social Network Analysis Methodology Process

This section describes the methods adopted to develop the discourse analysis for digital social networks. In this research, I adopted the exploratory sequential Mixed-Method design and applied it to a Phronetic netnography approach, and referred to

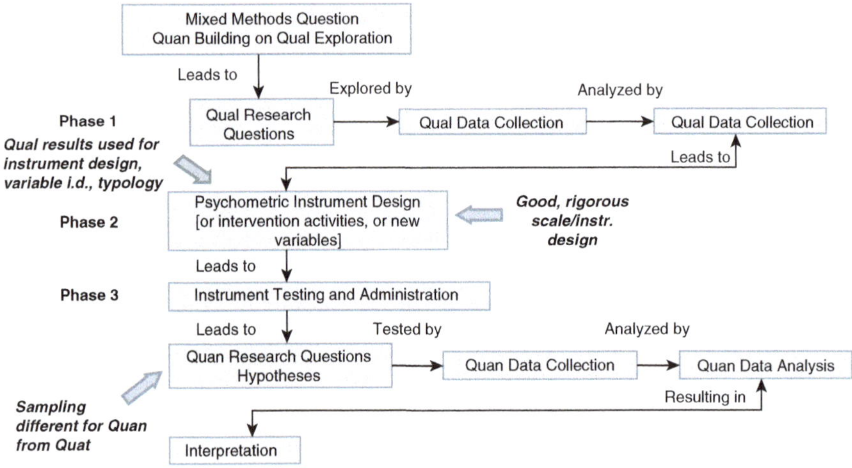

Fig. 10.7 Creswell and Plano Clark (2018) Mixed-Methods research design

it as a Phronetic Mixed-Method netnography approach (PMMN). According to Creswell and Clark (2018), the method evolved and addressed as the experimental intervention Mixed-Method design could be approached and conducted in a discourse (see Fig. 10.7).

The PMMN approach applied the concept map from Fig. 10.7 to the communication research by starting with the hypothesis and concepts that strengthened the study. Phase 1 involved collecting qualitative data from the social network site, http://www.wasatia.info, and the Facebook pages, Wasatia.info and Wasatia Vision and applying qualitative data analysis ending phase one of the analysis.

Phase 2 started with using the computer assistant software for qualitative analysis, the develops the datasets from the social network site, and the Facebook pages that from the text and visual data, collecting and dividing these into qualitative and quantitative datasets, the software that was used is Nivio10 for Windows. Phase three starts with quantitative questions that led to quantitative data collection and concluded with quantitative data analysis. The last phase, the interpretation, evolved from the analysis to examine the hypothesis of the study. Figure 10.8 illustrates the social network method applied.

10.6 The Phronetic Mixed-Method Netnography Integration

The intent of the integration in the exploratory sequential design emerges from integrating the social network site and the Facebook page. To generate a context-specific qualitative data collection and conveying this to link through a display using CAQCAS that summarizes how qualitative results informed each aspect of the quantitative results. This integration also involves interpreting at the end of the

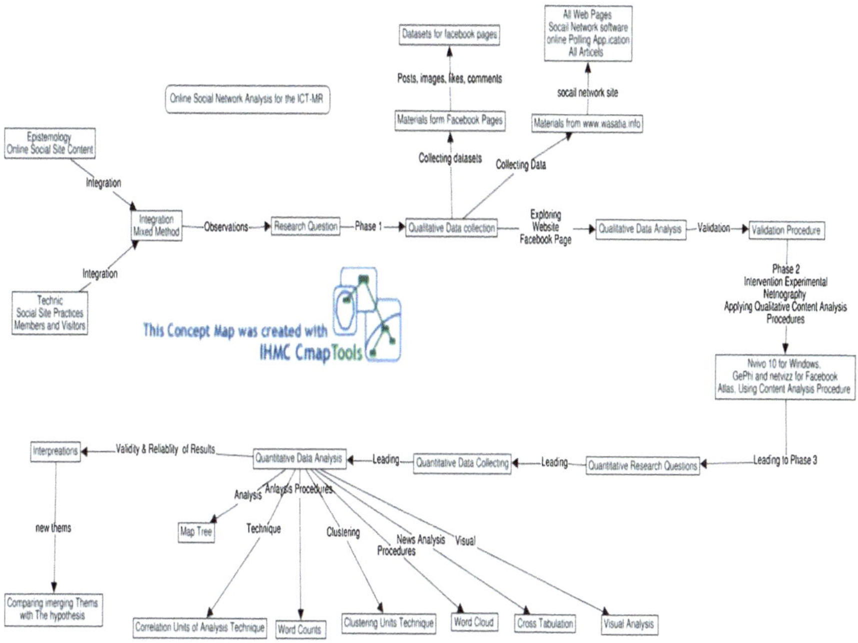

Fig. 10.8 Social network analysis methods used in the Internet Research ICT-MR

study, to the extent to which the method proves that it is the best method applied in the social network analysis and included the testing for the hypothesis of the research, this is illustrated in Chaps. 8 and 9.

10.7 The Experimental Discussion Phase (Social Change from A. to B)

The research illustrates a new research discipline signified as Internet Science Research according to the Internet Science Association. This type of research is known as analytics of netnography.[12] Netnographers applies the study of design to observe, collect data, and analyze these to interpret the resulting data and how the results related to the hypothesis.

In conducting the research, many conditions were expected to change and in different phases and simultaneously, some were predictable, and some were not, this led to demeanor the Phronetic research plan design by Flyvbjerg (Flyvbjerg 2006) and applied Phronetic social science (Flyvbjerg et al. 2012) an aspect of developing a method to conduct the ICT-MR Internet communication research that changes

[12] According to Kozinets (2015).

through time and in multiple phases. Kozinets (2015) defined phronesis as "*an approach through the application of practical judgment attentive to the specific context*" (p. 128). For example, after the surprise effect for the Palestinian students' trip to Auschwitz in March 2014, which was part of the Heart of Flesh Not Stone Project, it had unyielding reactions among Palestinian students who attended the trip, and a conflict incited within their community. According to Flyvbjerg (2001), the power relation was ultra dynamic, which involved using ICT as the element of power in the online research to shift powers between the participants and other participants. This kind of power dynamics resulted from applying phronesis, with power.

Tension arose within the Palestinian participants on the social site, and the tension created a problem of awareness among participants. The problem created a reason for the change, and the reason was interpreted with a question: Does empathy exist within the enemy? Another question arose through the tension: Why teach the Holocaust instead of Nakba?

The posted subject of the trip had 500 likes and some debates against the trip and against teaching the Holocaust to Palestinians. In this part of the research, the tension developed awareness, and the awareness called for social awareness and caused a social change of awareness about the enemy, this kind of conflict was referred by Fisher as the "*conflict spiral model*", according to Fisher, the model illustrates the vicious circle of action and reaction, in the research it is the change from animosity and hatred into understanding for the narrative of my own enemy (Fisher 2001, p. 38). Another encounter was that Wasatia was interpreted as the moderation of the west; this dialog happened with Hamas members, who debated that Wasatia for this NGO in Palestine or the West Bank was part of an ideology of brainwashing. Then, I had to debate using the Wasatia definition from the Quran and from famous Islamic scholars, such as Ibn Kathir (born c. 1300, died 1373), Abū Jaʿfar Muḥammad ibn Jarīr al-Ṭabarī (/ˈtɑːbəri/; Persian: محمد بن جریر طبری, Arabic: أبو جعفر محمد بن جرير بن يزيد الطبري) A224–310 AH; 839–923 AD), and Taqī ad-Dīn Ahmad ibn Taymiyyah (Arabic: تقي الدين أحمد ابن تيمية, January 22, 1263–September 26, 1328). This kind of debate also illustrated tension and resulted in members wondering whether the site could be trusted to teach the real moderate Islam.

After a month, Hamas ministry in schooling declared they wanted to teach Wasatia as an Islamic concept. The ICT-MR adopted the Applied Phronetic Social Science (Flyvbjerg et al. 2012); the shift of power made a social change within the participants, which was through developing tension, and then the tension created a problem that needed a reason to be solved, thereby further creating tension again. This Phronetic approach developed an influence for social change within the online community. In adopting the Phronetic approach, this ICT-MR platform was changed several times in several phases. The social site had to be integrated with two Facebook pages to enhance cooperation between participants and to attract more visits to the social site.

After Auschwitz trip by Palestinian Students (March-2014), a shift of change happened concerning studying the effect of ICT on the Palestinian community. The shift in the research changed to explore, collecting data, analyzing, validating and examine the effect of the Reconciliation process in both social sites and the two Facebook pages. Utilizing ICTs as the tool to develop an ego-centered social network that has at its core the adherence to expose the values of the Reconciliation process in the online society. According to Kozinets (2015). This approach was a humanist netnography.

10.8 The Theoretical Synthesis

The Internet communication research illustrated the birth of theoretical synthesis of a combination of approaches: Applied Phronetic mixed-method research design, netnography, and qualitative content analysis. Those approaches are researched within the discourse of the research. The benefits of this kind of synthesis are to explore, collect, analyze and interpret significant data provided by social networks such as Facebook pages, YouTube videos, and social network sites in this case (www.wasatia.info), where datasets were evolving. This new approach was referred to as the Phronetic Mixed-Method netnography approach used as a discourse for conducting the research. The approach developed the evaluation of experimental Mixed-Method netnography research design illustrated in the figure below, exploring the conduct of the research design within the discourse of the research (Fig. 10.9).

The empirical work for the discourse of the research, started by developing the social network site for the NGO Wasatia (http://www.wasatia.info). Then the social network site evolved and changed according to the development of the situation of the conflict. Different ICT applications were used for inducement of the Reconciliation process research, and on the social network site. The integration between the social network site and the Facebook pages developed had several impacts; the first influence was the integration of methods, qualitative method, and the quantitative method, which develops the mixed-method. The second influence was on the social network, more visitors were reading the materials of the site, and more members were registering to the site. The last count of members as of 28 December 2015 was 157 members of the social network. In the Facebook page

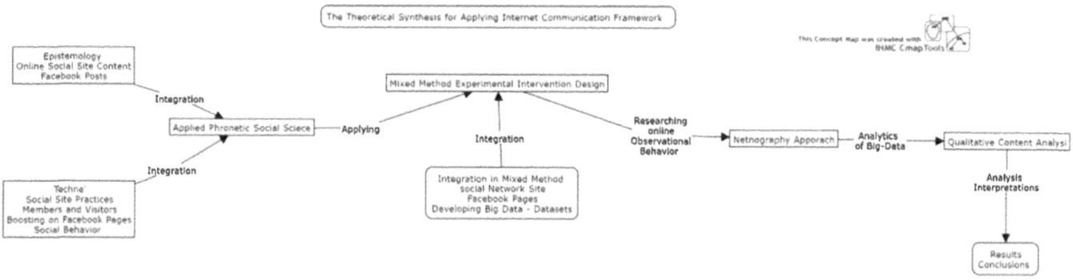

Fig. 10.9 The Theoretical Synthesis for the evaluation experimental mixed-method design

Wasatia.info, the last count as of 28 December 2015 the followers were 10,000[13] followers, and the Wasatia Vision had over 300 followers. Visitors to the site between the years 2013–2015 where more than 2Millon clicks on Wasatia.info, in 2016 the counts of click went down to more than 900,000 thousand clicks which illustrate the downfall for visitors to the site, as the trip to Auschwitz irrupted a debate that develops tensions the lead to problems and then leads to reason for social change, and as illustrated by (Bar-Siman-Tov 2004), social change develop a reason for change in societal believes that develop social conflict, this social conflict was address using ICT as a core element to dispute this conflict.

To collect, explore, analyze, validate, and interpret the big data that evolved from the datasets of the ICT-MR platforms, I adopted qualitative content analysis methods and procedures from several researchers (Bazeley and Jackson 2013; Kohlbacher 2012; Krippendorff 2013; Penna 2013; Saldana 2013a, b; Sandelowski 2000; Schreier 2012; Schulz 2012).[14] I explored a combination of methods to analyze and interpret social network sites.

10.9 The Online Interactive Reconciliation Workshops (IRW) (Qualitative Content Analysis (QCA), Interpretations, and Results)

The empirical work for conducting the Internet communication research was first by choosing the community for the research, which was the Palestinian community. Second by selecting Wasatia NGO in Palestine as the base for promoting the Reconciliation process through social network sites. Third, taking into consideration the Phronetic Research Plan as the abstract method to conduct the discourse for the research. Fourth, applying the synthesis of the three approaches, conducting Phronetic Mixed-Method Netnography approach PMMN, within the discourse of the study. Fifth, developing a Netnography design for the Wasatia NGO and collecting data that is evolving from datasets from both social network site http://www.wasatia.info and the Facebook pages, Wasatia.info and Wasatia Vision.

Then collecting, analyzing and interpreting data sets with the help of computer-assisted qualitative content analysis software CAQCAS, NVivo 10 for Windows.[15] The figure illustrates the discourse for the analysis of the research; this includes the procedure for discourse analysis, aggregative analysis, and closure analysis discourse as described by Krippendorff (2013).

---

[13] *see* Chap. 7.

[14] *The order is according to the importance and mostly referenced to in the communication research.*

[15] See Chap. 9.

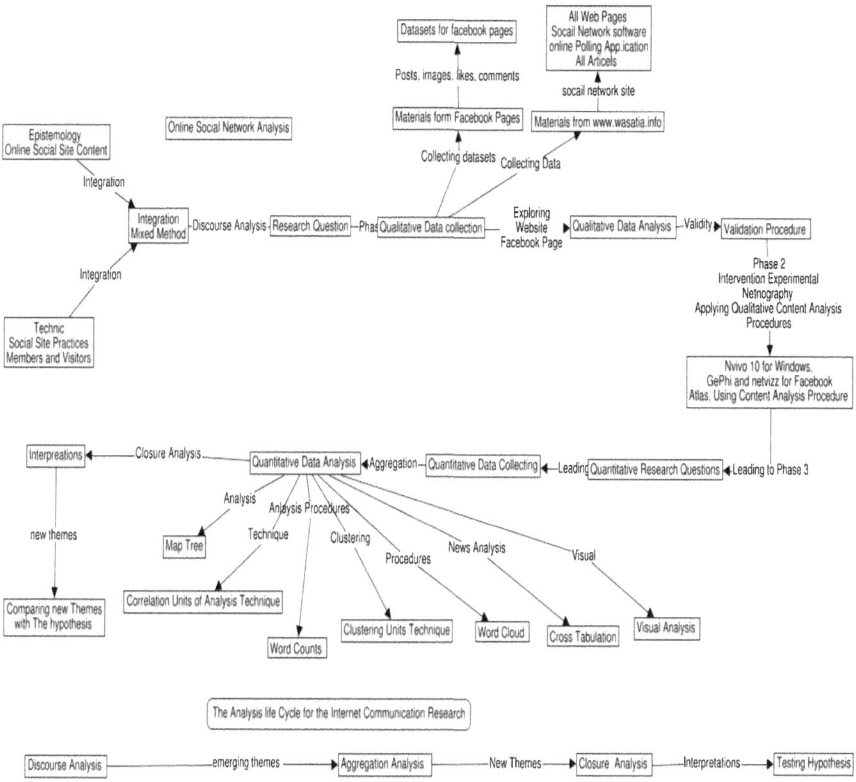

The figure illustrates the methodology used in the discourse of the ICT-MR community research. The discourse starts with research question then qualitative date collecting form resources. The validation process, then goes into the phase of quantitative analysis, exploring quantitative research question, then applying quantitatively research analysis techniques, such as MapTree, Correlation Units of Analysis Technique, Word Counts, Clustering Units Technique, Word Cloud, Cross Tabulation, Visual Analysis. The last phase is interpretations and then examining the results and new immerging themes with the hypothesis.

10.9.1 *IRW Practical Techniques for Conducting the Qualitative Content Analysis*

There are two Facebook Pages to be analyzed as mentioned before The Wasatia.info and Wasatia Vision, the reason to conduct a qualitative content analysis on the Facebook Pages are to investigate the practices of moral issues relating to ethical values and virtues. The reconciliation process is the ethical values as stated before. Therefore, the researcher examined both Facebook pages concerning the endorsements to the virtual community. In addition, the researcher explores the inferences that drove users to become members of the social network website. As from using the analysis techniques such as the words counts, word clouds, tree-maps and clustering of the interaction of users in the similarity of used terms, this illustrates that dataset from the Facebook page Wasatia.info and Wasatia Vision, the analysis includes impressions and interactions which were evident in the dataset.

Comparing between two case studies of Facebook Pages and how they promote the reconciliation process, the study illustrates that the integration between the social network website www.wasatia.info and the Facebook Page Wasatia.info, help promote the reconciliation process in the middle of conflict, but the without integration of the social network site, the Facebook Page – Wasatia Vision Facebook Page, did not promote the Reconciliation Process.

The techniques for analysis involved examining the contingencies and inferences for both the Facebook pages and the social network website. The practical techniques were applied for analysis to explore the correlations between terms in content used on the Facebook page.

10.9.2 Interactive Reconciliation Workshop QCA for the Facebook Page (Wasatia.info)

10.9.2.1 Word Counts Analysis Technique for Facebook Page

This method is used to determine ***the most frequently used terms*** in the Facebook pages and the social network website (http://www.wasatia.info). To also investigate indicators of the content used on the Facebook page and the website to determine the type of ethical values and virtues that the Facebook page contributes for the members and followers of the page. This method used to analyze the content using the word counts technique illustrated by (Krippendorff 2013).

The word count illustration involved gathering and counting all words used on the Facebook page. Table 10.1 shows the most used words, and it is only a sample of the word count table frequency, which contains 17,615 words used on the Facebook page (to check the whole table, please go to the accompanied CD created by NVivo 10 for Windows).

10.9.2.2 Word Clouds Analysis Technique Facebook Page (Wasatia.info)

The bigger the size of the word, makes it more impressive it had to indicate the use of the phrase in the content of the Facebook page; meaning, it was used more frequently, providing more impressions and likes. This type of technology was used to create an "*impressionistic content analysis*" according to Krippendorff (2013, p. 420) of the content used on the Facebook page, which indicated which words had

Table 10.1 Word counts on the Facebook page Wasatia.info

Word	Length	Count
Wasatia	7	529
الله Translate to (God)	4	438
الوسطية Translated Wasatia	7	440
Islam	5	428
Peace	5	358
God	3	302
Muslim	6	291
Muslims	7	299
Palestinian	11	268
Quran	5	254
Palestine	9	208
World	5	214
Islamic	7	204
Palestinians	12	204
Prophet	7	200
State	5	202
Moderation	10	193
Israel	6	183
الإسلام	7	174
Jewish	6	166
Religion	8	154
Good	4	132
Middle	6	136
الفلسطيني	9	130
Community	9	121
Moderate	8	119
Reconciliation	14	123
Right	5	122
Holocaust	9	105
Israeli	7	105
Jerusalem	9	108

more impressions on the users of the Facebook page and provided information on whether the Facebook page promoted the Reconciliation process (Fig. 10.10).

10.9.2.3 Correlation Units of Analysis Technique

Another type of analysis involved checking the association and correlation of the word "WASATIA" in the posts of the Facebook page and its relation to the comments. The word cloud visualization analysis was completed using NVivo for Windows to find the frequently occurring terms within the dataset; the query was

Fig. 10.10 Word cloud analysis technique for Wasatia.info Facebook page

created in the correlation to illustrate an association to most words on the Facebook page concerning the post and comments from the Facebook page dataset (Fig. 10.11).

10.9.2.4 Map Tree Analysis Technique

This technique created indicators for terms associated together, including nested used words. I examined if Wasatia was a term and a concept used in correlation with the posts and comments set up on the Facebook page. This type of technique showed that Wasatia was used by the Facebook page users, as illustrated in the word tree diagram in Fig. 10.12.

10.9.2.5 Clustering Units Technique

The researcher used clustering for Facebook pages because "*clustering in content analysis is based on intuitively meaningful similarities among units of analysis, and its resulting hierarchies resemble the conceptualization of text at various levels of abstractions*" (Krippendorff 2013, p. 206). Clustering provided vast analysis of the influence of users on the posts and comments on the Facebook page and their interactions with each other. It is not posted because the name of the users appears in each relation, so we do not have their consent.

Fig. 10.11 The correlation with the most used words on the Facebook page, the screenshot is date 20. May.2016

10.9.2.6 Interpretation for the Facebook Page (Wasatia.info)

The complete dataset of the Facebook page Wasatia.info consisted of 473 dataset rows acquired from the NVivo 10 for Windows. Table 10.2 illustrates a sample of the data for the interactions between users with posts published on the Facebook page. In addition, integrating materials derived from the website, http://www.wasatia.info. The dataset consisted of all comments, posts, likes, users, date, and times, as found in the Appendix.

Word Count Analysis

The word Wasatia, which is seven letters, was used 529 times in topics and posts from the content published on the Facebook page and from comments by users. This illustrated that the word was used by users on the Facebook page, and it had an influence on most users who were affiliated with the Facebook page. The word was used 529 times more than the other 17,615 words used on the Facebook page. Wasatia terms interacted with using the English and Arabic terms (see Appendix C).

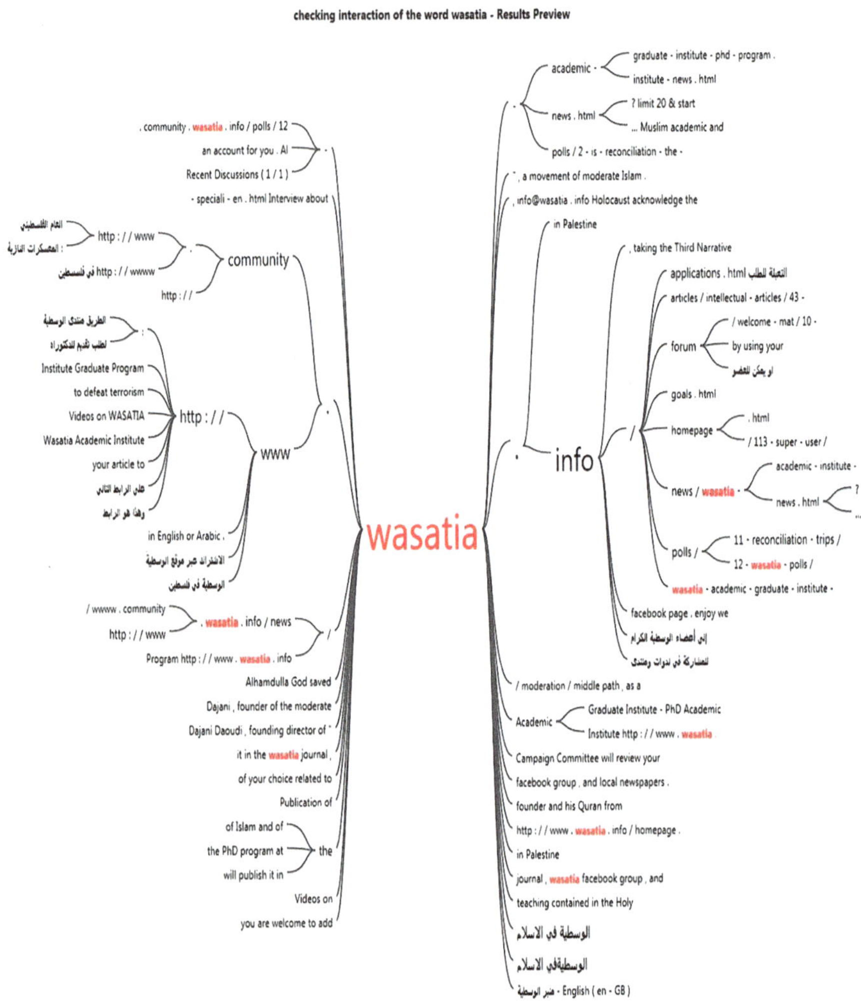

Fig. 10.12 Wasatia treemap of nested words, screenshot dated 20.May.2016

Word Cloud Analysis

The word cloud showed the word WASATIA in English, which had an influence on the Facebook page and was more often used by the users of the page in posts and comments. The word cloud also showed that the world *violence* had minimum influence on the page, as it was in small font. The object of the page was to expose the Reconciliation process for users of the page, thereby expecting them to become active members to promote the Reconciliation process within their communities. As a result, from both word counts frequencies and word cloud frequencies, it illustrated the influence and the intentionality of the users of the Facebook page.

Table 10.2 Datasets sample engineered from Wasatia.info Facebook collected 10. June.2016 Using Nvivo (posts are in Arabic and in English)

Post	Link	Likes	Time	Page likes
Online application to become a member at the Wasatia Ngo	Applications for the Ph.D. Program	2751	2/21/2015 8:32:36 PM	9996
To be part of the Wasatia Reconciliation interactive workshops, please register by clicking on the link below	Welcome to Wasatia Reconciliation Center "Registration"	2132	2/17/2015 2:34:31 AM	9996
http://www.community.wasatia.info/polls/11-reconciliation-trips/3-do-you-agree-by-encountering-with-the-suffering-of-the-other.html	Do you agree by encountering with the suffering of the enemy	1355	10/3/2014 4:03:48 PM	9996
http://ar.QnCAntara.de/content/nsht-lslm-lflstyny-mhmd-ldjny-ldwdy-mn-mtshdd-l-dy-slm-hrk-flstyny-wsty-lltsmh-wlhwr-wrdm	The Palestinian Movement for reconciliation Wasatia	1152	12/20/2014 9:51:24 PM	9996
Video; concerning the work of Professor Dajani and His Movement Wasatia	Inside the holy land: Professor Dajani Daoudi	1008	10/14/2014 3:30:18 AM	9996
http://www.wasatia.info/forum/	Wasatia first topic: what should we teach our kids about Wasatia.(1/1)	978	4/11/2015 5:29:58 PM	9996
An invitation for moderation by Professor Mohammad Dajani	A call for moderation by Dajani Daoudi	889	2/16/2015 1:52:03 AM	9996
VOTE your opinion, and what do you think. Click on the link to be part of the reconciliation polls http://www.community.wasatia.info/polls/12-wasatia-polls/2-is-reconciliation-the-answer-for-peace-for-the-israeli-palestinian-conflict.html	Is reconciliation the answer for peace in the Israeli-Palestinian conflict?	767	10/3/2014 2:41:37 PM	9996
Wasatia in Palestine www.wasatia.info		551	5/25/2015 10:36:48 PM	9996
Do you think the recognition of the Palestinian State by different countries in the world would advance the prospects of peace to reach a viable solution to the Palestinian-Israeli conflict?		439	10/22/2014 2:17:31 PM	9996
Teaching empathy and reconciliation	The Holocaust by Students supervised by Professor Dajani	349	10/16/2014 1:03:34 AM	9996

(continued)

Table 10.2 (continued)

Post	Link	Likes	Time	Page likes
https://m.facebook.com/story.php?story_fbid=724031317675286&id=453919494686471	The King of Jordan to support moderation in Islam	322	10/22/2014 7:50:43 AM	9996
Http://wwww.community.wasatia.info/news/wasatia-academic-institute-news.html	WASATIA Academic Institute News	322	2/9/2015 12:09:14 AM	9996
http://www.euronews.com/2014/10/04/sweden-becomes-first-european-nation-to-recognise-palestine/	Sweden becomes the first European nation to recognize Palestine	281	10/8/2014 7:33:16 PM	9996
Muslims, Jews, and Armageddon, by Professor Mohammed Dajani		243	2/15/2015 1:56:48 AM	9996
ترشيح البروفسور محمد الدجاني الداودي لجزائة نوبل لسلام	Do you agree with a Noble price for Professor Dajani	240	12/16/2014 12:59:47 AM	9996
Palestinian students trip to Auschwitz	Do you agree about conducting the trip	501	10/25/2014 9:26:00 AM	9996

Correlation Units of Analysis

The researcher analyzed the correlation between most words used in the post and comments on the Facebook page; as illustrated by the word cloud query, most words have a relation with the comments and the posts about WASATIA, which interpreted that the terminology of Wasatia was debated among the users of the Facebook page. The correlation appeared more with words associated with the word WASATIA.

TreeMap Analysis

Treemaps helped to analyze nested uses of words that were important for researching the hypothesis. I used this method to observe the words used on the Facebook page. From the word treemap, a justification of users interacting with each other was made, as it illustrated an association between users on the Facebook page. Using the word Wasatia in English demonstrated recurring themes of Wasatia terms.

Clustering

The clustering of users from Facebook page showed the interaction of users on the Facebook page, as illustrated in Fig. 10.13, ***the names of users are not shown***, part of the ethical consideration read Chap. 7. The figure proves user interaction with each other.

Clustering indicated users had an extensive interaction for using the word Wasatia; when users used this word, it meant they understood its terminologies, which showed intentions for understanding the Reconciliation process. In addition, the usage of that word on the Facebook page by users demonstrated that users were

Wasatia.info - Usernames Clustered by Word similarity

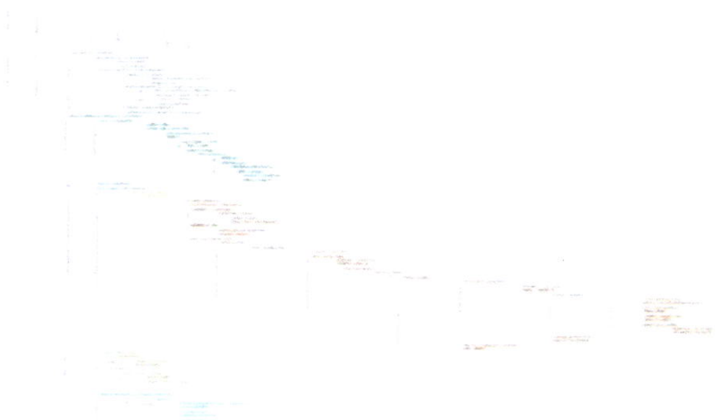

Fig. 10.13 Clustering Users that are using the same word similarities "WASATIA." It was engineered using Nvivo for Windows dated 20.July 2016

interacting with each other, which adhered to the Reconciliation process between the users who were affiliated with the Facebook page. The finding was that the influence was there, within the interactions of users and their deliberations.

10.9.3 Interactive Reconciliation Workshop QCA & Interpretation for the Facebook Page (Wasatia Vision)

Wasatia Vision was another Facebook page used for the NGO Wasatia in Palestine.

10.9.3.1 Word Counts Analysis

I used this technique to create indicators that would imply whether the Facebook page exposed the Reconciliation process or promoted some other values (see Table 10.3).

From the word count procedure, the analysis interpreted that the frequently occurring terms were mainly regarding Islam, and the most used words were associated with the religion of Islam.

10.9.3.2 Word Clouds Analysis

Searching for the recurring terms in Wasatia's Facebook page (WASATIA Vision), it can be concluded that *Islam* and *Peace* were used most frequently by Muslims, which demonstrated that Wasatia Vision was conducive to the spread of Islamic teachings rather than Wasatia's Circle values (see Fig. 10.14).

Through analyzing Wasatia Vision and Wasatia.info Facebook pages, I realized that both pages were different with varying interactions regarding Wasatia values and teachings of Islam. The conclusions indicated that the page utilized information regarding Islam rather than discussions surrounding Wasatia's Circle values. This interpretation was an essential distinction between both Facebook pages.

10.9.3.3 TreeMap Units for Analysis Technique

I examined if the Wasatia Term was used on the Facebook page. Figure 10.15 illustrates that Wasatia is mentioned on the Facebook page, but it is not the focus of the participants in the Reconciliation process. The TreeMap indicates that Wasatia is used within the content of the Facebook page. However, it has no impressionistic use, as illustrated in the word cloud before. This finding demonstrates that the Wasatia Vision Facebook page contributes to the meaning of Wasatia in Isla, but is

Table 10.3 Word Count for used word in Wasatia Vision Facebook page dataset collected in 25. July.2016

Word	Length	Count	Weighted percentage (%)
Islam	5	172	0.42
Peace	5	138	0.34
Muslims	7	134	0.33
Muslim	6	121	0.30
الوسطية Wasatia	7	120	0.30
God	3	119	0.29
الله God	4	109	0.27
Wasatia	7	107	0.26
Palestinian	11	104	0.26
Quran	5	99	0.24
World	5	96	0.24
Palestinians	12	93	0.23
Prophet	7	85	0.21
Islamic	7	84	0.21
Moderation	10	78	0.19
Jewish	6	77	0.19
State	5	77	0.19
Israel	6	73	0.18
Time	4	73	0.18
Believe	7	67	0.16
الإسلام	7	67	0.16
Allah	5	66	0.16
Palestine	9	64	0.16
Religious	9	62	0.15
Religion	8	61	0.15
Good	4	59	0.15
University	10	57	0.14
Right	5	55	0.14
هذه	3	54	0.13
Middle	6	52	0.13
كما	3	52	0.13
Man	3	51	0.13
Jerusalem	9	48	0.12
Jews	4	48	0.12
Moderate	8	47	0.12
Group	5	46	0.11
Holocaust	9	46	0.11
Israelis	8	46	0.11
Reconciliation	14	46	0.11

Fig. 10.14 Wasatia's Circle values data analyzed 25.July.2016

not focused of the followers of the page; mostly the page focuses on the Islamic terminologies, the teachings of Islam, and what Wasatia would mean in the Islamic world.

10.9.3.4 Clustering Units for Analysis

I clustered for the frequently occurring terms that showed the relations of terms being used between users; this technique illustrates that the frequently occurring terms words written by users, datasets analyzed and collected [20. July.2016], the photo is not published for any providing of consent from users. The clustering illustrates that users are interacting within the discussions in the *post*, ***but names are not seen because of the ethical*** *consideration* issues discussed in Chap. 7; it just illustrates that they were interacting with each other on terms and religious concepts. The dataset collected and analyzed [20. July.2016].

The Wasatia Vision was done to promote the Wasatia as a moderate concept, and all the posts were to develop a more understanding of moderation in Islam. On the contrary, Wasatia Vision analysis was still interpreting impressions that the page presented Wasatia terminologies but also represented the teaching of Islamic terms. It focused on Islamic ideologies and teachings, as well as the concerns of the users on the page who were not from the Reconciliation process. They were more focus on Islamic religious terms.

Fig. 10.15 Treemap of Wasatia Vision illustrates the topic discusses in Arabic language datasets analyzed 27. June. 2016

10.9.4 Users' Interactions in the Facebook Pages

This ICT-MR illustrated the interactions by clustering users who used the same word frequently, as in Fig. 10.16. In addition, ICT-MR illustrated where particular terms occurred in the content of the dataset of the Facebook page wasatia.info. regarding similar words used by the users which illustrate users have done debates using similar words.

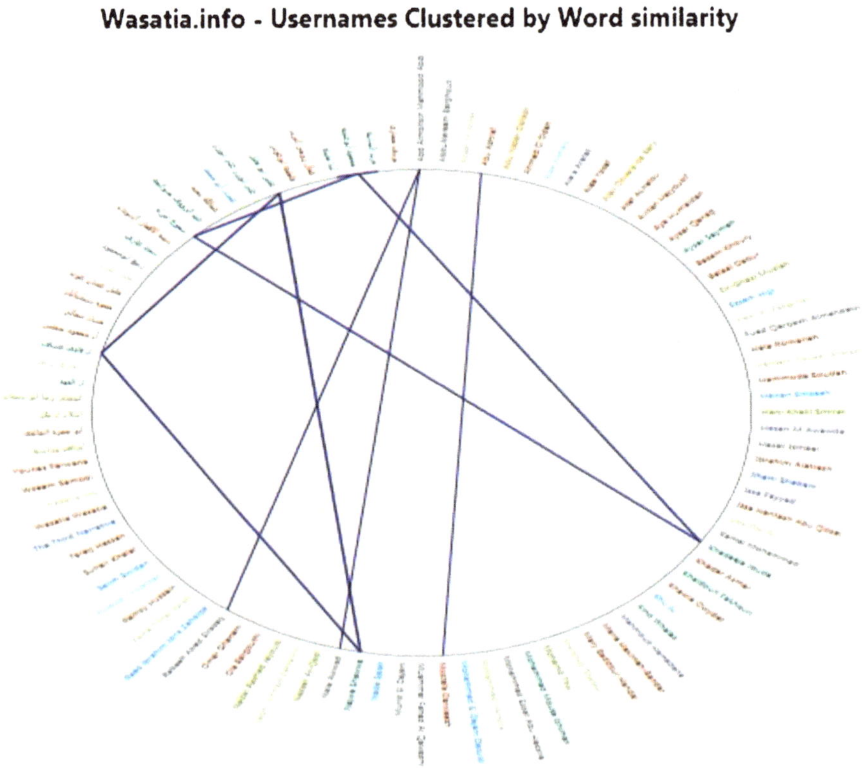

Fig. 10.16 Clustering interactive users, as part of the ethical consideration, the names of participants are not shown in the diagram. Datasets analyzed 20.July.2016

The Wasatia nested TreemMap, as shown below, illustrates that the term Wasatia was being used by the users and it develops a nested occurrence (Fig. 10.17).

10.9.5 Analyzing Online Interactive Reconciliation Articles (OIRA)

ICT-MR gathered over 250 web news articles about interactive reconciliation workshops and embedding them in the social network site http://www.wasatia.info with news-web-links. Most circulated OIRA on the Internet was the Auschwitz Trip (2014), with Palestinian students. The trip combined Palestinian students from different universities, supervised by the Heart of Flesh Not the Stone project team (HoF), which combines Israelis – Palestinians – Germans scholars working together to conduct Interactive Reconciliation Workshop. It took place from March 24 until 30th 2014. The visit was publicized by the news media, utilizing ICT platforms as a tool for publishing the event. The scope of this section is to illustrate, the practical procedure for analyzing the online news articles that covered the event, and to

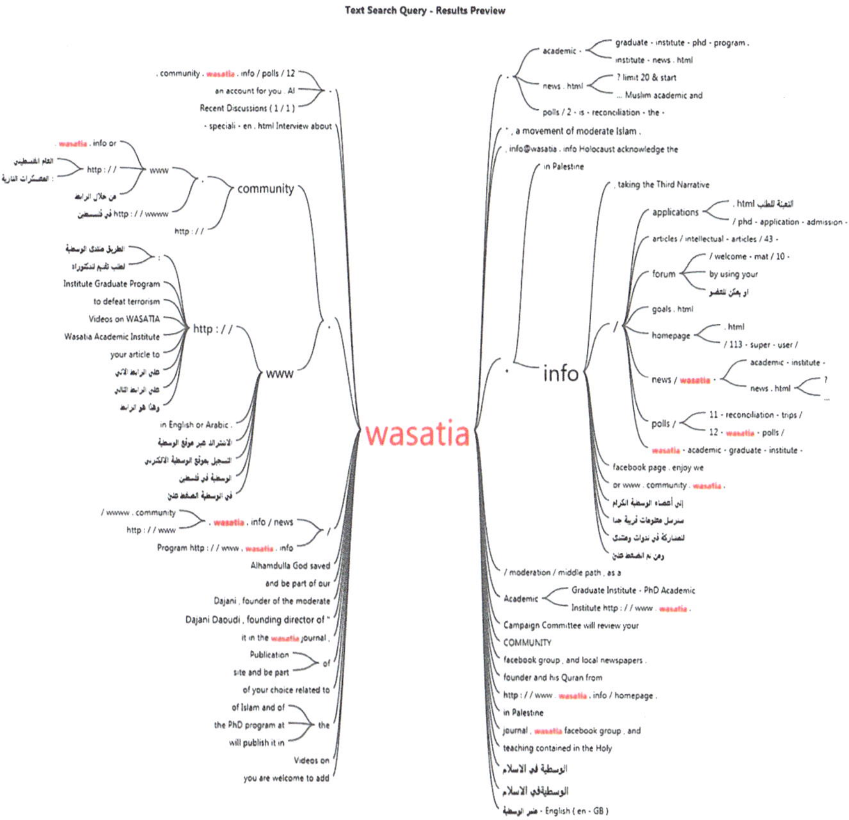

Fig. 10.17 Wasatia nested Treemap datasets collected and analyzed [20.July.2016]

observe the focus of the OIRA, that utilized ICT to publicize the trip on online social media, such as Facebook and other online news magazines.

The analysis and interpretation are conducted on the online news article section of the social network website, to analyze and interpret the influence of the IRW on the visitors and members, to expose visitors and members to the OIRA and to investigate whether the IRW and OIRA expose virtues of the reconciliation process. In addition, interpret what was the primary focus of online news agencies covering the workshop, and to explore the inferences that the online news media developed on WASATIA Ngo. and the heart of flesh, not a stone project.

This communication research gathered online 250 agencies that published online about the IRW – Auschwitz Trip 2014 – the most IRW was the Auschwitz trip (2014) more described in Chap. 5. All materials were gathered from the online social network website http://www.wasatia.info/news.html. The techniques for analysis are to examine the contingencies, inferences, and frequencies for online articles and news articles about the interactive conflict reconciliation workshops.

The analysis would Identify frequently occurring terms in the content of the online articles and news articles for the interactive conflict reconciliation workshops

(IRW) including the search for content depending, on how it is being coded. Implementing cross tabulation for coded terms used in the online articles and to explore and examine correlations between terms in the IRW.

10.9.5.1 Word Count Analysis Technique for IRW – Auschwitz Trip Workshop –

In analyzing the online article and online news articles about the IRW Auschwitz trip that was conducted in 2014. It was essential to search for the most frequent words used and their terminologies in the news articles regarding Auschwitz trip workshop. Using NVivo Software for Windows, a query was used to generate results for all used words within the news articles regarding IRW workshop and how many times there were used (see Table 10.4).

The table demonstrates there are about 1100 words used more than once in the news articles regarding Auschwitz. The table shows that the term "Wasatia" terminology is used for more than 566 times. The word "Wasatia" was used more than any other word written in the news articles that were gathered from 250 Internet news articles. This illustrates that the interactive conflict resolution workshop "trip to Auschwitz" which was conducted by the Project Heart of Flesh Not Stone and had Wasatia Ngo participates in it, had the most observations by the news articles. The word count analysis illustrated that WASATIA was the most focus concerns in the interactive conflict resolution workshop in the news articles, and journalists were interested in Wasatia as an NGO and the Wasatia values and their interactivity to change the community.

10.9.5.2 Word Cloud Analysis Technique – Auschwitz Trip Workshop –

Another aspect of analyzing the news articles regarding the IRW workshop was to search the particulate terms that were used in the news articles, as illustrated in Fig. 10.18. The most 1000 common words, used by the news articles regarding the IRW workshop "Auschwitz trip by Palestinian Students," are illustrated in Fig. 10.18.

The impressionistic analysis illustrates that the news articles covered news about the Palestinians. In this case, it is the Palestinian students who went to Auschwitz, so we can conclude that the focus of the news was also on Palestinian students not on the importance of the trip. The word WASATIA has the most number of occurrences which interprets that WASATIA as a word, concept and as presenting Wasatia NGO received more focus from news agencies even though the Wasatia's values were not at the core. However, the use of WASATIA, definitions, concepts, and principles was present in all news articles regarding the interactive conflict resolution workshops. Whether news articles wanted to share the views of Wasatia, I interpreted the influence of the interactive conflict resolution workshop via exposure by the online news articles directed towards the Reconciliation process which was

Table 10.4 Word Count for Online News Articles, data analysis dated 20.July.2016

Word	Length	Count
Palestinian	11	676
Wasatia	7	566
Islam	5	499
Peace	5	450
الوسطية Wasatia in English	7	440
الله God	4	438
University	10	411
Israel	6	409
Palestinians	12	408
Students	8	404
Muslims	7	374
Muslim	6	342
Holocaust	9	337
World	5	320
God	3	315
Israeli	7	311
Auschwitz	9	308
Trip	4	299
Jewish	6	278
Palestine	9	269
State	5	265
Quran	5	264
Jews	4	252
2014	4	246
Islamic	7	231
Moderation	10	219
Prophet	7	211
Time	4	209
Believe	7	194
Group	5	189
Jerusalem	9	189
Israelis	8	186
Middle	6	182
Religion	8	177
Even	4	174
Reconciliation	14	174
الإسلام	7	174
هذه	3	168
Good	4	167
كما	3	164
Allah	5	163
Just	4	162

Table 10.4 (continued)

Word	Length	Count
Religious	9	161
Facebook	8	159
Right	5	159
Moderate	8	155
Conflict	8	154
Life	4	154
Visit	5	154

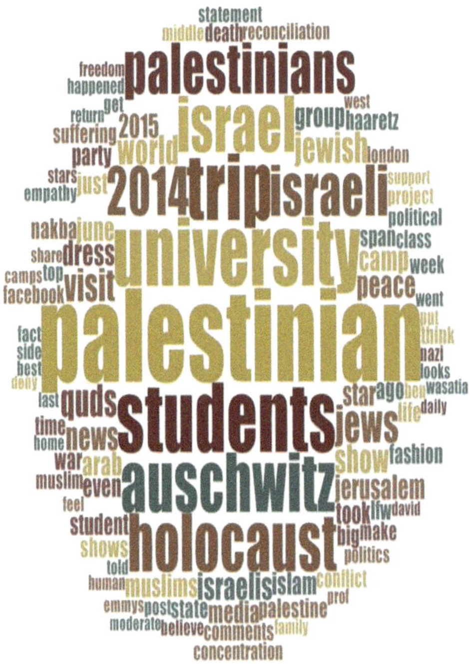

Fig. 10.18 The most used words utilized in the news articles about the interactive conflict resolution workshops "Auschwitz trip by Palestinian Students" – analyses dated 20. July 2016

eminent to the readers of the online news articles. I concluded from the analysis that the Reconciliation process was highly recommended by many news articles in the digital news agencies, using the online social network (Fig. 10.19).

10.9.5.3 Cross Tabulation: IRC – Auschwitz Trip Workshop –

News articles about Auschwitz trip by Palestinian Students, the sample is taken from the social network site www.wasatia.info, the analysis is for News Articles that wrote about the Auschwitz trip by Palestinian Students. Figure 10.20 shows the

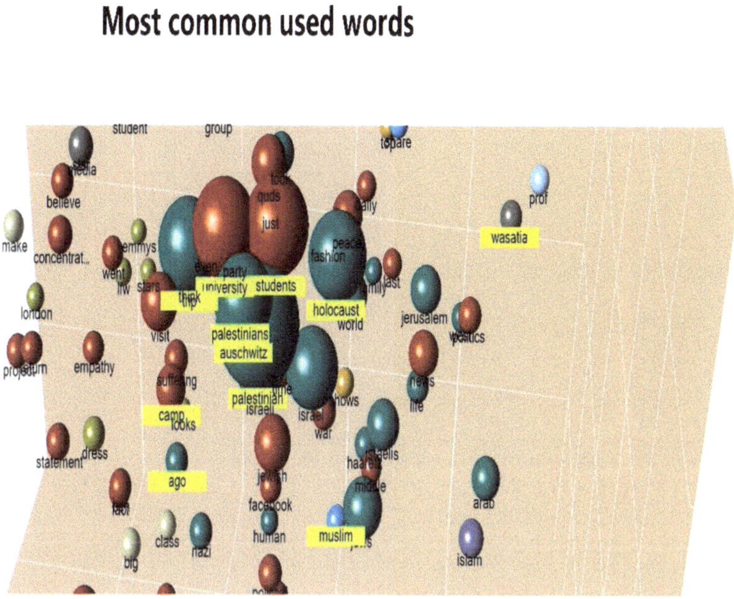

Fig. 10.19 Most commonly used words in the news articles about Auschwitz Analytics Dataset collected [20, July 2016]

percentage of coverage of the terms Wasatia in the news articles that were published about the Auschwitz trip by Palestinian Students. Using this technique involved developing indicators of coverage for terms used by a journalist to cover the Auschwitz trip by Palestinian Students. I concluded that the news journalists covered in their content were more about the Auschwitz trip by Palestinian Students than about the Wasatia as NGO and the Reconciliation process. However, I also found the Reconciliation process was mentioned in the article, which did help promote the Reconciliation process in the digital world.

10.9.6 *Analysis and Interpretations for Online Interactive Reconciliation Articles (OIRA)*

The techniques for analysis are to examine and explore the contingencies, inferences, and frequencies for the articles published online on the social network site and boosted in the Facebook Pages. The practical techniques applied for analysis are to interpret, where the appropriate term occurs in the content of OIRA written by various professionals in the field of moderation, reconciliation, and democracy to give a more understanding definition for the virtues of the reconciliation process

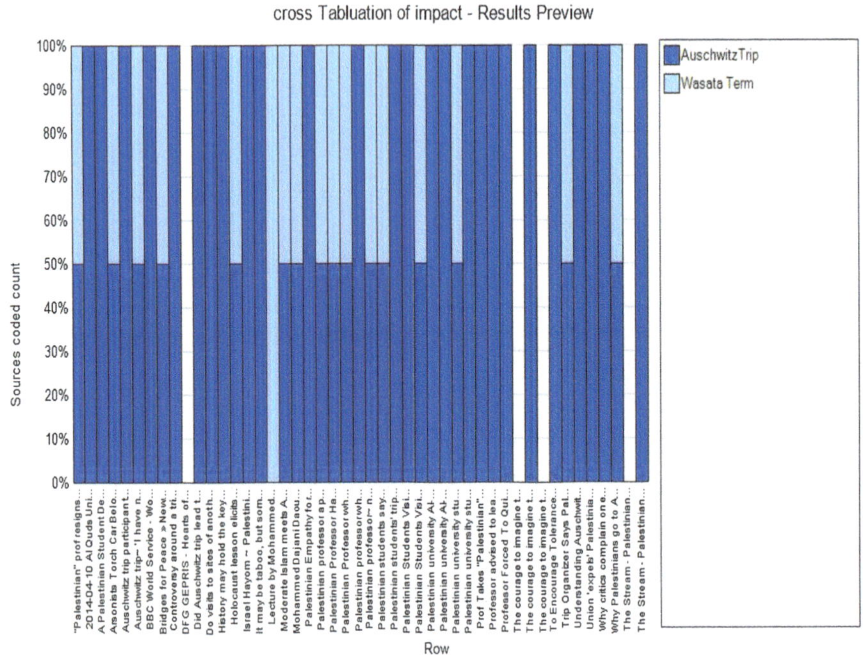

Fig. 10.20 Cross-tabulation chart for terms used in newspapers, Analytics dated 20.June.2016

to the visitor and members of the website. Other techniques for analysis involved identifying frequently occurring terms in the content of the OIRA. Search for content depending, on how it is being coded, including Cross tabulation for coded terms used in OIRA. The importance of the analysis was to investigate whether the OIRA give a more affirmative understanding to the concepts of moderation, reconciliation, and democracy to visitors and members.in addition to investigating the correlations of the OIRA with the members and visitors.

As part of the ICT-MR objectives adhere to the Reconciliation process to visitor members in site and advance them to become active members in Wasatia's work. The analysis of Wasatia articles proves whether the content adheres to that purpose. The terms to check if they occur in the context of the promotion of the Reconciliation process:

1. Reconciliation
2. Moderation
3. Wasatia
4. Moderate
5. Moderates
6. Peace
7. Peaceful

1		A : moderation term	B : peace term	C : reconciliation term	D : wasatia term
2	1 : ~How Do You Say Moderate in Palestinian~ Wasatia~ By Gil Troy	9.93%	2.03%	0%	3.21%
3	2 : About Wasatia Articles	0.37%	0.29%	0%	1.43%
4	3 : Arabic Articles	0%	0%	0%	0.71%
5	4 : Articles	0%	0%	0%	1.07%
6	5 : Articles About Wasatia	1.47%	0.58%	0%	2.14%
7	6 : Auschwitz Correspondance Letters (Palestinian Student Aschwitz Trip)~	0%	2.61%	3.85%	1.79%
8	7 : BIG DREAM~SMALL HOPE A Peace Vision Mohammed S. Dajani Daoudi	0%	41.45%	50%	1.07%
9	8 : Dr. Mohammed Dajani ["Moderation in all things." Cleobulus, Greek philosopher, 6th Century B	2.21%	0.87%	0%	3.93%
10	9 : Goals	1.1%	0.87%	7.69%	1.43%
11	10 : Intellectual Articles.	0.74%	0.58%	0%	1.07%
12	11 : Interview with Professor Dajani on Auschwitz	0.37%	0.29%	7.69%	1.43%
13	12 : Palestinian Students Say Auschwitz Trip was "for Educational Purposes Only" Written by Abd	0.74%	0%	15.38%	1.79%
14	13 : Religious peace between Jews and Muslims~ an interview with Mohammed S. Dajani Daoudi	2.94%	2.32%	0%	4.64%
15	14 : Seeking Moderation and Conciliation Mohammed Dajani Daoudi	6.62%	2.03%	0%	5%
16	15 : The Christian Science Monitor	0.74%	0.58%	0%	1.79%
17	16 : The Holy Books as Light and Guide to Humanity By Mohammed S. Dajani Daoudi	0.37%	1.74%	0%	0.71%
18	17 : The Role of Wasatia to promote education in the Palestinian culture~	2.21%	2.03%	7.69%	5.36%
19	18 : WASATIA~ The Middle Road	8.82%	2.03%	0%	12.5%
20	19 : Why Palestinians Should Learn About the Holocaust By MOHAMMED DAJANI DAOUDI and	0.37%	1.45%	0%	1.07%
21	20 : www.wasatia.info~images~centrismandmoderation.pdf	18.38%	4.93%	0%	6.07%
22	21 : www.wasatia.info~images~wasatiathespiritofislam.pdf	42.65%	33.33%	7.69%	41.79%

Fig. 10.21 Tabulation analysis for terms occurs in news articles; Analytics dated 20.6.2016

The referencing of the terms of the Reconciliation process indicates that those particular terms are occurring in OIRA – refer to the NVivo Appendix – Wasatia articles and Wasatia reference analysis on the CD.

To analyze which term of (Wasatia, moderation, reconciliation, peace), has more coverage and more use of words that provide inferences in OIRA. This kind of analysis is called cross tabulation analysis between the OIRA and the terms, Wasatia, moderation, reconciliation, and peace. The purpose is to create indicators to examine if the OIRA on the social network site and to examine if they expose visitors and members to the Reconciliation process.

Figure 10.21 illustrates the cross-tabulation analysis technique between the ORIA and the terms Wasatia, moderation, reconciliation, peace and provides percentages of inferences toward the Reconciliation process.

Figure 10.22 illustrates the evident influence of the terms Wasatia, moderation, reconciliation, peace. It demonstrates the comparison of the coverage of those terms in the OIRA. This also indicates that the Reconciliation process is being explored and exposed through the content of OIRA in the social network site www.wasatia.info.

The chart illustrates the comparisons for the influence of the terms (Moderation, peace, reconciliation, Wasatia) in the IROA exploring the concepts and their terminology. As illustrated by the chart the term Wasatia, has more coverage and referenced more in the OIRA than the other terms, this also interprets that the OIRA in the social network site promote toward Wasatia terminologies as well promotes moderation reconciliation that is part for paving the for social change towards the Reconciliation process.

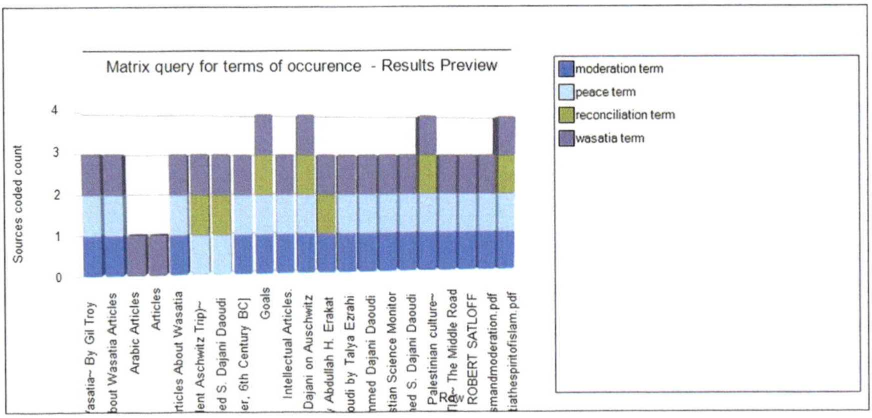

Fig. 10.22 Chart of comparison for the coverage of terms, analytics dated 20.June.2016

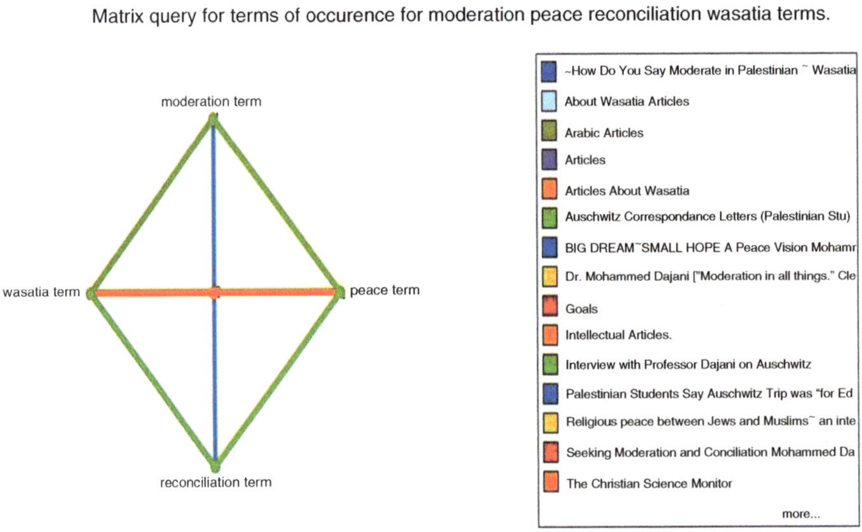

Fig. 10.23 The indicators for the correlation between terms in the articles published online in the social network website with posts in the Facebook Page, Analytics dated 15. July.2016

Applying online matrix analysis for the social network site to assist interpreting in depth the OIRA in the social network site, and its correlation with provisions for the Reconciliation process that are as posts published on the Facebook Page. The correlation and association of the Reconciliation process terms and the published articles indicate that members and visitors are exposed to those terms respectively.

Figure 10.23 illustrates, **the exposure for the terms in depth, that provides the existence of correlations of terms in all OIRA, that proves that visitors and**

members are exposed for social change towards the Reconciliation process. The figure also illustrates that all terminologies **are transdisciplinary of each other**. The qualitative data structure (QDS) is presented in Figure below, which shows the results for the correlation of terms of the Facebook Page and the Social Network Website in the ICT-MR communication research. The figure shows each connected dot, comprehend to an association between articles related to the Reconciliation process terms in the online materials.

Applying qualitative content analysis to the OIRA in the social network website investigates the correlation between the terms used in the Facebook Page in the comments and the published articles in the social network website. It proves whether OIRA adheres to social change towards the Reconciliation process using the Facebook page for that purpose, it also interprets that members and visitors are exposed to those terms respectively. Figure 10.23 interprets that the term moderation leads to the expose the Wasatia terminologies as well as disclosures of terms such as peace and reconciliation terminologies. Finally, this proves that OIRA are exposing visitors and members for social change towards the reconciliation process, as evident from the qualitative content analysis techniques.

The correlation coefficients are a descriptive analysis, that indicate the strength or the weakness of relationship in the research. The Fig. 10.23 indicates the correlation between most used terms in the Online Interactive Reconciliation Articles in the online social network website, with the datasets from the Facebook page. We discover the coefficient correlation from the results of the Matrix Correlation technique done by Nvivo software for windows.

The interpretations of the Fig. 10.23 are as follows:

R would signify the correlation between the **Most used terms in the Datasets in the Facebook page** with the **published online articles** in the social network website.

R_1: Moderation is in correlation with the terms Wasatia and Peace, in online datasets from qualitative content analysis results of Matrix Correlation Techniques.

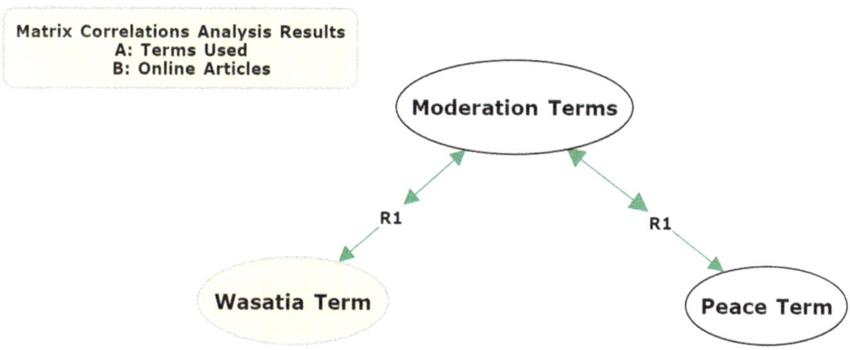

R_2: **Moderation is in correlation with reconciliation** from the qualitative content analysis results.

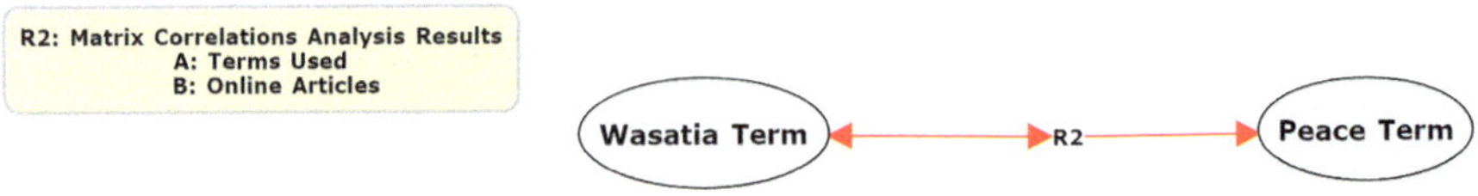

R_{3}: **Wasatia terms in correlation with reconciliation** and **with the term peace.**

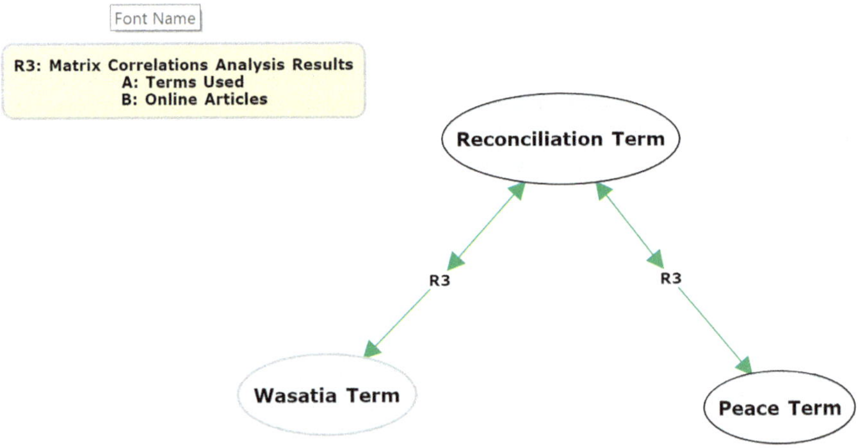

R_4: **The Wasatia term is in correlation with Peace Terms**.

$$H_0 = R1 + R2 + R3 + R4.$$

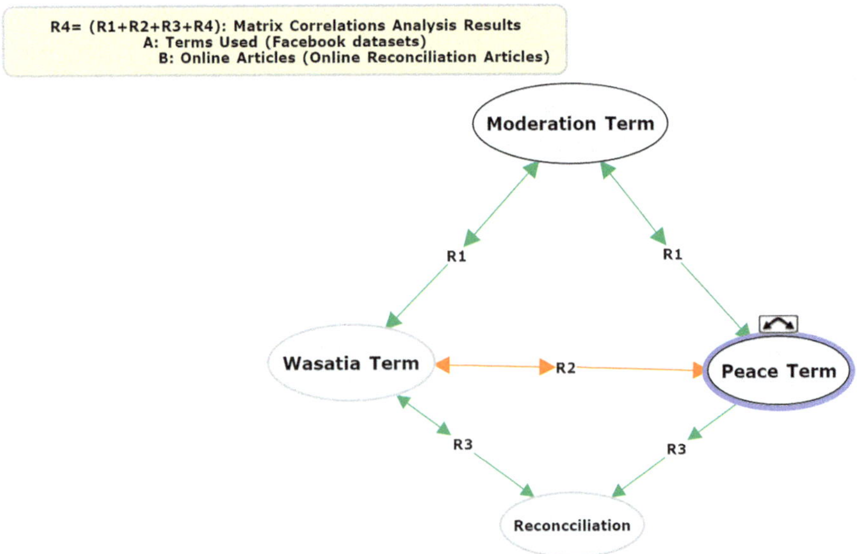

They hypotheis as invistigated is being approved by the matrix analysis ***that ICT can foster the reconciliation process for both visitors of the online social network and the Facebook Page, they illustrate the Wasatia Cycle described in Chap. 5.***

10.9.7 Visual Content Analysis Methods

Concerning the ethical consideration for conducting such an analysis. The photos used for the analysis procedures are owned by Wasatia NGO in Palestine; therefore, they can be used in the analysis procedure. According to the ethical consideration stated in Chap. 7, there is no human body harm for such an analysis method, and it is only conducted for the purposes of the research. That gives the right to conduct the analysis on photos published on the social network site www.wasatia.inof, and the Facebook pages, Wasatia.info and Wasatia. Vision.

The methods of analysis applied in the discourse of analysis were accredited to a pre-conference workshop for the AOIR2017 in Estonia, which I attended. I am a member of the Association of Internet Researchers. The workshop was done by Maria Schreiber, introducing visual social media methods, and Uta Russmann, introducing corpus/data. Figure 10.24 illustrates the method for visual analysis, as demonstrated in the workshop.

According to Russmann, the methods for analysis were conducted through measured principles, and these included perception, image management, and integration.

1. Perception can be explored by three aspects, perspective, broadcasting, and mobilization.
2. Image management is analyzed through personalization, privatization, and celebrities.
3. Integration is analyzed by hybridity, shared content, and campaign references.

Online social platforms, such as Facebook pages, Twitter, and Instagram, were visually analyzed by conducting the following principles:

1. Interactivity: Number of likes and posting, which is done by the owner of the page and comments by followers.
2. The content of captions by the poster, including measurement of emotions, posting with intrinsic values or posting without intrinsic values.
3. Negative and positive tonality and attitudes.
4. Reciprocity, where reactions are related to previous comments or related to other comments or these are separated and have a new dilemma.
5. Comments by followers, including the coding scheme.

Some samples to conduct the visual analysis methods were taken as snapshot photos from the social network site http://www.wasatia.info (see Fig. 10.25 for the first photo).

To conduct the visual analysis, I follow the principles stated before to conduct a visual analysis for the images of the professor, the first part is perception, as we can

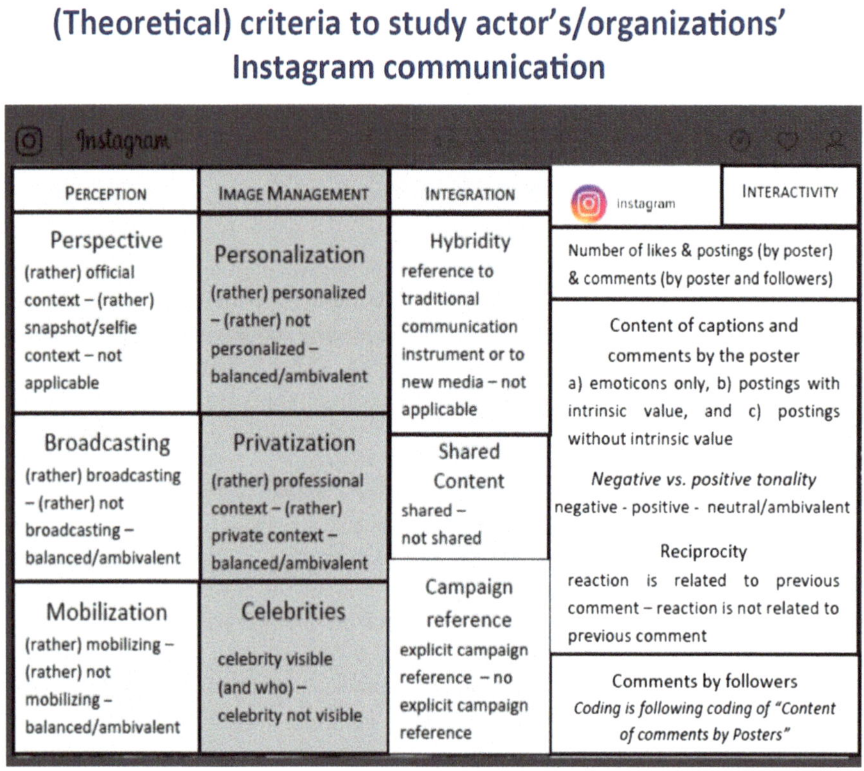

Fig. 10.24 Visual analysis methods. (Source: Pre-Conference workshop at the AOIR-2017 in Estonian work accredits to Uta Rossman workshop name, The date of the Workshop dated October 2017)

see the image in the picture is a person visiting Auschwitz camp. The perspective in the picture was taken as part of the trip to Auschwitz, which illustrates a personal experience. Broadcasting, the image illustrate a professor in the Nazi camp known by Auschwitz Mobilization; the image illustrates a trip to Auschwitz camp in Poland.

Image Management; can be analyzed by applying those criteria's:

1. Personalization: This is a personal picture of the professors; he is the only one in the picture.
2. Privatization: This picture is not private, it is in a professional context, and the visit to the Auschwitz camp was with his students.
3. Celebrities: none celebrities, does not make him famous.
4. Integration: can be analyzed and interpreted through the following criteria's:
5. Hybridity: Traditional communication aspect, Palestinian usually does not visit Auschwitz, and it is a Taboo, in the Palestinian cultures.
6. Shared Content: the picture is widely shared as it is in the first photos of a slide-show on the social network site.

Fig. 10.25 Professor Dajani in Auschwitz, the with the consent of Professor Dajani to use this photo for analysis

Fig. 10.26 Meeting with students at Wasatia office – screenshot dated July.6.2013

7. Campaign reference: The professor does this visit with his Palestinian and the picture is a proof and reference to the visit.

The second photo is in Fig. 10.26.

Using the principles stated before to conduct the visual analysis and they are as follows:

Fig. 10.27 Workshops with Palestinian students – screenshot dated 2. Feb.2014, consent approved by Wasatia NGO

1. Perception: The view is in the Wasatia office in Jerusalem with a scene of a lecture.
2. Perspective: An official picture of lecturing students, It was taken at his lecture in the Wasatia office.
3. Broadcasting: A lecture with students.
4. Mobilization: There is no mobility in the picture
5. Image Management: can be analyzed by applying those criteria's:
6. Personalization: The picture is not a personal picture it is a professional one.
7. Privatization: The picture is not private; it is in an open space, a lecture room.
8. Celebrities: none celebrities, none are famous.
9. Integration: can be analyzed and interpreted through the following criteria's:
10. Hybridity: Traditional communication aspect, a professor giving a lecture to a student at his office.
11. Shared Content: the picture is widely shared, as it is in the slideshow on the social network site.
12. Campaign reference: it is not referencing any events.

The third picture is presented in Fig. 10.27.

Using the principles stated before to conduct the visual analysis and they are as follows:

1. Perception: The view is a photo taken with students and his professor with a guest, who presented his workshop.

2. Perspective: The picture is an official picture, it includes students and his guest, and they were doing a workshop.
3. Broadcasting: Palestinian students with a famous guest, introducing his book and his movie, "*Lost Stories from the Holocaust's Long Reach into Arab Lands*."
4. Mobilization: The place is at the university.
5. Image Management: can be analyzed by applying those criteria's:
6. Personalization: The picture is not a person, includes a professor, students, and a guest.
7. Privatization: This picture is not private, it is in a professional context. It was a lecture or a workshop with students.
8. Celebrities: the person in the picture is Professor Dajani and Robert Satloff who presented his new movie about Arabs saving Jews in the time of the Nazi-Germany.
9. Integration: can be analyzed and interpreted through the following criteria's:
10. Hybridity: Traditional communication aspect, Palestinians taking a workshop with Robert Satloff a famous professor and the director of the Washington Institute for Middle East studies.
11. Shared Content: the picture is widely shared as it is in the first photos of a slide-show on the social network site.
12. Campaign reference: references the workshops done, as part of Wasatia empirical work toward moderation reconciliation in the community.

10.10 The Validity of the Analysis of the ICT-MR

Content analysis is valid if the inferences from the qualitative data analytics which resulted from the online materials, measures and investigates the research question and examining the hypothesis. In our QLCA, we measure whether the content of the ICT-MR indorses toward the Reconciliation process and changes the members from the intentionality less for moderation and reconciliation to intentionality's that are more for moderation and reconciliation. As the growth of followers on the Facebook page and the registration in the social network sites are increasing, shows that our measure to prove the hypothesis is realistic. Krippendorff (2013) stated, "*Validation provides compelling reasons for taking the results of scientific research seriously*" (p. 329).

10.10.1 Evidence of Sampling Validity

Krippendorff (2013) defined this kind of validity: "*it is in concerns to the degree of text accurately represents the population of phenomena in whose place it is analyzed*" (p. 335). The evidence in the analysis of the ICT-MR depended on the material that was taken from the social network site http://www.wasatia.info and the

Facebook page/wasatia.info and the Wasatia Vision; all were in correlation with each other. The samples collected for analysis were generated from the influence of the social network site and the Facebook pages, and these developed the datasets for analysis and interpretations. According to Krippendorff (2013), the *"text becomes available by their source and choice and contains intentional biases in representing the phenomena of their interest"* (p. 335). I used a qualitative analysis software that created the QDS for the ICT-MR Internet communication research.

10.10.2 Evidence of Semantic Validity

Evidence on semantic validity "*is the extent to which the categories of an analysis of texts corresponds to the meaning these texts have within the chosen context*" (Krippendorff 2013, p. 338). This ICT-MR illustrated the semantic validity in collecting datasets from the social network site with contents published online by Wasatia NGO, and all content promoted the Reconciliation process that the NGO Wasatia endorsed. The datasets had to mean and contributed to the inducement of the Reconciliation process, as the data were part of the work completed by the NGO Wasatia.

10.10.3 Structural Validity

Structural validity is "*the structural correspondence between available data or established theory and the modeled relationships*" (Krippendorff 2013, p. 335). The theory was to endorse and advance the community from intentionality that was for hate and enmity toward intentions that were more for the Reconciliation process.

10.10.4 Functional Validity

Functional validity involved analyzing the functionality correspondence between the data and the behavior in contrast that it frequently shares underlying correspondence within the data (Krippendorff 2013). The functional validity in ICT-MR was presented by the use of CAQCAS, which was the NVivo 10 software for Windows that was utilized to conduct analysis and interpretations. The CAQCAS assisted in the interpretation and indicated that the data promoted the Reconciliation process, as these originated from articles and books promoted by Wasatia NGO.

10.10.5 Correlative Validity

Correlative validity *is an epistemological fact* (Krippendorff 2013, p. 336). This was illustrated in the analysis of the newspapers regarding Auschwitz trip by Palestinian Students, where the news agencies on the Internet wrote only about the trip, but Wasatia terms and terminology were always mentioned. This indicated the correlative validity of the data that was used in the Auschwitz trip by Palestinian Students, but either term could not distinguish the other.

10.10.6 Predictive Validity

Predictive validity derived from the matrix query for terms analysis technique in the previous section, which proved the content indicated the material analyzed interpreted that the articles promoted the Reconciliation process in the ICT-MR Internet communication research. The prediction was valid, as the analysis of the content of the ICT-MR validated the phenomena studied, which was the hypothesis for this research.

10.11 Reliability of the Analysis of the ICT-MR

Reliability means that different analysts can use the data generated for content analysis: "*The data are to be trusted and has a ground of reasoning and have been generated with all conceivable precautions*" (Krippendorff 2013, p. 270). This ICT-MR used the CAQCAS software NVivo 10 for Windows to produce data from the site and from the Facebook pages, which satisfies that the data, generated from the content analysis, were reliable. As Krippendorff (2013) stated, there are three types of reliability, and these are the following:

1. Stability,
2. Replicability, and
3. Accuracy.

10.11.1 ICT-MR Stability

ICT-MR in the content analysis represents stability and replicability, thus representing reliability. For example, the analysis of Wasatia articles illustrates the coding procedure yields the same results on repeated trials as it appears in the previous sections of analysis techniques. The tree-maps, word clouds, clustering, correlation units for analysis, cross-tabulation analyses, and charts that illustrate the same results of interpretation. The material endorses the Reconciliation process for the members and visitors of the social network site and the Facebook pages.

10.11.2 ICT-MR Represents Replicability

The use of the software for analysis can illustrate the replicable process. NVivo 10 for Windows created the same procedure that could be used by different analysts working under varying conditions and yields the same results. Such procedures of analysis involved analysis where the particular term occurred. This analysis technique is shown in the word TreeMap when analyzing the datasets from the Facebook pages and the social network site. Moreover, I identified frequently occurring terms. This was shown when analyzing which common terms occurred on the Facebook page as well in the news articles and in the online-published articles and materials that are on the social network site. I also conducted a cross-tabulation analysis. This process was illustrated in the published online articles analysis and could be used as a process with different analysts yields the same results.

10.11.3 The Accuracy of Analysis in ICT-MR

I used the software for generating results for interpretation and finding indicators, inferences, and frequencies, which could permit different analysts to reproduce the process. In addition, if the analyst is working under varying conditions and different locations or using different but functionally equivalent measuring instruments such as the CAQCA, develops an accurate analysis. In the ICT-MR, the NVivo 10 for Windows was the instrument for qualitative content analysis for the research that developed accuracy in analysis from the dataset in both Facebook pages and the social network site.

10.12 Conclusion

From conducting the qualitative content analysis, different techniques were utilized, such as word counts, word trees, word clouds, treemaps, clustering units, referencing, cross-tabulations, matrix query, graphs, and charts.

The Internet communication research proves that ICTs can have a role in promoting Moderation, reconciliation democracy within and between nations that are in the middle of the conflict. The ICT-MR was successfully developing an ego-centered social network that adheres to the values of the Reconciliation process. Implementing validating procedures, and a follow-up of interpretations for the

Qualitative and Quantitative data using the Experiment -Intervention Mixed-Method Research Design approach for (Creswell and Plano Clark 2018).

The results illustrate that Internet communication research successfully reached a considerable amount of people within the Palestinian community, whether they are members or followers and encourages social change towards the Reconciliation process. The research demonstrates that social networks and its applications are a tool for a deliberation process. Therefore, the online social network platforms applied deliberative democracy over the Internet, which could be used in a different aspect of deliberation. The research utilized ICTs for exposing online members and visitor to social change towards the Reconciliation process in the Palestinian community in the middle of the conflict. The research applies different methodologies as an in a tribulation process in depth and introduces the theoretical synthesis in applying Mixed Method – Netnography – Qualitative Content Analysis.

The purpose from applying the theoretical synthesis is to investigate the inducement to social change towards the Reconciliation process from the online material of the social network site www.wasatia.info, and the dataset generated from both Facebook pages (Wasatia.info and Wasatia vision).

By applying the theoretical synthesis to conduct the ICT-MR communication research, the method investigates the online materials and demonstrates an understanding of what online material endorses such as the IRW and OIRA. The analysis proved that ICT-MR communication research exposed visitors and members online and offline to social change towards the Reconciliation process and promoted the ideologies and concepts of moderation, reconciliation, and democracy in the social network website www.wasatia.info and the Wasatia.info Facebook Page. Conversely, the other Facebook Page Wasatia Vision promoted the religion of Islam and understanding of Islamic terminologies, regardless of Reconciliation process values, as well as promoted toward Wasatia as a religious concept.

The ICT-MR Internet communication research proves interactions between users on the Facebook page (wasatia.info), illustrating an interactive reconciliation workshop dialog, which was present between participants and visitors and they have deliberated on Reconciliation process virtues.

The ICT-MR was measured for its validity and reliability; the conclusion is that it yields both aspects of validity and reliability for the analysis and interpretation conducted. The validity and reliability were measured using methods introduced by (Krippendorff 2013, pp. 271–354) as it complies with Krippendorff validity and reliability concepts.

The ICT-MR communication research was the interactive reconciliation workshop that exposed social change towards the reconciliation process in the Palestinian community. In addition, the research proved that the members in the online social network, were interactively active and deliberated on virtues of moderation, reconciliation, peace, democracy, this proves that members and visitors attitudes are more exposed toward the Reconciliation process, which has more realistic to change their intentions. According to Fisher (2001), interactive conflict resolution workshops induces a cognitive change in participants way of thinking. The ICT-MR communication research developed a cognitive change in the direction of the realization of the reconciliation process using ICTs in applying the IRW. Therefore, the interactive participants in the online social network became members in the social network website and started requesting to become active in the interactive reconciliation workshops that are done by Wasatia NGO in the community. Finally, this illustrates the change of intentionalities to more for the reconciliation process within the Palestinian community and illustrates move evident to proof the hypotheses of the communication research.

Chapter 11
Research Findings and Recommendations

The figure illustrates the categories researched in this chapter:

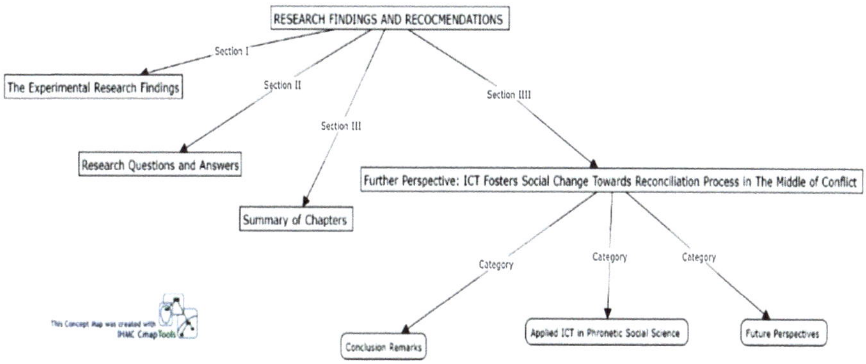

11.1 The Experimental Research Findings

This ICT-MR developed interdisciplinary research with many possible theories, methods, and application related to the analysis of social networks, such as social network site, like Facebook, Twitter, YouTube, relating toward democracy, peace, reconciliation, and moderate Islam within the Palestinian community. The objectives of the ICT-MR Internet communication research are to preserve a habitation such as fostering an online community, where Palestinians can participate and interact within the advancement for the understanding for social change towards the Reconciliation process in the Palestinian community. In addition, the researcher used Internet communication technology for the purpose of exposing the concepts

I. M. AlDajani, *Internet Communication Technology (ICT) for Reconciliation*, Lecture Notes in Social Networks, https://doi.org/10.1007/978-3-030-41203-6_11

of moderation, reconciliation, and democracy in the midst of the conflict. **The added value from the experiment was to develop an online social network website in the Palestinian community, where different public Palestinian groups could share, interact, post, and access to different materials to gain information about moderation, reconciliation, and democracy, resulting in social change towards the reconciliation process in the middle of the Israeli-Palestinian conflict.**

The achieved results of the ICT-MR Internet communication research:

1. The development of a data mine for terminologies and information about moderation, reconciliation, and democracy, to expose social change toward the Reconciliation process in the community.
2. Fostering media pluralism in an online social network website and the Facebook pages, by developing a more understanding for the virtues of the Reconciliation process utilizing interactive reconciliation workshops and online interactive reconciliation articles in reconciliation forums, reconciliation chats, reconciliation polls, reconciliation blogs, reconciliation videos in You-Tube and finely reconciliation Facebook pages.
3. Investing in the public's minds to become an online social network literate and change an aspect of the Palestinian community to a virtual network literate community that has the mind to interact with each other online, through the online social network to expose the ideologies and terms of Reconciliation process between each other. For example, the visits to the online social network website according to a statically online software provided by the server as illustrated in Chap. 7, over two million visitors in the year of 2014, and then went down in the 2015 to nine hundred visitors, due to a trip to Auschwitz that was done by the Heart of Flesh not Stone Project, that was condemned by some important members of the Palestinian community. But on the contrary a lot of Palestinians over one thousand participants in the online social network www.wasatia.info and Facebook Page Wasatia.info. Participants and visitors expressed that they are being pleased with such visits, and would like to participate in more visit to Auschwitz Nazi Camp, they more expressed the wanting to understand the narrative of their enemy for wanting to change their notion from animosity and hatred into understanding of the real reality of the other, this is part of reconciliation, and proof that internet communication technology can pave the way for reconciliation in the middle of conflict.
4. ICT-MR constructed data analysis system for social networks using CAQCAS such as NVivo 10 for Windows that implemented Qualitative Content Analysis on the Internet communication research as implement and practiced in the previous chapters, to investigate and research the datasets and themes that emerged from conducting the Netnography and examine the data's correlations with the hypothesis of the communication research.
5. ICT-MR Internet communication research as a contribution for new researchers demonstrates how to use the Phronetic Netnography Mixed-Method Approach applied to research, theoretically and the practice was sufficient.

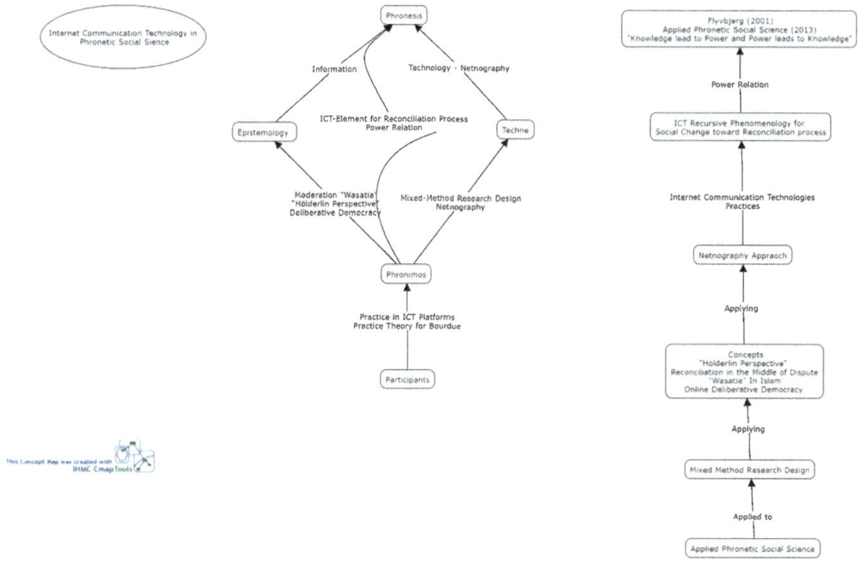

Fig. 11.1 Synthesis of theoretical ideas for social change towards the reconciliation process

6. ICT-MR Internet communication research explores and examines big data by applying a mixed-method – Netnography – Qualitative content analysis to conduct Internet communication research. Developing the synthesis of theoretical ideas that includes three approaches (Mixed-Method Research Design, Netnography, Qualitative Content Analysis).
7. The theoretical framework to conduct communication research is illustrated in Fig. 11.1. This applies to the discourse of the Internet communication research analysis, illustrating new terms, such as Phronetic Mixed-Method Netnography Approach[1], developing new methods for online social networks analysis[2].
8. Figure 11.2 illustrates the results of PMMN see Chap. 7, as applied to the social network site and the Facebook pages, which illustrates that members and visitors were exposed to Reconciliation process values. (described in detail in Chap. 10)

The figure indicates that the terms and concepts of the Reconciliation process related to the green lines, and have been exposed in IRW and OIRA, the terms are (moderation, Wasatia, Reconciliation, Peace). The blue line indicates that two specific terminologies are exposed in the several articles together (moderation, Reconciliation) and both terminologies are also part of the Reconciliation process, the orange line indicates the terminologies that were exposed together in several articles (Wasatia, Peace). As a result, the analyst can comprehend that the social

[1] See Chap. 10.

[2] See Chap. 8.

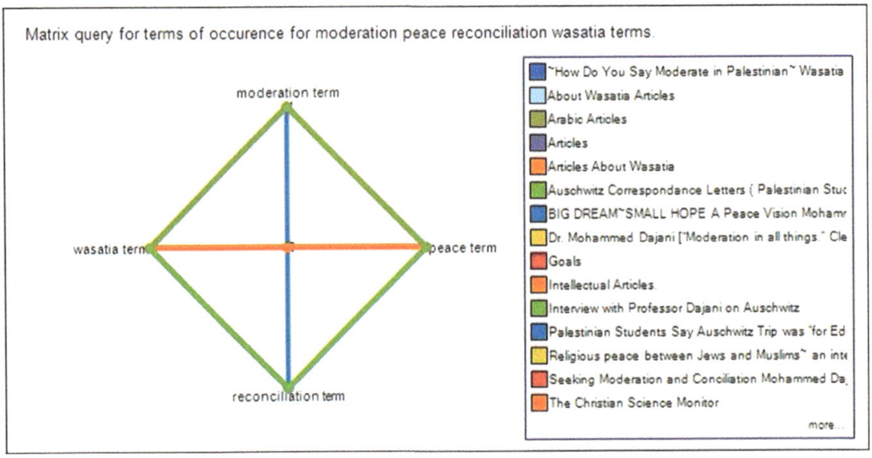

Fig. 11.2 Results of PMMN

network site fosters the terminologies and concepts indicated in the Reconciliation process in the middle of the Israeli Palestinian conflict.

Altering the Palestinian community to a better understanding of the Reconciliation process and adhering the understanding of Wasatia NGO Goals utilizing ICT for those goals. Which are as follows:

(a) To bring a broader and more rational understanding of Islam to Moslems as well as to non-Moslems.
(b) To clarify the distortions to which Islam has been subjected at home and in the West.
(c) To educate Palestinians on taboo topics such as the Holocaust taking a new humanistic approach.
(d) To seek answers for the profound religious, political, social, and economic crises inflicting the Palestinian society.
(e) To strive and work for ending the Israeli military occupation through negotiations and peaceful means.
(f) To spread and promote Islamic tolerant concepts, values, and principles within the Palestinian community.
(g) To encourage the practice of moderation among Palestinians in order to mitigate religious radicalism and bigotry and reduce political extremism.
(h) To bring a message of peace, moderation, justice, coexistence, tolerance, and reconciliation to the Palestinian community through vocal civic leaders.
(i) To teach creative and critical thinking and open-mindedness.
(j) To empower the potential for leadership in their society. The goal of dialogue and education is to deconstruct mythologies and distortions and misinterpretations and to promote knowledge and empathy for the other.
(k) Wasatia NGO addresses all aspects of life, the way we eat, the way we dress, the way we spend money. Moderation is a value shared with the various thinkers

and philosophers as well as all faiths and therefore could become a fruitful foundation for dialogue to achieve peace and reconciliation.

ICT-MR Internet communication research meets the "Heart of Flesh, Not Stone" project's objectives by achieving the necessary goals in promoting social change towards the Reconciliation process and developing an ego-centered social network that has at its core the values of the Reconciliation process The Internet communication research developed a methodological discourse for analysis according to expert and scholars in the academic research, the figure below illustrates the findings to conduct internet research methodology. (see also Chap. 9 for details.)

According to Hollenbeck, "*Reconciliation is a relationship state in which parties no longer harbour animosity toward the other and agree to behave in a manner that does not threaten the other basics human rights*" (Hollenbeck 2001, p. 76). in the social network online website and the Facebook Pages, the users and visitors became members, and all members have an understanding that they are members will their fellow Palestinians, and their Israeli neighbors that want moderation, reconciliation, and democracy, and have all those members registered in the site, proves that we have such a reconciliation process that is in its start, there is 250 members in the online social network and over 9000 members in the Facebook Page Wasatia.info.

The other aspect is the IRW – Auschwitz Trip -, after the visit to Auschwitz there was a breakdown, where the Professor, the leader of the Palestinian team, resigned his job from pressure onto him by the administration of the university in the following year, and his car was torched by an anonymous. In the online social media Facebook Page Wasatia.info, you find that Palestinians still want to go to Auschwitz concentration camp in Poland and understand their enemy without wanting to harm their enemy, this is fertile and a start for the reconciliation process. This kind acceptance and willing to reconcile, in accordance to Kreisberg, reconciliation "*refers to the process but which parties have experienced an oppressive relationship or a destructive conflict with each other move to attain or to restore a relationship that they believe to be minimally accepted*" (Kreisberg 2001, p. 48).

11.2 Research Questions and Answers

Questions and answers are a type of evaluating the process to the internet communication research to tackle each part of the communication research.

The communication research questions and answers are a constructed type of evaluation according to (Corbin and Strauss 1990) investigating the following aspects:

1. Methods to conduct the research.
2. Exploring methods and procedures to conduct the research.
3. Type of coding methods in the research.
4. The research procedures.
5. Empirical findings.

The question and answers are as follows;

What are the methodologies used in Internet communication research?

The mixed-method experimental intervention research design which was conducted as a discourse mechanism to apply ICT in a phronetic research procedure, the concept is discussed in Chap. 1. In addition to a netnography approach and qualitative and quantitative approaches. Recording behaviors and impressions of visitors and partisans in the online social network. The methodology included validity and reliability process and constraints. The methodology also included collecting data, investigating, analyzing and interpreting the datasets as they emerged from the research. (See Chap. 1).

What are the procedures used to collect data in Internet communication research?

Data collection methods and procedures used validation and analysis procedures discussed in Chap. 1 as illustrated by (Creswell and Clark 2018). The Analysis and the datasets are in an interrelated process. The datasets were collected from the social network site, www.wasatia.info, and the two facebook pages, Wasatia.info and Wasatia Vision. The social network site was integrated with the Facebook Page Wasatia.info, where posts were boosted and targeted to a specific audience. (see Chap. 2, Figs. 2.9, 2.10 and 2.11).

What are the generating concepts for coding that were adopted in the research?

1. Concept-driven
2. Data-driven
3. Combining concept driven and data-driven coding.

What are the types of coding used in the research?

4. Open Coding.
5. Axial Coding.
6. Selecting Coding.

Chapters 8 and 9 explores the coding frames and coding procedures in details.

How was the procedure implemented?

The selecting of the datasets was through downloading all web pages and articles from the social network website and converting them into.pdf files for articles and jpeg for images, the use of the software of Nvivo software assisted in the process of collecting, analyzing and interpreting the data.

The major categories in the process were www.wasatia.info, Wasatia.info Facebook Page, and Wasatia Vision Facebook Page. The behavior and impression of the participants in the social network site and the Facebook pages, where more observed to terms such as Wasatia in Islam, Peace, Reconciliation, Democracy, Wasatia NGO, Auschwitz-trip, those terms were used as sub-categories for investigation, analysis, and interpretations. (See Chaps. 7, 8 and 9 for more details). Also, see the photo below that explains the discourse analysis for the ICT-MR.

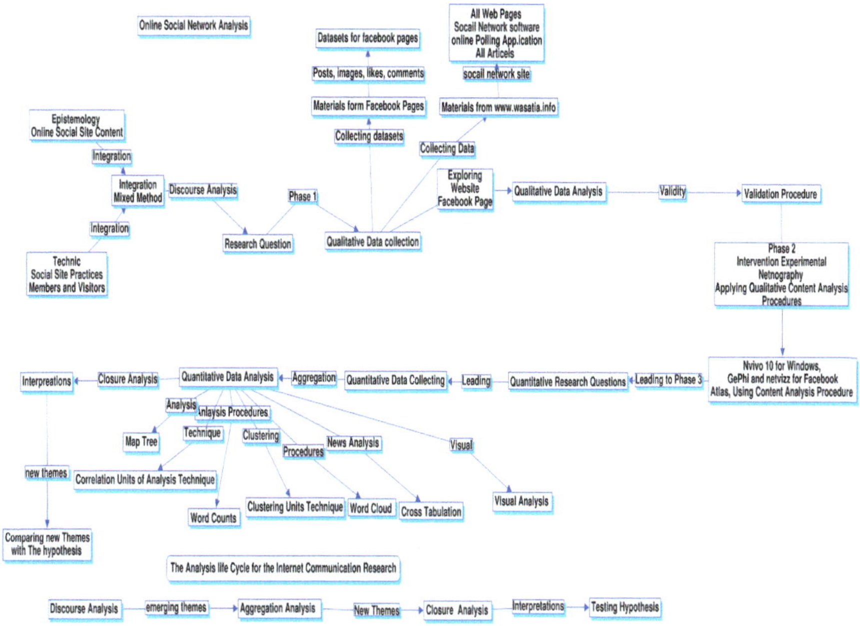

Discourse Analysis for online social network

Where the hypotheses of the research pertaining to relations among categories?

Yes, the hypothesis was to investigate and explore the impact of the online social networks for social change towards the reconciliation process, which was defined as a process from moderation to reconciliation leading to democracy. The terms that were investigated were Wasatia as NGO, Wasatia in Islam, Wasatia as a moderation, Reconciliation, Democracy, peace, those terms were correlated with each other, as the analysis illustrates in Chaps. 9 and 10, which relates to the main hypothesis in the research,

How were the discrepancies accounted for using online social media? How did they relate to examining the hypotheses?

The hypothesis concentrated on exploring the ICT for Reconciliation in the positive behavior and outcome but did illustrate the disadvantages arising from online social media, but it was not the focus of the study. The study was to understand how much focus can social media impact for social change towards the reconciliation process in a society that is in conflict, and if the impact would be changed from animosity to more empathy and reconciliation and understanding the narrative of the other. Whether the impact was positive or negative, the hypothesis stands correct and supported by conducting internet communication research. The research proves that ICTs can be used as a tool for social change, in the positive and in a negative way, in that regards ICT can be a tool for social change for the reconciliation process as in this case of the study. ICT develops a mechanism for the reconciliation process in the middle of the conflict, as stated in previous chapters.

What are the theoretical and empirical findings in the research?

There were generated some concepts, and they are as follows:

- The phenomenology of ICT in Applied Phronetic social science, this concept is illustrated in Chap. 2 of the research, it is an evolved concept from applying Flyvbjerg approach in social science referred to it as the applied Phronetic social science, also discussed in details in Chap. 2.
- Applied Phronetic Internet research theory, it is illustrated in Chap. 2.
- The mixed-method research design for internet communication research also explained in Chap. 2 of the research and in Chap. 9.
- Phronetic Netnography Approach for online social network analysis explained in Chap. 7.
- Theoretical synthesis emerged from conducting the research; the synthesis was evolved from conducting a mixed-method research design, netnography, and qualitative contact analysis procedures and methods. This approach was referred to it as the Phronetic Mixed-Method Netnography approach to conduct an academic internet communication research. The concept is more explained in Chaps 2 and 7.
- Applied Wasatia Cycle, in Chap. 5, exploring the relation between moderation, reconciliation, and democracy in a systematic social change to develop social transformation towards the reconciliation process by applying the Phronetic social science using Internet communication technologies.
- Qualitative and quantitative social network approach for analyzing digital social media. Such as the social network website and the Facebook Pages.
- ICT for social change towards the reconciliation process in the middle of the conflict.

What are the political intentions for consciousness-raising from conducting internet communication research?

The research subjugated the exposure of Reconciliation process values and terminologies in the Palestinian society from the use of ICT platforms for that purpose. The research applies ICT-mechanism for reconciliation within the online debates of the participants and deliberations. ICT developed a virtual community that adheres to the values of the Reconciliation process. The exposure for the value of the Reconciliation is illustrated by the integration methods between the Facebook page wasatia.info with the social network site www.wasatia.info. The influence was vast and developed an ego-centered social network that has at its core the values of the Reconciliation process. The Wasatia social network site has over (157 members), and the Facebook page has over 10,000 followers as of [28 April 2016[3]]. The social network site had over two million in [2014–2015] and went down in [2016] in conclusion to the of the Palestinians trip to Auschwitz. Figure 11.3 clarifies the integration in the model on the social network site and the Facebook page influencing the Palestinian community as a discourse of social network analysis.

The advancement in the community towards the Reconciliation process was within the adapted concepts by the WASATIA NGO in Palestine (West Bank) using

[3] See Chap. 8.

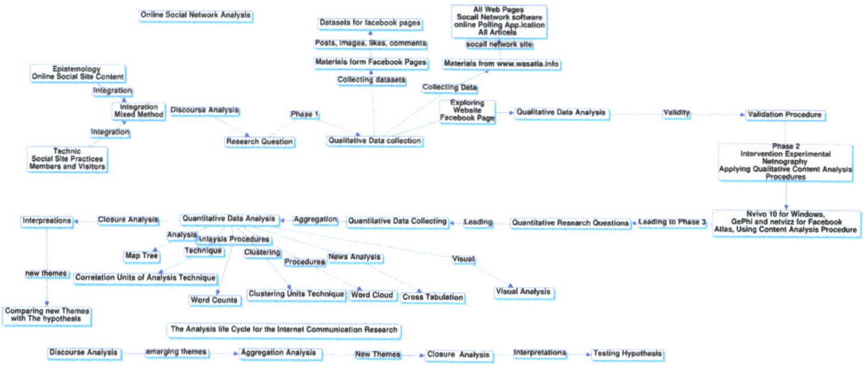

Fig. 11.3 The analytics procedure in internet communication research

Internet communication technologies, and the influence was identified and investigated by the synthesis of theoretical ideas developed within the research discourse.

The Internet communication research ICT-MR endorsed the Reconciliation process by introducing terminologies, definitions, and values that are predefined by WASATIA NGO. In addition, utilizing Internet communication network technologies that encouraged the participation in debating on crucial issues that are considered as taboos in the community. The ICT-MR endorsed interactions between participants online and offline, which can be observed in the social network site www.wasatia.info, and the Facebook page Wasatia. Info and Wasatai Vision. Visitors interacted, debated and some3 became followers of the WASATIA Facebook page and then integrated and registered as members of the WASATIA social network site www.wasatia.info and became ready to be active in workshops in the community. ICT-MR created a haven for members to share, contribute their thoughts to the society about the terms of moderation, reconciliation, and democracy to promote the values of the Reconciliation process[4].

The Wasatia.info Facebook page has up to *10,000 followers*[5]; this interprets that the Facebook page of Wasatia NGO, tends to get more awareness and more followers. The attitude of the members in the Facebook page is astonishing as they interact and developed deliberation among participants with each other, and discuss specific topics such as Auschwitz trip by Palestinian Students, that was conduct in (April 2014) by the Heart of Flesh, not stone Project. Members also discussed the goals of Wasatia NGO and its intentions toward the society. This is illustrated in the Fig. 11.4 below.

According to Fig. 11.4, which is an example of one of the posts on the Facebook page of WASATIA.INFO on [10, November 2014], illustrates that the post reached (186,560) and got (1000) likes, comments, and shares, (950) likes and (117) Shares and that post clicks are (4687). This means the tendency of healthy attitudes toward

[4] See Chap. 4.

[5] *According to* Chap. 6,

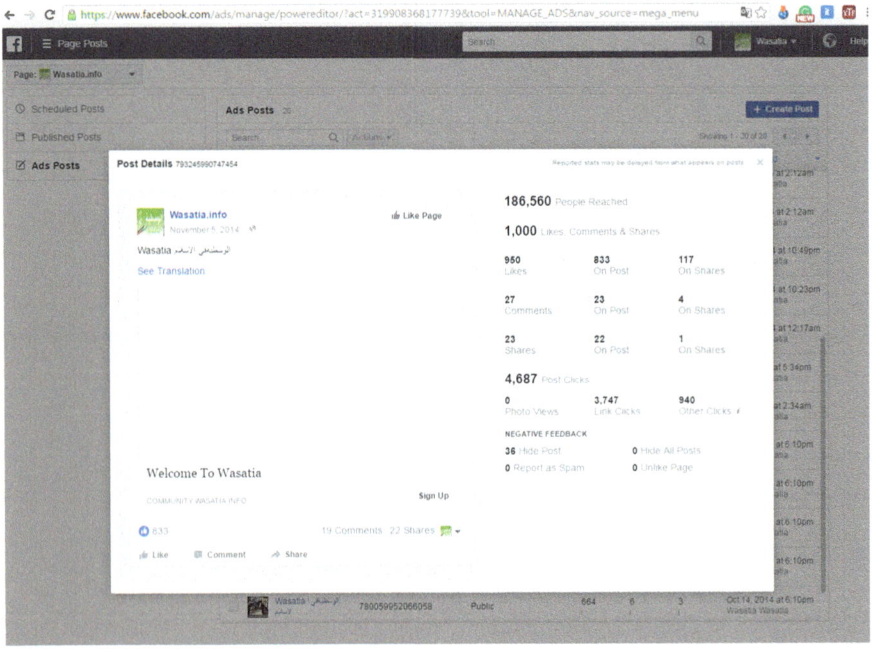

Fig. 11.4 Wasatia welcomes page members and its goals, post dated November 5, 2014

Wasatia NGO was definite at the time because many followers of the page read the posts and interacted and develop a deliberative debate between each other, whether they liked or commented on the post. This meant they were interested in what Wasatia.info as NGO should offer the community more studies and contribute to social change towards the reconciliation process. On another aspect, attitudes could change, according to a deliberative debate between participants on the trip to Auschwitz (2014); when boosting and promoting a post about the trip. On [October 25, 2014], the post reached (54,368), with (550) likes, (43) shares, (13l) comments were some against, and some were deliberating whether the trip was for the interest of the community or not. The post also had 632 post clicks, which illustrates interested in the trip. The Fig. 11.5 shows the change of attitudes toward Wasatia.info as an NGO has shifted.

After the Auschwitz trip by Palestinian Students, the NGO posted its goals of Wasatia.info to promote Reconciliation process and its aims; again, the results were not as expected (see Fig. 11.6).

As shown in Fig. 11.6, the positive attitudes have declined toward Wasatia NGO with only 107 likes; on post 103, there were only two comments, showing a broad decline in Wasatia NGO, according to its high number of followers which is at that time 10,000 followers the momentum for moderation and reconciliation is low. I can acknowledge that attitudes can always change, but the benefits that social media network of Wasatia NGO can be a tool to reach homes and individuals in the

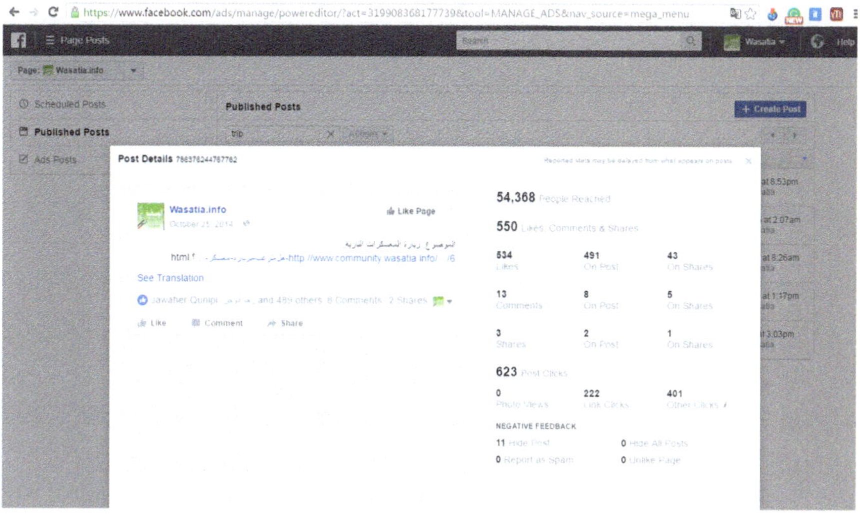

Fig. 11.5 Auschwitz trip by Palestinian Students Post user interactions, dated October 25, 2014

Fig. 11.6 After Auschwitz trip by Palestinian Students user debated against Wasatia, as the post illustrated in the figure the change of attitude post Accessed [May, 25, 2015]

Palestinian society. The ICT-MR platforms brought more awareness and understating of the Reconciliation process. ICT-MR developed an ego-centered social network that in its core values the Reconciliation process, this is a proof for our

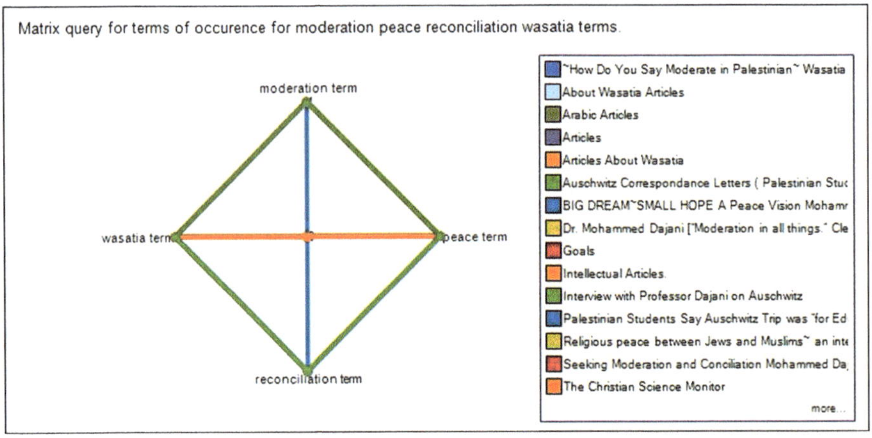

Fig. 11.7 Results of the synthesis of theoretical ideas, Analytics acquired [20 Oct 2016]

hypothesis, that ICT can develop social change towards Reconciliation process in a society that is in conflict[6].

Did the online social network deliberation correlate with the virtues of the reconciliation process?

Figure 11.7 demonstrates the synthesis of theoretical ideas between the correlations of Moderation, Peace term, Wasatia term, and Reconciliation term, which is more used in the social network site. Those terms were a result of the Content Analysis of the social network site and the datasets from Facebook pages – Wasatia.info – Wasatia Vision. The investigation was to check which terms are in the correlation between the Facebook Page Wasatia.info and the material in the social network site. The investigation discovered that the terms more used Moderation, Peace, Wasatia, Reconciliation. The correlation is illustrated in the figure below and shows the correlation between the term used and the articles that are published on the social network site[7].

The analysis illustrated in the previous chapter confirms that online deliberative democracy is achievable. The two Facebook pages, promote Wasatia values as a religious value and as a reasonable term in an online debate. The Wasatia.info Facebook page promoted reconciliation and moderate Islam, but the Wasatia Vision Facebook page promoted Wasatia as a religious concept and promoted Islam as a faith and religion, this shows that the online deliberation varies upon the participants and what they want to debate on, no control on the deliberative debates. This illustrates the validity and reliability of the analysis conducted in Chap. 9. The results illustrate that the discourse of analysis achieved the goals of analyzing the two Facebook pages that were under the supervision of the Wasatia NGO. In

[6] See Chap. 3.

[7] *see* Chap. 8 *for more analysis on the figure.*

addition, the results were different, but the intentions from both Facebook pages were to promote the Reconciliation process. This also illustrates that the synthesis of the theoretical approaches such as that Mixed-Method research, Netnography and Content analysis, provided the discourse to explore, investigate, analyze, and interpret the Internet communication research and developed a new term which is referred to as the Phronetic Mixed-Method Netnography Approach that helped examine the hypothesis.

Did the use of ICTs change intentionalities From Hatred and Animosity to adapt towards the virtues of the Reconciliation process? "What is Before and After"

Before deploying the ICT-MR project, there was no online deliberative democracy exposed in the Palestinian community, and thus no investigation on topics that are taboos in the Palestinian community.

In (Chap. 7) explores the integration between the social network site http://www.wasatia.info and the Wasatia social Facebook page (Wasatia.info). This kind of integration illustrates a tool to increase the exposure of the Reconciliation process, by using the OIRA and IRW in the online social network website and boosting those OIRA posted on the Facebook page. This technique assures that all IRWs of Wasatia Ngo are published on the social network site and are made available on the Facebook page. By using social networks to publish and promote articles, activities, workshops, to endorse and expos develop social change towards the Reconciliation process in the Palestinian community. In addition, the integration between both platforms helps recruit new members for the NGO and promotes those members to become active members in the NGO's workshops and activities in the society. The visitors and participants accepted that Wasatia NGO, and considered it as a reliable source for the promotion of moderate Islam, reconciliation, and democracy within the Palestinian community. With the increasing number of members registering in the social network site www.waatia.info and showing concerns to become really active members in activities done by Wasatia NGO in Palestine, this proves that ICT exposed the Reconciliation process and even intentionality's where drawn toward the Reconciliation process. This proves that ICT's developed the ego-centered network that acts toward having a more understanding of the Reconciliation process and advancing new members for a change of their intentionality's to more to adapt the Reconciliation process.

Explain how does the Wasatia Cycle concept in phronetic social science transforms the conflict into the objective side of the conflict.

According to (Rasmussen 2001, p. 110), he stated that the assumption for improving attitudes, emotions, trust, and communication, among parties, whether in conflicts or not "*will facilitate a more collaborative and integrative approach to dealing with the objectives of the side of the conflict*" (Rasmussen cited from Fisher 1995, p. 47).

The Wasatia Cycle concept portraits the use of the ICT to improve virtues in community, attitudes, emotions, trust, forgiveness, acknowledgment, recognition,

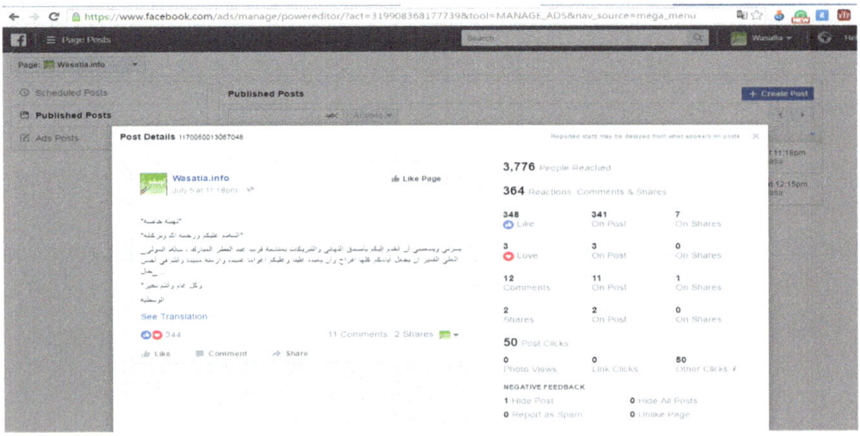

Fig. 11.8 Visitors have clicked on posts, post-dated July 5, 2015

communication among conflicted parties in a protracted conflict. Wasatia Cycle model would facilitate better collaboration and a more integrative approach within dealing with both communities or nations that are in conflict by applying IRW and OIRA.

The ICT-MR platform exposes the visitors and members to the Reconciliation process by published materials that are promoting toward tolerance and the acceptance of the other, empathy, truth, justice, acknowledgment, recognition in different online deliberative debates. Figure 11.8 shows that visitors are clicking on the posts and in contrast, they are being transferred to the site and reading articles that are published on the site (Fig. 11.9).

Wasatia does its annual celebration for its foundation every year, and more than 200 participants arrive to be active within the celebration. Figure 11.10 also illustrates the integration of the community in the Wasatia office in Jerusalem.

Figure 11.10 demonstrates that the articles are being read by the followers and members of the Wasatia NGO, and published articles are being viewed by visitors of the social network site.

Wasatia's social network works toward encouraging the Reconciliation process, in accordance affecting the community and its leaders. One of the examples of the effect of the synthesis of Internet communication research is illustrated in the figure below. The following illustration, the King of Jordan, speaking in front of the European Parliament and talks about moderate Islam. Wasatia NGO, using the ICT-MR had spread the voice of moderates such as His Royal King of Jordan Abdel Ben Al- Hussein the second. The post reached 1182 of Wasatia community that is on Facebook. The comments were astonishing for the support of moderate Islam and for support of the Wasatia values in the Palestinian Society. The post also illustrates that the Palestinian has an affection toward moderation and that they are open to becoming more moderate (Fig. 11.11).

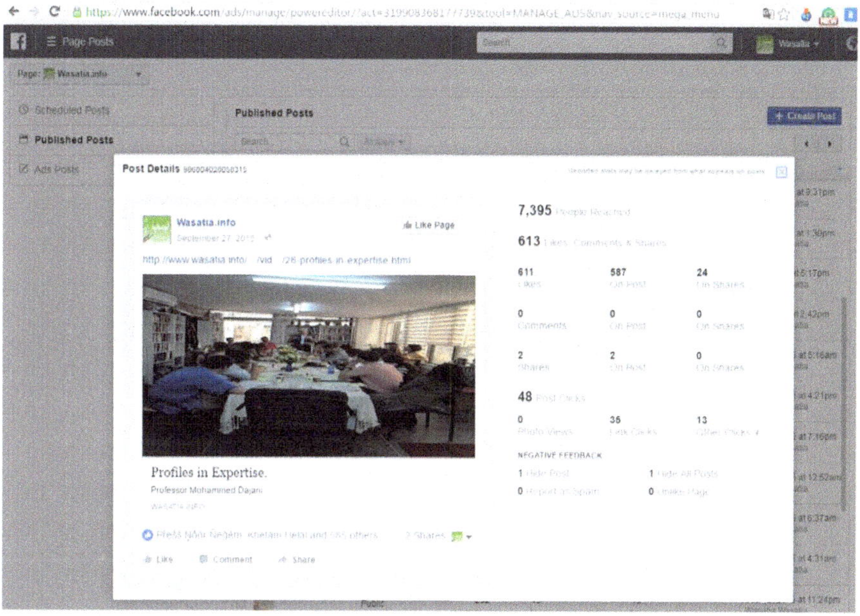

Fig. 11.9 The integration of the community in the Wasatia office in Jerusalem screenshot dated September 27, 2015

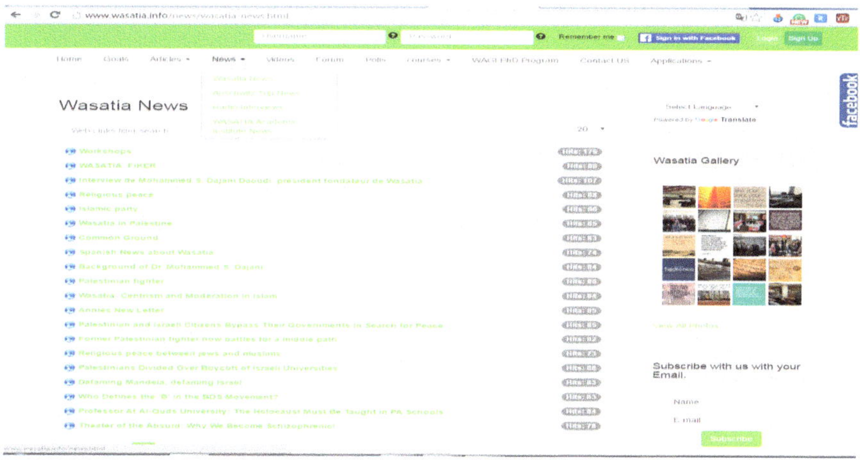

Fig. 11.10 Wasatia Articles published online Accessed [20.May.2015]

The post illustrates that the endorsement for online deliberation for moderation in Islam was also supported by His Royal Majesty King Abdulla the second. The figure also shows online deliberation and support for "Wasatia" as they support the message of the kind of Jordan on moderation in front of the European Parliament.

Fig. 11.11 King of Jordan the second encourages moderation. Screenshot accessed [June.25.2015]

According to Fisher (2001), Interactive Reconciliation Workshops must develop a space where participants can participate in conduct that gives space for the exposure of attitudes, emotions, trust, forgiveness, acknowledgment, recognition, communication among conflicted parties in a protracted conflict. ICT-MR communication Research was the Interactive Reconciliation workshop for the advancement of the reconciliation process, which created a better a virtual space for collaboration and a more integrative approach to the reconciliation process.

11.3 Summary of Chapters

11.3.1 Summary of Chapter 1

This chapter contained research about Phronetic research plan developed by Flyvbjerg (2006) and the concept of applying phronesis to communication research (Flyvbjerg et al. 2012) in Applied Phronetic Social Science, as it was crucial to understanding the discourse of any research that wants to develop social change. The idea was to understand ways in which to develop the intellectual discourse for an investigation and the concept of steps and principles needed to do so. This was introduced in this chapter and referred to as The Phenomenology of ICT in Phronetic social science. The chapter applied a new terminology to the Mixed-Method research design when using Phronetic approach, referring to it as applied Phronetic mixed-method experimental intervention research design, the Mixed-Method experimental research design was introduced by (Creswell and Clark 2018). This design was the roadmap for applying the empirical work discourse to gain results from conducting the Internet communication research. The chapter introduced qualitative content analysis methods and procedures, as these were conducted in the discourse of the research.

11.3.2 Summary of Chapter 2

This chapter discussed the reason and the importance to study reconciliation, as it represented a broad concept that could affect the discourse of conflicts, whether these were cultural, social, and economic conflicts. Reconciliation represented the road taken to resolve and adjust the transformation of the conflicts for animosity to empathy and understanding the other. The chapter introduced the concepts for reconciliation, emphasizing reconciliation during conflicts, as Leiner and Fläming (2016) explained. In addition, the chapter introduced the evolvement of media from simple media, such as radio and TV, to high technological media, such as Internet communication technologies (ICT). Furthermore, ICT was discussed regarding its involvement in changing the discourse of a political change, such as the Arab Spring. In addition, the chapter introduced the research project, which was referred to as ICT-MR to study moderation, reconciliation, and democracy in the middle of a conflict. The chapter explores, Why reconciliation in the middle of the conflict, Methods for Reconciliation, ICT for Reconciliation, and Reconciliation in the Arabic language.

11.3.3 Summary of Chapter 3

The chapter discussed the hypothesis for the communication research and researched the influence of ICT on societies and nations. The chapter generated an understanding for exploring whether ICT could foster moderation, reconciliation, and democracy in the midst of conflict, which is referred to it as the Reconciliation Process. The chapter explores ICT in transformation towards social change in content, structure, profiles, and groups, and primary Issues. It discussed the disadvantages of using ICT and introduced in-depth the Internet communication research discourse considering the deliberative process, thereby exploring ICT as a means of intrusion in the midst of conflict. The chapter contained research about ways in which ICT could be a tool for peacebuilding and an instrument of culture, social, and economic change, thereby introducing different theories from different professionals in the field of ICT for education, development, and social justice.

11.3.4 Summary of Chapter 4

This chapter introduced "Wasatia" as a concept and a movement to describe the foundations of the NGO "Wasatia" in Palestine by explaining the "Wasatia" concept as a moderate notion of thought in Islam and introduces the founder of the "Wasatia" Movement. In addition, the chapter explored definitions and terminologies of moderation as an Islamic term and how it was introduced in society. The chapter introduces the "Wasatia" Movement in Palestine and its role to develop the notion of the reconciliation process. Also, introduced the theoretical idea of the "Wasatia" Cycle and the adherence of ICT for the discourse for the Reconciliation process. In addition, I discussed ways in which ICT adhered in the development of moderation reconciliation and democracy in society.

11.3.5 Summary of Chapter 5

This chapter explored ICT in democracy and how it develops into e-government to affect the social capitals for social change. It describes new terminologies used for social change towards a democratic approach in societies, such as e-democracy, e-service, and e-governments. The chapter also discusses deliberative democracy and online deliberation, as it was an essential part of the discourse for the Reconciliation process. The chapter introduced definitions for deliberative democracy in different aspects, especially definitions introduced by Gutmann and Thompson (2004), as well as many other professors in the field of deliberative democracy, such as Cohen (1998). I described the goals, characteristics, and values of deliberative democracy to achieve a reasonable deliberative process. The chapter emphasized the discourse of deliberative democracy over digital social networks, including how deliberative democracy could apply to the Reconciliation process. Moreover, it included ways in which deliberative democracy could operate over the Internet in theory and in practice. I discussed the advantages and disadvantages of deliberative democracy, as well as using deliberative democracy over the Internet. Introducing a study conducted by Pietro Speroni di Fenizio (Coimbra University of Portugal), Alois Paulin (Faculty of IS, Novo Mesto, Slovenia), Cyril Velikanov (foundation Euractiv/Politech, Brussels; Finizio 2011). Finally, the chapter introduced a model for the process of deliberative democracy on the Cloud.

11.3.6 Summary of Chapter 6

This chapter introduced the discourse for the netnography approach, a term adopted by Kozinets in 2002. The chapter explored the implementation of principles of netnography, as these are described by Kozinets (2015), and applied these to the Internet communication research in this ICT-MR. The chapter presented the ethical considerations that should be adapted to any Internet communication research according to the Association of Internet Research – AoIR (Ethics Working Committee, 2012, pp. 4–5). In addition, I presented the netnography for the social network site adopted by this Internet communication research. The chapter introduced a new research approach, as Creswell (2014) mentioned four types of research tradition in his book in 2014. I wanted to add a fifth approach and refer to it as the Mixed-Method netnography approach (MMNA). The approach was applied to the communication research ICT-MR considering the 12-principle process for researching a netnography by Kozinets (2015). The chapter was used to describe ways in which to apply Internet research on an online community and explore the tools that were used to study the virtual society in adopting the new approach: the PMMN.

11.3.7 Summary of Chapter 7

This chapter explored qualitative content analysis theories and methods for social network analysis. Those methods were used in theory for social network analysis. These derived from a combination of different methods used by different professors in the field of qualitative content analysis. The qualitative content analysis procedure introduces the work of Weitzman and Levkoff (2000), Sandelowski (2000), Kohlbacher (2012), Schreier (2012), Schulz (2012), Krippendorff (2013), Saldana (2013a, b), and Bazeley and Jackson (2013). The chapter also enlightened one on using software for analysis, such as NVivo 10 for Windows, when applying the method of social network analysis to the PMMN. The chapter introduced the Internet research communication framework design for conducting the analysis in practice.

11.3.8 Summary of Chapter 8

This chapter introduced practical techniques for social network analysis, which was described in theory in the previous chapter and applied in this chapter using qualitative analysis software, NVivo 10 for Windows, thereby introducing the coding structure techniques and procedures. In addition, the chapter illustrated the logical path and evaluation techniques applied to qualitative content analysis model, as introduced in Chap. 8. In addition, the chapter illustrated the empirical work and the results from the analysis procedure of the social network site, http://www.Wasatia.info, and analyses of the developed datasets from two Facebook pages, and http://Wasatia.info, as followed by 10,000 followers, and Wasatia-Vision, as followed by 258 members, as of December 31, 2015.

11.3.9 Summary of Chapter 9

Introduces the evaluation of research, including the aim and the purpose of the research. The philosophical and theoretical foundation for conducting the research, the adopted social network analysis methodologies. Also, evaluate the use of the PMMN. In addition introduces the experimental discussion phase, the synthesis theory conducted in the discourse of the research. The empirical work evaluation that describes techniques of analysis and interpretation by using qualitative content analysis for a different specialist in the field. In addition to exploring and investigates the hypothesis, interpreting new emerging themes from the analysis. The chapter also explores the validity of the research and measure the reliability and stability of the research.

11.3.10 Summary of Chapter 10

The Chapter introduces the conclusion from conducting internet communication and whether ICT develops social change towards reconciliation process in the middle of the conflict. In addition, introduces the experimental research findings, research questions, and answers, a summary of chapters. Also introduces a further perspective for a study to be explored in different research, and to study whether ICT can develop social change towards the reconciliation process in a different aspect of the social conflict. The chapter also includes conclusion remarks and exploration for applied ICT in Phronetic Social Science and future perspectives.

11.4 Further Perspective: ICT Fosters Social Change Towards Reconciliation Process in the Middle of Conflict

11.4.1 Conclusion Remarks

ICTs, such as social network sites and network platforms (e.g., Facebook, Twitter, and YouTube), can have a significant role in steering conflicts in the direction of peace or war. As a result, from the communication research, ICTs can encourage the

Reconciliation process[8] within society and in the middle of the conflict. The Internet communication research illustrated that the change of intentionalities of members in "Wasatia"'s social network site and Facebook pages were directed toward the Reconciliation process. The social network site developed an aspiration of having a democracy in the Palestinian society, which affected the community for social change toward moderation and reconciliation. This ICT-MR successfully developed an ego-centered social network that was willing to adopt the Reconciliation process and encourage it in the society. This aspect could be shown with the increase of registered new members on the social network site, http://www.wasatia.info. In addition, the objective of those members was to be active in the Palestinian society, thereby exposing the society to the Reconciliation process. Those members have the tendency to accept and understand the Reconciliation process and work toward implementing it in their daily lives. For this reason, the hypothesis that was examined within the discourse of the research—ICT could foster and promote the Reconciliation process in the middle of conflict—was attainable.

11.4.2 Applied ICT in Phronetic Social Science

The concept for the theory developed in Chap. 2 could be applied in any ICT research to observe the social change in the society. An example for applying ICT in Phronetic social science, such as the phenomenology of ICT for social change toward reconciliation process, could force a social change toward a reconciliation process; this aspect was taken from Habermas' "*The Abstract of the Force for a Better Argument*" (Flyvbjerg 2001, p. 96). Figure 11.12 illustrates the concept of applied ICT in Phronetic social science.

Figure 11.12 shows the discourse of knowledge and technology as one part of the other to develop applied phronesis using ICT to develop a communicative relationship between these and apply the Habermas appeal for the force of a better argument. This discourse involved applying reconciliation terminology, philosophy, and techniques that would lead to social change toward a reconciliation action process. Once applied to the community, it would have its effect, negatively and positively, which should be respected equally. There was a combination of tools considered to apply the phenomena, and these included methods for reconciliation, netnography approach, types of social change effect, and ICT application (see Table 11.1). Applied ICT for social change toward a reconciliation process can be emphasized in Table 11.1. Table 11.1 illustrates the roles of netnography and ICT for social change toward the reconciliation process, as well as how it can be achieved, researched, and applied.

[8] See Chap. 5.

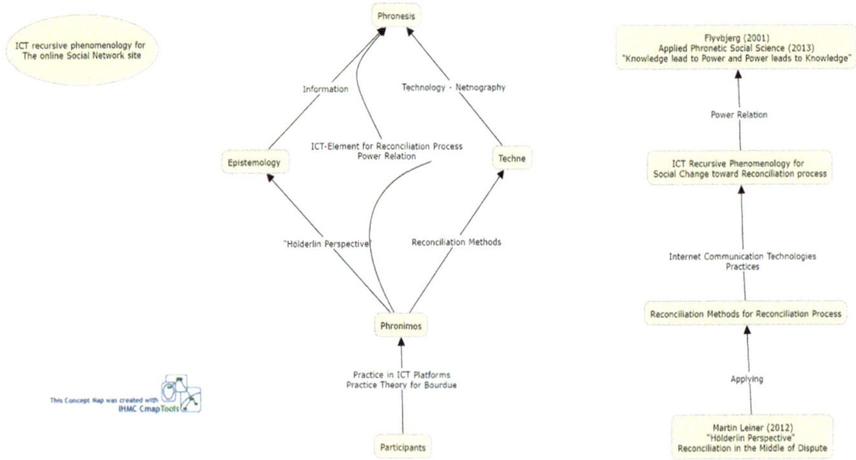

Fig. 11.12 Applied phronetic internet research phenomenology

11.4.3 Future Perspectives

ICT-MR illustrates an understanding of how ICT can influence society through ICT, as it is clarified in Fig. 11.13. Figure 11.13 presents ICT as the core toward advancing the discourse for the development of a tolerant society that can be directed for peace and coexistence to develop a path for reconciliation within societies or nations. ICT can be used to work toward social change for the Reconciliation process.

As a result, one can understand that ICT can be an instrument one may use to influence conflict transformation in social transformation to navigate the way for moderation, reconciliation, and democracy in the society. As Hizkias Assefa (2015) stated,

> "Mediation and reconciliation work is about a profound quest for justice and social transformation. But at the same time, they are about service, solidarity, about exploring and rediscovering the human spirit that has been lost or shattered by human conflict, cruelty, ignorance, and greed" (Assefa 2015).

ICT-MR demonstrates the work of WASATIA NGO to endorse and promote the reconciliation process within the society in the middle of the conflict. ICT can be a tool one may use to develop a bridge through social networking, social justice, and multicultural understanding between communities and nations in conflict. ICT can act as a mediator in the conflict, pushing toward another level of tolerance and acceptance of the other.

The Internet communication research developed synthesis of theoretical ideas, such as the following approaches: Applied Phrenetic Social Science (Flybjerg 2013) in adopting experimental intervention Mixed-Method research design (Creswell and Clarck 2018), netnography (Kozinets 2015), and qualitative content analysis

Table 11.1 Methods used for the reconciliation process

Methods for the reconciliation process	Netnography	Social change	Examples of ICTs
Apology Education Published meetings Truth and reconciliation commissions Public trails Reparations payments The work of NGOs Joint projects	Introspection Investigation Information Interview Inspection Interaction Immersion Indexing Interpretation Iteration Instantiation Integration	Provide information	Internet connectivity Mobile phones and personal data assistant Geographic information systems. Forums Radio Chat Videos
Writing a common history Published meetings Truth and reconciliation commissions Public trails Reparations payments The work of NGOs Joint projects	As the above	Help people process information	Social network sites, and portals Data visualization tools Online dispute resolution tools
Education Public trails The work of NGOs Joint projects	As the above	Improve decision making	Virtual command center Games and simulations Online dispute resolution tools
Education The work of NGOs Joint projects	As the above	Reduce scarcity	Mobile phones Hand held portable device, such as iPad, tablets
Apology Published meetings. Truth and reconciliation commissions Public trails Reparations payments The work of NGOs Joint projects	As the above	Support relationship	Social network tools Online collaboration tools Mobile phones Virtual reality Telecenters
Apology Truth and reconciliation commissions Public trails Reparations payments Joint projects The work of NGOs	As the above	Help people understand each other	Translation software Blogs Social network tools Multimedia

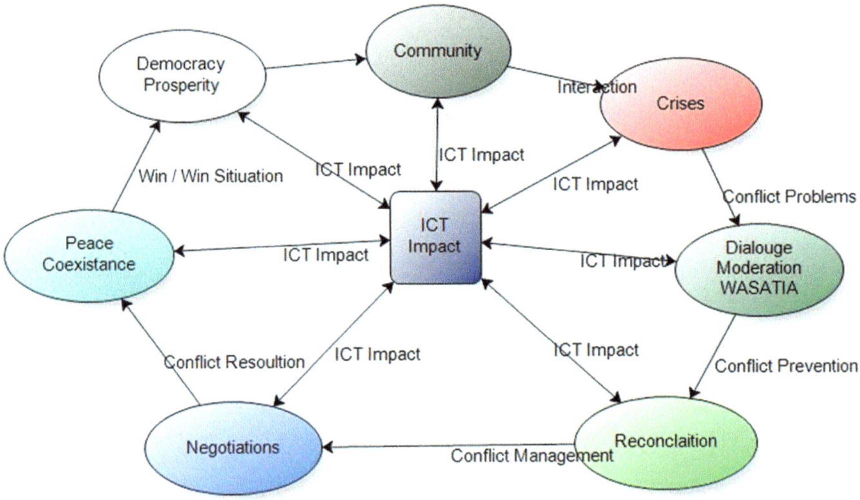

Fig. 11.13 Model of ICT tolerance

(Bazeley and Jackson 2013; Kohlbacher 2012; Krippendorff 2013; Saldana 2013a, b; Schreier 2012; Schulz 2012). This approach was signified as PMMN for Internet communication analysis. This type of research was interdisciplinary research that contributed to assisting other researchers in the field of Internet research science, which was a new discipline of researching the Internet and virtual communities to understand ways in which these influenced real communities. This interdisciplinary study included philosophy, social science, religion, political science, communication science, and computer and information sciences. The phenomena of the research hypothesis were that ICT could foster the Reconciliation Process in the midst of the conflict. The focus of the research was on developing a virtual community in the midst of a conflict that endorsed moderation, reconciliation, and democracy in the Israeli/Palestinian conflict. The goal was to explore and examine the influence of ICT within the scope of this virtual society and the effect of social changes towards the reconciliation process in the middle of the conflict, and the attempt to transform the conflict from a practical understanding of the reconciliation process into a real application in the community.

Finally, the Internet communication research, ICT-MR, developed an ego-centered social network within the Palestinian community that adopted the values of the Reconciliation process in the middle of the Israeli/Palestinian conflict. Malcolm (2005) stated the following:

> The media is the most powerful entity on earth. They have the power to make the innocent guilty and to make the guilty innocent, and that is the power. Because they control the mind of the masses. (Malcom 2005)

Bibliography

Books

Achcar, G. (2013). The people want: A radical exploration of the Arab uprising, translated by G. M. Goshgarian. Berkeley: University of California Press.

Aristotle. (1976). *The Nicomachean ethics* (abbreviated as N. E.), translated by J. A. K. Thomson, revised with notes and appendices by H. Tredennick, Introduction and Bibliography by J. Barnes. Penguin: Harmondsworth.

Bar-Siman-Tov, Y. (Ed.). (2004). *From conflict resolution to reconciliation*. New York: Oxford University Press.

Barakat, Z. (2017). *From heart of stone to heart of flesh: Evolutionary journey from extremism to moderation*. München: Herbert Utz Verlag GmbH.

Bazeley, P., & Jackson, K. (2013). *Qualitative data analysis with NVivo* (2nd ed.). London: SAGE.

Bekerman, Z., & Horenzsyk, G. (2009). Computer supported collaborative Intercultural Education. In C. Vrasidas, M. Zembylas, & G. Glass (Eds.), *ICT for education, development and social justice* (pp. 201–213) Charlotte: Information Age Publishing.

Benkler, Y. (2006). *The wealth of networks: How social production transforms markets and freedom*. New Haven: Yale University Press.

Bernhard, B. (1952). *Content analysis in communicative research*. New York: Free Press.

Bidgoli, H. (2006). *Handbook of information security: Information warfare, social, legal, and international issues and security foundations* (2nd ed.). New Jersey: Wiley.

Bloomfield, D., Barnes, A. T., & Huyse, L. (Eds.). (2003). *Reconciliation after violent conflict: A handbook*. Stockholm: International Institute for Democracy and Electoral Assistance.

Boellstrorff, T., Nardi, B., Pearce, C., & Taylor, T. (2012). *Ethnography and the virtual world: A handbook of methods*. Princeton: Princeton University Press.

Bohman, J., & Rehg, W. (Eds.). (1997). *Deliberative democracy: Essays on reason and politics*. Cambridge, MA: The MIT Press.

Bourdieu, P. (1977). *Outline of a theory of practice* (R. Nice, Trans.). New York: Cambridge University Press.

Bryman, A. (2006). *Mixed-methods*. London: SAGE.

Cammaert, B., & Carpentier, N. (2007). *Reclaiming the media: Communication rights and democratic media roles*. Chicago: University of Chicago Press.

Castells, M. (2010). *The rise of the network society. With a new preface* (2nd ed.). Chicester: Wiley.

I. M. AlDajani, *Internet Communication Technology (ICT) for Reconciliation*, Lecture Notes in Social Networks, https://doi.org/10.1007/978-3-030-41203-6

Couldry, N. (2012). *Media, society, world social theory and digital media practice*. Cambridge: Polity Press.

Creswell, J. W. (2014). *Research design qualitative quantitative mixed-methods approach* (4th ed.). California: SAGE.

Creswell, J., & Plano Clark, V. (2011). *Designing and conducting mixed-method research* (2nd ed.). California: SAGE.

Creswell, J., & Plano Clark, V. (2018). *Designing and Conducting Mixed-Methods Research*. London: SAGE. Available at Amazon [Kindle Edition].

Cronin, C. (2008). *Between naturalism and religion by Jürgen Habermas*. Cambridge: Polity. [German, 2005b].

Dajani, M. (2008a). *WASATIA: From theory to practice*. Jerusalem: Wasatia Publications. (in Arabic)

Dajani Daoudi, M. (2009). *WASATIA: The Spirit of Islam*. Jerusalem: WASATIA Publishing.

Dajani Daoudi, M. (2012). *The future of Jerusalem*. Jerusalem: Wasatia Publishing.

Dajani, M., & Schwartz, R. (2002). *YES PM: Years of Experience in Strategies for Peace Making, Looking at Israeli-Palestinian People-to-People Activities, 1993–2002*. Jerusalem: Israel Palestine Center for Research and Information (IPCR I).

Dryzek, J. (2002). *Deliberative democracy and beyond: Liberals, critics, contestations*. Oxford: Oxford University Press.

Elster, J. (1998). *Deliberative democracy* (Cambridge studies in the theory of democracy). Cambridge: Cambridge University Press. https://doi.org/10.1017/CBO9781139175005.

Fishkin, J. (2009). *When the people speak: Deliberative democracy and public consultation*. Oxford: Oxford University Press. [Kindle Edition] Available at Amazon.

Flybjerg, B. (2013). *Mega projects and risks*. Cambridge: Cambridge University Press.

Flyvbjerg, B. (1991). *Rationality and power: Democracy in practice*. Chicago: The University of Chicago Press.

Flyvbjerg, B. (2001). *Making social science matter*. Cambridge: Cambridge Press.

Flyvbjerg, B. (2012). Making Social Science Matter. In G. Papanagnou (Ed.), *Social science and policy challenges: Democracy, values, and capacities* (pp. 25–56). Paris: UNESCO Publishing. Available at SSRN. https://ssrn.com/abstract=2278218.

Flyvbjerg, B. (2013). *Mega projects and risks*. Cambridge: Cambridge University Press.

Flyvbjerg, B., Landman, T., & Schram, S. (Eds.). (2012). *Applied Phronesis in Real Social Science*. Cambridge: Cambridge University Press. [Kindle Edition] Available at Amazon.

Gastil, J., & Levine, P. (2005a). *The deliberative democracy handbook: Strategies for effective civic engagement in the twenty-first century*. New York: Wiley.

Gould, M. (2013). *Social media gospel*. Minnesota: Liturgical Press.

Green, J. C. (2007). *Mixed-methods in social inquiry*. San Francisco: Jossey-Bass.

Gutmann, A., & Thompson, D. (2004). *Why deliberative democracy?* Princeton: Princeton University Press. [Kindle Edition] Available at Amazon.

Gutmann, A., & Thompson, D. (2009). *Why deliberative democracy?* Princeton: Princeton University Press.

Haberman, F. W. (1991). *Noble lectures: Peace*. London: World Scientific Publishing.

Harold, I. (1951). *The bias of communication*. Toronto: University of Toronto Press.

Huxley, A. (1958). *Brave new world revisited*. New York: Harper & Row Publishers.

Junca, D., & Ribeaute, D. (2006). *Comment rédiger les bibliographies*. Paris: Adventure Works Press.

Kawasaki, G. (2014). *The art of social media*. New York: Penguin Group.

Kozinets, R. (2015). *Netnography redefined* (2nd ed.). California: SAGE.

Krippendorff, K. (2013). *Content analysis: An introduction to its methodology* (3rd ed.). Pennsylvania: SAGE.

Langford, D. (Ed.). (2000). *Internet ethics*. London: Macmillan Press Ltd.

Leiner, M., & Flämig, S. (Hg.). (2012). *Latin America between conflict and reconciliation*. Gottingen: Vandenhoeck & Ruprecht.

Leiner, M., & Schließer, C. (Eds.). (2018). *Alternative approaches in conflict resolution*. Palgrave Macmillan.

Lenhardt, C., & Nicholsen, W. (1990). *Moral consciousness and communicative action*. Cambridge, MA: MIT Press. [Original in German, published in 1983].
Lysack, C. L., Krefting, L., & Gliner, J. A. (1994). Qualitative methods in field research: An Indonesian experience in community based practice. *The Occupational Therapy Journal of Research, 14*(2), 93–111. https://doi.org/10.1177/153944929401400203.
Mcleod, S., & Lehmann, C. (2012). *What school leaders need to know about digital technologies and social media*. San Francisco: Wiley.
Miall, H. (2003). *Conflict transformation: A multi-dimensional task.* Berghof Research Center for Constructive Conflict Management – Edited version, August 2004 (First launch March 2001).
Muḥammad Sulaymān al-Dajjānī Dāwūdī. (2016). In M. S. Dajani Daoudi, M. Leiner, & Z. Barakat (Eds.), *Teaching empathy and reconciliation in the midst of conflict*. Jerusalem: Wasatia Press.
Murphy, C. (2010). *A moral theory of political reconciliation*. Cambridge: Cambridge University Press.
Newman, M. (2010). *Networks: An introduction*. Oxford: Oxford University Press.
Nissenbaum, H. (2010). *Privacy in context: Technology, policy, and the integrity of social life*. Stanford: Stanford University Press.
O'Malley, M., Leiner, M., Summe, D., & Knoeppfler, N. (Hrsg.). (2017). *Thüringen: Braucht das Land Versöhnung? Kritisches Jahrbuch der Philosophie, Band 17* (17th ed.). Jena: Thüringischen Gesellschaft für Philosophie e.V.Vorsitzender: Klaus-M. Kodalle.
Pool, I. (Ed.). (1959). *Trends in content analysis*. Chicago: University of Illinois Press.
Pool, I. (1983). *Technologies of Freedom*. Cambridge, MA: Belknap Press.
Ritzer, G. (2008). *Classical sociological theory* (5th ed.). New York: McGraw Hill Higher Education.
Saldana, J. (2013a). *The coding manual for qualitative researchers* (2nd ed.). London: SAGE. [Kindle Edition] Available at Amazon.
Sandel, M. (2009). *Justice: What's the right thing to do*. Oxford: Oxford University Press.
Schreier, M. (2012). *Qualitative content analysis in practice*. Los Angeles: SAGE. [Kindle Edition] Available at Amazon.
Selian, B. A. (2002). *ICTs in support of human rights, democracy, and good governance*. Geneva: International Union Telecommunication.
Shapiro, I. (2005). *The flight from reality in the human sciences*. Princeton: Princeton University Press.
Sirimarco, E. (2005). *The civil rights movement*. New York: Marshall Cavendish.
Steiner, J. (2012). *The foundations of deliberative democracy: Empirical research and normative implications*. Cambridge: Cambridge University Press.
Tashaakkori, A., & Teddlie, C. (2003). *Handbook of mixed-methods in social and behavioral sciences*. Thousand Oaks: SAGE.
Tashaakkori, A., & Teddlie, C. (2010). *The handbook of the mixed-method in social and behavioral sciences* (2nd ed.). Thousand Oak: SAGE.
Vrasidas, C., & Glass, G. (Eds.). (2005). *Preparing teachers to teach with technology* (Current perspectives on applied information technologies). Charlotte: Information Age Publishing.
Vrasidas, C., Zembylas, M., & Glass, G. (2009). *ICT for education development and social justice*. Charlotte: Information Age Publishing.
Warschauer, M. (2003). *Technology and social inclusion: Rethinking the digital divide*. Cambridge: MIT Press.

Articles in Journals & Chapters in Books

Andrews, M. (2000). Forgiveness in context. *Journal of Moral Education, 1*, 29.
Atchison, C. (2014). *An overview of NVivo*. Accessed 10 Sep 2018. Available at qsrinternational.com/Resource/NVivo10/NVivo-10-Overview.pdf.

Bailey, M. T. (1992). Do physicists use case studies? Thoughts on public administration research. *Public Administration Review, 52*, 47–54. Aristotle. (1976). *The Nicomachean ethics*. Harmondsworth: Penguin.

Bar-Tal, D., & Bennink, G. H. (2004). The nature of reconciliation as an outcome and as a process. In Y. Bar-Siman-Tov (Ed.), *From conflict resolution to reconciliation* (pp. 11–38). Oxford: Oxford University Press. (also published in Politika, 2002, Issue No 9, pp. 9–34 in Hebrew).

Basset, E. H., & O'Riordan, K. (2002). Contesting the human subject research model, ethics and information technologies. *Ethics of Internet Research, 4*, 233–247.

Bohman, J. (1998). The coming age of deliberative democracy. *The Journal of Political Philosophy, 6*(4), 400–425.

Boneva, B., Kraut, R., & Frohlich, D. (2001). Using email for personal relationships. *American Behavioral Sciences., 43*(3), 530–549.

Botes, J. (2001, Autumn/Winter). Conflict transformation: A debate over semantics Botes. *The International Journal of Peace Studies, 6*(2), 1–4.

Bourdieu, P. (1983). Ökonomisches Kapital, Kulturelles Kapital, Social Kapital. *Soziale ungleichheiten, 2*(2), 183–198.

Brantmeier, E. (2009). ICT for peace and reconciliation. In C. Vrasidas, M. Zembylas, & G. Glass (Eds.), *ICT for education, development and social justice* (pp. 213–236). Charlotte: Information Age Publishing.

Buchstein, H. (1997). Bytes that bite. *The Internet and Deliberative Democracy, Constellations, 4*(2), 248–263.

Cohen, J. (1998). Democracy and liberty. In J. Elster (Ed.), *Deliberative democracy* (Cambridge studies in the theory of democracy) (pp. 185–231). Cambridge: Cambridge University Press. https://doi.org/10.1017/CBO9781139175005.010.

Corbin, J., & Strauss, A. (1990). Grounded theory research: Procedures, canons, and evaluative criteria. *Qualitative Sociology, 13*(t), 3–20.

Cuban, L. (2003). Review of oversold and underused computers in the classroom. *Language Learning & Technology, 7*(3), 42–45.

Dahlberg, L. (2007). The internet, deliberative democracy, and power: Radicalizing the public Sphere. *International Journal of Media and Cultural Politics, 3*(1), 47–64.

Dajani, M. (2008b, Summer). Big dream/small hope: A peace vision. *Crosscurrent, 2*, 191–219; ("Big dream/small Hope: A peace vision," *in* Educating toward a culture of peace, edited by Yaacov Iram Greenwich, Connecticut: Information Age Publishing, 2006), pp. 39–56.

Dajani, M. (2009, December). The Arab peace initiative: Lost in the translation. *Crosscurrent, 59*(4), 532–539.

Dajani Daoudi, M. (2011). Palestinian refugees. In E. Matthews, D. Newman, & M. S. Dajani Daoudi (Eds.), *The Israel-Palestinian conflict: Parallel discourses* (pp. 36–47). London/New York: Routledge.

Dajani Daoudi, M., & Dajani, A. M. (2011). Jerusalem: Where to? In H. G. Brauch & others (Eds.), *Coping with global environmental change. Disasters and security: Threats, challenges, vulnerabilities, and risks* (pp. 459–483). New York/Heidelberg/Berlin: Springer.

Dangwal, K. L., & Srivastava, S. (2015). Fostering peace education through ICT. *International Journal of Research and Analytical Reviews., 2*(2), 64–73.

During, J., & Sherif, J. (2003). The effects of virtual community on changing worldwide business and culture. *Journal of Academy of Business and Economics, 2*(1), 202–212.

Edwards, R. (2002). The moderator as an emerging democratic intermediary. *The Role of the Moderator in Internet Discussions about Public Issues, 7*(1), 3–20.

El-Gantri, R., & El-Mufti, K. (2018). Not without dignity, views of Syrian refugees in Lebanon on displacement. *International Center for Transitional Justice.* Date published: 6/9/2017.

Ess, C., & Jones, S. (2004). Ethical decision-making and internet research: Recommendations from the aoir ethics working committee. In *Reading in virtual ethics issues and contverseries* (p. 357). Hershey: Information Science Publishing.

Ethics Working Committee. (2012). Recommendations from the Association of Internet Research Ethics Committee. *Ethical Decision-Making and Internet Research, 2*, 19.

Fisher, R. J. (1995). Pacific, impartial third-party intervention in international conflict: A review and an analysis. In J. A. Vasquez et al. (Eds.), *Beyond confrontation: Learning conflict resolution in the post cold war era*. Chicago: University of Chicago.

Fisher, R. (2001). Social – psychological processes in interactive conflict analysis and reconciliation. In M. Abu-Nimer (Ed.), *Reconciliation, justice, and coexistence* (pp. 25–45). Lanham: Lexington Books.

Fishkin, J., & Habermas, J. (1995). The voice of the people: Public opinion and democracy, deliberative politics. In Estlund (Ed.), *Democracy, deliberative politics* (pp. 33, 114–115, 120–121). New Haven: Yale University Press.

Fishkin, J., & Laslett, P. (2006). Debating deliberative democracy. *Philosophy, and Society, 3*, 11–13.

Fishkin, J. S., & Luskin, R. C. (1999). The quest for deliberative democracy. *The Good Society, 9*(1), 1, 4–8.

Fishkin, J. S., & Luskin, R. C. (2000). The quest for deliberative democracy. In M. Saward (Ed.), *Democratic innovation: Deliberation, association, and representation*. London: Routledge.

Fishkin, J. S., & Luskin, R. C. (2005). Experimenting with a democratic ideal: Deliberative polling and public opinion. *Acta Politica, 40*, 284–298.

Fishkin, J. S., & Luskin, R. C. (2006, January). Broadcasts of deliberative polls: Aspirations and effects. *British Journal of Political Science, 36*, 184–188.

Fishkin, J. S., Luskin, R. C., & Jowell, R. (2000). Deliberative polling and public consultation. *Parliamentary Affairs, 53*, 657–666.

Fishkin, J. S., Baogang, H., Luskin, R. C., & Siu, A. (2010). Deliberative democracy in an unlikely place: Deliberative polling in China. *British Journal of Political Science, 40*, 435–448.

Flew, T., & Young, G. (2005). *From E-government to online deliberative democracy* (pp. 1–11). Creative Industries Faculty, Queensland University of Technology, Brisbane, Australia, National Forum.

Flyvbjerg, B. (2002, June). Bringing Power to Planning Research One Researcher's Praxis Story. *Journal of Education and Research, 21*(4), 353–366.

Fox, S. (2004, January 1). The new imagined community: Identifying and exploring a bidirectional continuum integrating virtual and physical communities through the Community Embodiment Model (CEM). *Journal of Communication Inquiry, 28*(1), 47–62.

Gadamer, H. G., Krajewski, B., & Heinemann, R. (1992). The expressive power of language: On the function of rhetoric for knowledge. *PMLA, 107*(2), 345–352.

Gastil, J., & Levine, P. (2005b). The deliberative democracy handbook. In *The deliberative democracy handbook*. New York: Wiley.

Gehrlein, W. (2006). Condorcet's Paradox. *Theory and Decision, 15*(2), 29–35.

George, A. (1959). Quantitative and qualitative approaches to content analysis. In I. de Sola Pool (Ed.), *Trends in content analysis* (pp. 1–32). Chicago: University of Illinois Press.

Gibson, J. (2006). The contribution of truth to reconciliation: Lessons from South Africa. *Journal of Conflict Resolution, 3*(50), 409–432.

Gret, B., & Moor, J. (1998). On morality and computer ethics. In *Proceedings of CEPE98*. London School of Economics.

Habermas, J. (1996). Contributions to a discourse theory of law and democracy. In addition, see his Discourse Ethics. In *Moral consciousness and communicative action*, translated by Lenhardt and Shier. In *Between facts and norms* (p. 94). Cambridge, MA: MIT Press.

Hart, W. (1998). Intercultural computer-mediated communication. *The Edge: E-Journal of Intercultural Relations, 1*(4). www.interculturalrelations.com/Resource/The Edge.htm

Hattotuwa, S. (2004). Untying the Gordian knot: ICT for conflict transformation and peacebuilding. *Dialogue, 2*(2), 39–68.

Hölderlin, F. S. W. (2004). Briefe und Dokumente in zeitlicher Reihenfolge. *Bremer Ausgabe, 6*, 169.

Hollenbeck, M. E. (2001). The attainment of justice through restoration, not litigation. In M. Abu-Nimer (Ed.), *Reconciliation, justice, and coexistence* (pp. 65–85). Lanham: Lexington Books.

Hoven, J. V. D. (2000). The internet and varieties of moral wrongdoing. In D. Langford (Ed.), *Internet ethics* (pp. 127–158). London: Macmillan Press ltd.

Hsieh, H. F., & Shannon, S. (2005). Three approaches to qualitative content analysis. *Qualitative Health Research, 15*(9), 1277–1288. https://doi.org/10.1177/1049732305276687.

Hu, G., Pan, W., Lu, M., & Wang, J. (2009). The widely shared definition of e-government. *The Electronic Library, 27*(6), 968–985.

Jankovic, I. (2012). Deliberative democracy: History of an idea. *Theoria, Beograd, 55*, 21–52.

Janssen, D., & Kies, R. (2005). Online forums and deliberative democracy. *Acta Politica, 40*(3), 317–355.

Jentz, N. K., & Zimmerman, M. (1996). Evaluation of 37 AIDS projects: Successful approaches and barriers to program effectiveness. *Health Education Quarterly, 23*(1), 80–97.

Johnson, D. (2000). Democratic values and the internet. In D. Langford (Ed.), *Internet ethics* (pp. 181–200). London: Macmillan.

Kaefer, F., Roper, J., & Sinha, P. (2015). A software assisted qualitative content analysis of news articles: Example and reflections. *Forum Qualitative Sozialforschung/Forum: Qualitative Social Research, 16*(2), 1–5. Art. 8. http://nbn-resolving.de/urn:nbn:de:0114-fqs150283

Karefe, J. H. (2012). Participation through mobile phones: A study of SMS use during the Ugandan general elections 2011. In *Proceedings of the fifth international conference on information and communication technologies and development* (pp. 249–258).

Kneopffler, N., & O'Malley, M. (2016). Wasatia: A way to enduring peace in Palestine. In M. Dajani Daoudi, M. Dajani Daoudi, M. Leiner, & Z. Barakat (Eds.), *Teaching empathy and reconciliation in the midst of conflict* (pp. 19–29). Jerusalem: Wasatia Press, Jena Center for Reconciliation Studies, Ethik-Zentrum Jena.

Kofi, A. (1999, September 9). *The work of the organization.* REPORT from the UN Secretary-General. New York: UN. (A/54/1)(1).

Kohlbacher, F. (2012). The use of qualitative content analysis in case study research. *Forum Qualitative Sozialforschung / Forum: Qualitative Social Research, 7*(1), 1–30. From http://nbn-resolving.de/urn:nbn:de:0114-fqs0601211.

Kozinets, R. V. (2002). The field behind the screen: Using Netnography for marketing research in online communities. *Journal of Marketing Research, 31*(1), 61–72.

Kriesberg, L. (2001). Changing forms of coexistence. In M. Abu-Nimer (Ed.), *Reconciliation, justice and coexistence* (pp. 47–64). Lanham: Lexington Books.

Lambourne, W. (2001). Justice and reconciliation: Post-conflict and peacebuilding in Cambodia and Rwanda. In *Reconciliation justice and coexistence*. Oxford: Lexington Books.

Langer, R., & Beckman, S. C. (2005). Sensitive research topics: Netnography revisited. *Qualitative Market Research: An International Journal, 8*, 23.

Larrauri, H. P. (2013). New technologies and conflict prevention in Sudan and South Sudan. In F. Mancini (Ed.), *International Peace Institute – New Technologies and Conflict Prevention in Sudan and South Sudan* (pp. 71–86).

Leiner, M. (2016a). What is reconciliation and why it is needed? In M. Daoudi, M. Daoudi, M. Leiner, & Z. Barakat (Eds.), *Teaching empathy and reconciliation in the midst of conflict* (pp. 30–45). Jerusalem: Wasatia Press, Jena Center for Reconciliation Studies, Ethik-Zentrum Jena.

Leiner, M. (2016b). Thinking differently about identity and harmony. In P. Tolliday, M. Palme, & D. Choon Kim (Eds.), *Asia-Pacific between conflict and reconciliation* (pp. 183–203). Gottingen: Vandenhoeck &Reprecht GmbH & KG.

Leiner, M. (2018). Conclusion: From conflict resolution to reconciliation. In M. Leiner & C. Schliesser (Eds.), *Alternative approaches in conflict resolution* (pp. 175–183). Palgrave Macmillan.

List, C., Luskin, R., Fishkin, J., & McLean, I. (2013, January). Deliberation, Single-peakedness, and the possibility of meaningful democracy: Evidence from deliberative polls. *The Journal of Politics, 75*(1), 80–95. https://doi.org/10.1017/S0022381612000886.

Luskin, R. C. (2003). The heavenly public: What would the ideal democratic citizenry be like? In M. B. MacKuen & G. Rabinowitz (Eds.), *Electoral Democracy*. Ann Arbor: University of Michigan Press.

Lysack, K. (1994). An Indonesian experience in community-based practice, the occupational therapy. *Qualitative methods in field research, 14*(20), 93–110.

Mančić, Z. (2012). Deliberative democracy and the internet: Whether online deliberative democracy can replace classical democracy? *Philosophy and Society, XXIII*(2), 168–186.

Mason, J. (2002). Qualitative researching. In J. Saldana (Ed.), *The coding manual for qualitative researchers* (2nd ed., p. 12). SAGE Publications. [Kindle Edition] Available at Amazon.com.

Mathews, D. (1994). Politics for people. In *Finding a responsible public voice*. Urbana: University of Illinois Press.

Mayring, P. (2000). Qualitative content analysis. *Forum: Qualitative Social Research., 1*(2), 20.

Moaz, I. (2004). Social cognitive mechanisms in reconciliation. In Yaacov Bar–Simon-Tov (Ed.), *From conflict resolution to reconciliation* (pp. 225–237). New York: Oxford University Press.

Nadler, A. (2010). Intergroup reconciliation: Definitions, processes, and future directions. In L. Tropp (Ed.), *Oxford handbook of conflict* (pp. 1–5). New York: Oxford University Press.

Nussbaum, S. (2012). Social media is changing the way we live and learn. In *What school leaders need to know about digital technologies and social media* (pp. 67–73). San Francisco: Jossey-Bass.

O'Connor, C., & Skulnick, R. (2009). Social networking applications, social justice, and multicultural understanding. In M. Z. C. Vrasidas (Ed.), *ICT for education development and social justice*. Charlotte: Information Age Publishing.

Pavličić, P. (2010). Condorcet paradox. *Philosophy and Society, XXIII*(2), 278–279.

Rasmussen, J. L. (2001). Toward Integrating Relationship Building and Reconciliation into Official Peace Negotiations. In M. Abu-Nimer (Ed.), *Reconciliation, justice, and coexistence* (pp. 101–127). Lanham: Lexington Books.

Rehm, M. (2000). An artistic approach to the virtual community. *Consumer Science Research Journal, 2*, 29.

Richards, T. A. (1994). Using computers in qualitative analysis. In N. Denzin & Y. Lincoln (Eds.), *The handbook of qualitative research* (pp. 445–462). Thousand Oaks.

Richmond, O. P., & Tellidis, I. (2014, October-December). Emerging actors in international peacebuilding and statebuilding: Status quo or critical states*? Global Governance: A Review of Multilateralism and International Organizations., 20*(4), 563–584.

Saldana, J. (2013b). *The coding manual for qualitative researchers* (2nd ed.). London: SAGE. [Kindle Edition] Available at Amazon.

Sandelowski, M. (2000, August). Whatever happened to qualitative description? *Research in Nursing and Health, 23*(4), 334–340.

Sawilowsky, S. S. (2009). New effect size rules of thumb. *Journal of Modern Applied Statistical Methods, 8*(2, Article 26), 597–599. https://doi.org/10.22237/jmasm/1257035100. Available at: http://digitalcommons.wayne.edu/jmasm/vol8/iss2/26

Sirianni, C., & Friedland, L. (2003). Civic innovation, conflict, and politics: Response for the good society symposium on civic innovation in America. *The Good Society, 12*(1), 74–82.

Skinner, B. F. (1985). Quoted in Bourdieu's theory of practice as theoretical framework. In *Repatriation to France and Germany, mir-Edition* (p. 7). © Springer Fachmedien Wiesbaden. https://doi.org/10.1007/978-3-658-05700-8_2.

Sundqvist, J. (2011). Reconciliation as a societal process: A case study on the role of the Pentecostal movement (ADEPR) as an actor in the reconciliation process in post-genocide Rwanda. *Svensk Missionstidsskrift, 99*(2), 157–195.

Sundqvist, J. (2017). *Beyond an instrumental approach to religion and development: Challenges for church-based health care in Tanzania*. (Doctoral communication research). Uppsala: Acta Universitatis Upsaliensis.

Van der Meer, T. G. L. A., Gelders, D., & Rotthier, S. (2014). E-democracy: Exploring the current stage of e-government. *Journal of Information Policy, 4*, 489–506.

Viera, A. J., & Garrett, J. M. (2005). Understanding interobserver agreement: The kappa statistic. *Family Medicine, 37*(5), 360–363.

Walther, M. (2014). *Repatriation to France and Germany* (mir-Edition) (pp. 7–16). © Springer Fachmedien Wiesbaden. https://doi.org/10.1007/978-3-658-05700-8_2.

Weitzman, P. F., & Levkoff, S. E. (2000). Combining qualitative and quantitative methods in health research with minority elders: Lesson from a study of dementia caregiving. *Fields Methods, 12*(3), 195–208. https://doi.org/10.1177/1525822X0001200302

Zembylas, M., MacIssac, C., & Vrasidas, C. (2004). Of nomads, polyglots, and global villagers: Globalization, information technologies, and critical education online. In *Distance education and distributed learning*. Greenwich: Information Age.

Internet Resources

Assefa, H. (2015). *Assefa Quotes*, [Online on 2015]. Accessed 20 June 2016. Available at: http://www.doonething.org/heroes/pages-a/assefa-quotes.htm

Bonnemann, T. (2009). *What is deliberation*? By TIMin DELIBERATION, DICTIONARY. [Online 4, March, 2009]. Accessed 10 May 2015. Available at www.intellitics.com/blog/2009/03/04/what-is-deliberation

Channel, S. (2012). *The truth on Israel Palestine conflict*. Accessed 9 May 2016. Available at https://www.youtube.com/watch?v=Nc8EjQEpZ3s

Cole, R., & Crawford, T. (2007). *Building peace through information and communications technologies*. Accessed 24 Jan 2016. Available at http://apo.org.au/node/16132

Dajani, M. (2007). *Wasatia Booklet*. Ramallah: Modern Press. Accessed at 9 Sept 2015. Available at http://www.wasatia.info/articles/about-wasatia.html

Dajani, M. (2008). Wasatia: The middle road. *Wasatia The Middle Road*. Accessed 22 May 2015. Available at http://www.wasatia.info/articles/about-wasatia/16-wasatia-the-middle-road.html. p. 5.

Dajani, M. (2010). Wasatia-the-middle-road article: الوسطية تحديات الواقع الفلسطيني. Accessed 12 Oct 2014. Available at http://www.wasatia.info/articles/about-wasatia/16-wasatia-the-middle-road.html

Dajani, M. (2013). Wasatia: Centrism and moderation in Islam. *Wasatia Articles*, p. 20. Accessed 14 July 2015. Available athttp://www.wasatia.info/articles/about-wasatia/32-wasatia-centrism-and-moderation-in-islam.html

Dajani, M. (2014). *Wasatia NGO*. Accessed 15 Jan 2016. Available at www.wasatia.info/aboutWasatia

Dajani, M. (2015). *www.wasatia.info*. Accessed 22 Feb 2016. Available at http://www.wasatia.info/articles/intellectual-articles/63-palestinian-authority-urged-to-release-journalist.html

EDUCAUSE. (2006). *Seven things you should know about Facebook*. [Online 10 Sep 2006]. Accessed 22 Feb 2016. From https://net.educause.edu/ir/library/pdf/ELI7017.pdf

Erakat, A. H. (2014). *About Wasatia Articles*. [Online 28, April 2014]. Accessed 20 June 2014. From http://www.wasatia.info/articles/about-wasatia/17-palestinian-students-say-auschwitz-trip-was-%E2%80%9Cfor-educational-purposes-only%E2%80%9D-written-by-abdullah-h-erakat.html

ESTRIN, D. (2015) *For Palestinians social media is a source of news and anger*. Associated Press, [Online Oct. 24. 2015]. Accessed 20 May 2016. Available at http://www.businessinsider.com/ap-for-palestinians-social-media-is-source-of-news--and-anger-2015-10?IR=T.

Finizio, P. S. (2011). *Online mass deliberation process, "Mass Online deliberation, how to make it happen."* Accessed 20 Aug 2013. Available at https://www.youtube.com/user/pietrosperoni/videos

Flyvbjerg, B. (2006). *Five misunderstandings about case-study research.* Accessed 6 May 2015. Available at from http://qix.sagepub.com/content/12/2/219.short

Godwin, L. (2016). *Medium.* Accessed 14 Jan 2017. Available at: https://medium.com/@RevolutionaryId/godwins-law-as-an-online-discussion-grows-longer-the-probability-of-a-comparison-involving-5a6eb7365f7a

Haines, W. (2018). Iep.utm.edu. *Internet Encyclopedia of philosophy.* Consequentialism. Accessed 4 Apr 2018. Available at https://www.iep.utm.edu/conseque/

Jorgic, B. (2013). *Kenya tracks Facebook, Twitter for election "hate speech."* Accessed 20 Apr 2014. Available at http://www.reuters.com/article/2013/02/05/kenya-elections-socialmedia-idUSL5N0B4C4120130205, 05, 2-5.

Leiner, M. (2015a). https://de.wikipedia.org. (n.d.). *Wikipedia.* Accessed 15 Sept 2015. From https://de.wikipedia.org/wiki/Martin_Leiner#cite_note-10

Leiner, M. (2015b, January 13). *Hearts of flesh not stone: AICGS.* Accessed 17 Mar 2015. Available at: https://www.aicgs.org/site/wp-content/uploads/2015/02/Leiner_HeartsWashington_2-3-2015.pdf

Malcolm, X. (2005). *Media's Media's Influence on society.* Accessed 25 Dec 2015. Available at: http://www.csun.edu/~sm60012/GRCS-Files/Final%20Projects/MediaInfluenceonSociety-GRCS.htm

Maroun, F. (2016a). *Peace starts with facing the harsh reality of hate.* Accessed 2 Aug 2016. Available at: https://www.gatestoneinstitute.org/7920/palestinians-peace

Maroun, F. (2016b). [Online 25, April, 2016]. *Gate stone institute.* Accessed 25 Apr 2017. From http://www.gatestoneinstitute.org/7920/palestinians-peace

Owais, S., & Kiat, H. (2013). *Freelancers: 8 awesome reasons why you should join online forums.* Accessed 23 May 2015. From http://www.hongkiat.com/blog/freelancers-why-join-online-forums/

Penna, S. E. Q. (2013). *Qualitative data analysis with NVIVO.* [Online 7 June 2013]. Accessed 20 May 2014. Available at from www.youtube.com: https://www.youtube.com/watch?v=0YyVySrV2cM

Perera, M. (2015). *Reconciliation in Sri Lanka: The role of ICT.* [Online 30, June 2015]. Accessed 20 Jan 2016. Available at https://www.linkedin.com/pulse/reconciliation-sri-lanka-role-ict-mahesh-perera

Rosensteel, S. (2013). *Why online video is vital for Your 2013 content marketing objectives.* Accessed 23 May 2015. Available at http://www.forbes.com/sites/seanrosensteel/2013/01/28/why-online-video-is-vital-for-your-2013-content-marketing-objectives/

Schiller, T. (2017). Direct democracy. *Encyclopædia Britannica.* Accessed 9 Sept 2018. Available at : https://www.britannica.com/topic/direct-democracy.

Schulz, D. (2012). *YouTube – analyzing your interviews.* Accessed 4 May 2015. Available at Southampton Education School. Available at http://www.southampton.ac.uk/~sesvideo/

Siegfried, K. (1952–1953). The challenge of qualitative content analysis. *The Public Opinion Quarterly, 16*(4), 631–642. Accessed 17 July 2016. Available at http://www.jstor.org/stable/2746123

Special meeting of the Security Council Counter-Terrorism Committee and its Executive Directorate conference held on 16–17 December 2015. *Preventing Terrorist from Exploiting the Internet and social media to recruit terrorists and incite terrorist acts,* New York: Available on https://www.un.org/sc/ctc/news/2015/11/19/special-meeting-of-the-counter-terrorism-committee-and-technical-sessions-of-the-counter-terrorism-committee-executive-directorate-on-preventing-and-combating-abuse-of-ict-for-terrorist-purposes-new/. Accessed on 20 Jan 2016.

Taies, N., & Marc, A. (2009). *Disarmament, demobilization, and reintegration.* Accessed 4 Apr 2014. Available at http://siteresources.worldbank.org/EXTSOCIALDEVELOPMENT/Resources/244362-1164107274725/DDRFinal3-print.pdf

UNCTED. (2015). *ICT for peace foundation*. Accessed 10 Jan 2016. Available at http://ict4peace.org/?p=4037#sthash.btC3G5bo.dpuf

UNDP. (2008). *Action for cooperation and trust (ACT): Building lasting relationships islandwide*. Nicosia: UNDP-ACT.

van der Meer, T. G. L. A. (2016). Public Frame Building: The Role of Source Usage in Times of Crisis. *Communication Research, 45*(6), 956–981.

Venn, J. (1880 Published in July). Venn diagram *http://en.wikipedia.org/wiki/Venn_diagram*. Accessed 9 May 2014. From http://en.wikipedia.org/wiki/Venn_diagram

Vox. (2016). *The Israel-Palestine conflict: A brief, simple history*. Accessed 20 June 2016. Available at https://www.youtube.com/watch?v=iRYZjOuUnlU

Wasatia.info "Homepage" [Online 3 Feb 2015] Accessed 13 Mar 2015. Available at http://www.wasatia.info/

Wasatia.info "Our Goals" [Online 3 Feb 2015]. Accessed 13 Mar 2015. Available at http://www.wasatia.info/goals.html

Wasatia.info "About Wasatia" [Online 3 Feb 2015]. Accessed 13 Mar 2015. Available at http://www.wasatia.info/articles/about-wasatia.html

Wasatia.info "Articles" [Online 3 Feb 2015]. Accessed date 13 Mar 2015. Available at http://www.wasatia.info/articles/about-wasatia.html

Wasatia.info "Videos" [Online 3 Feb 2015]. Accessed 20 July 2015. Available at http://www.wasatia.info/homepage.html

Wisegeek. (2018). *www.wisegeek.com: What is an internet forume*. Accessed 26 Apr 2018. Available at: http://www.wisegeek.org/what-is-an-internet-forum.htm

World Conference against Racism, R. D. [Online 31, Aug 2001]. *World conference against racism, racial discrimination, xenophobia, and related intolerance*. Accessed 26 May 2015. Available at http://www.un.org/WCAR/durban.pdf

Printed by Printforce, the Netherlands